Animal Disease in Relation to Animal Conservation

Animal Disease in Relation to Animal Conservation

*(The Proceedings of a Symposium held at
The Zoological Society of London
on 26 and 27 November 1981)*

Edited by

MARCIA A. EDWARDS

and

UNITY McDONNELL

*The Zoological Society of London, Regent's Park,
London NW1, England*

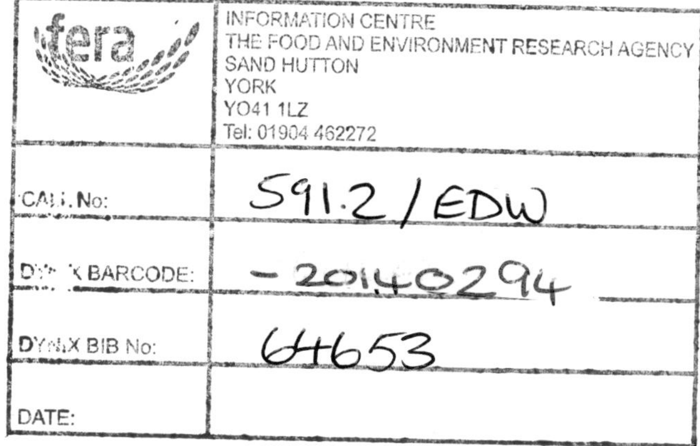

Published for
THE ZOOLOGICAL SOCIETY OF LONDON
BY
ACADEMIC PRESS
1982

ACADEMIC PRESS INC. (LONDON) LTD
24/28 Oval Road, London NW1 7DX

United States Edition published by
ACADEMIC PRESS INC.
111 Fifth Avenue, New York, New York, 10003

British Library Cataloguing in Publication Data

Animal disease in relation to animal conservation.
  —(Symposia of the Zoological Society of London
ISSN 0084-5612; no. 50)
  1. Wildlife conservation—Congresses
  2. Vertebrates—Diseases—Congresses
  I. Edwards, M. A.   II. McDonnell, U.
  III. Series
  639.9'6   QL81.5

ISBN 0-12-613350-6

Printed in Great Britain at the Alden Press
Oxford London and Northampton

Contributors

BARNES, A. M., *Chief, Plague Laboratory, Center for Infectious Diseases, Department of Health and Human Services, PO Box 2087, Fort Collins, Colorado 80522, USA* (p. 237)

HENDERSON, W. M., FRS, *Yarnton Cottage, High Street, Streatley, Berks RG8 9HY, England* (p. 287)

JONES, D. M., *Institute of Zoology, The Zoological Society of London, Regent's Park, London NW1 4RY, England* (p. 271)

KAPLAN, M.,* *Director General, Pugwash Conferences on Science and World Affairs, 11a Avenue de la Paix, 1202 Geneva, Switzerland* (p. 121)

LAINSON, R., FRS, *The Wellcome Parasitology Unit, Department of Parasitology, Instituto Evandro Chagas, Fundação Serviços de Saúde Pública, Caixa Postal 3, 66.000 Belém, Pará, Brazil* (p. 137)

MOLYNEUX, D. H., *Department of Biology and Centre for Development Studies, University of Salford, Salford M5 4WT, England* (p. 29)

NELSON, G. S., *Department of Parasitology, Liverpool School of Tropical Medicine, Pembroke Place, Liverpool L3 5QA, England* (p. 181)

PLOWRIGHT, W., CMG, FRS, *Department of Microbiology, A.R.C. Institute for Research on Animal Diseases, Compton, Newbury, Berks RG16 ONN, England* (p. 1)

ROSS, J., *Agricultural Science Service, Ministry of Agriculture, Fisheries and Food, Worplesdon Laboratory, Tangley Place, Worplesdon, Guildford, Surrey GU3 3LQ, England* (p. 77)

GORDON SMITH, C. E., CB, *Dean, London School of Hygiene and Tropical Medicine, Keppel Street (Gower Street), London WC1, England* (p. 207)

SMITH, G. R., *Nuffield Laboratories of Comparative Medicine, Institute of Zoology, The Zoological Society of London, Regent's Park, London NW1 4RY, England* (p. 97)

STECK, F., *Vet.-Bakteriologisches Institut, Universität Bern, Postfach 2735, Länggass-Strasse 122, CH-3001 Bern, Switzerland* (p. 57)

---

*Martin Kaplan is a consultant to the influenza laboratory of the WHO Collaborating Centre for Collection and Evaluation of Data on Comparative Virology, Veterinary Faculty, University of Munich.

*The Zoological Society of London is much
indebted to the Ministry of Agriculture, Fisheries and Food
for the generous grant which enabled this Symposium to be held*

Other Participants in Discussions

HRH The Prince Philip, Duke of Edinburgh, KG, KT (President, World Wildlife Fund)

BARR, M., *Ministry of Agriculture, Fisheries and Food, Room 1013, Tolworth Tower, Surbiton, Surrey KT6 7DX, England*

BRAY, R. S., *Imperial College of Science and Technology, Field Station, Silwood Park, Sunninghill, Ascot, Berks SL5 7PY, England*

CHAPMAN, M., *2 Thornton Close, Easingwold, York YO6 3JL, England*

COX, F. E. G., *Department of Zoology, King's College London, Strand, London WC2R 2LS, England*

DAVIES, J. I., *Ministry of Agriculture, Fisheries and Food, Burghill Road, Westbury-on-Trym, Bristol BS10 6NJ, England*

FIELDING, M. J., *11 Braxfield Court, St Anne's Road West, St Anne's, Lancs FY8 1LQ, England*

GALLAGHER, J., *no address given.*

JEWELL, P. A., *Physiological Laboratory, University of Cambridge, Downing Street, Cambridge CB2 3EG, England*

JONES, G. W., *Ministry of Agriculture, Fisheries and Food, Room 1013, Tolworth Tower, Surbiton, Surrey KT6 7DX, England*

KEYMER, I. F., *Ministry of Agriculture, Fisheries and Food, Veterinary Investigation Centre, Jupiter Road, Norwich, Norfolk, England*

KHALIL, L. F., *Commonwealth Institute of Parasitology, 395a Hatfield Road, St Albans, Herts AL4 OXU, England*

KINGDON, J., *The Elms, Islip, Oxon OX5 2SD, England*

LITTLE, T. W. A., *Ministry of Agriculture, Fisheries and Food, Central Veterinary Laboratory, Bacteriology Department, New Haw, Weybridge, Surrey KT15 3NB, England*

MACPHERSON, C. N. L., *Department of Zoology and Applied Entomology, Imperial College of Science and Technology, Prince Consort Road, London SW7 2BB, England*

MULLANEY, R., *Department of Veterinary Pathology and Microbiology, Veterinary College of Ireland, Ballsbridge, Dublin 4, Ireland*

ORMEROD, W. E., *Department of Medical Protozoology, London School of Hygiene and Tropical Medicine, Keppel Street, London WC1E 7HT, England*

PAYNE, J. M., *Director, ARC Institute for Research on Animal Diseases, Compton, Nr. Newbury, Berks RG16 ONN, England*

SCOTT, M. E., *Department of Pure and Applied Biology, Imperial College of Science and Technology, Prince Consort Road, London SW7 2BB, England*

THOMPSON, H. V., *Ministry of Agriculture, Fisheries and Food, Worplesdon Laboratory, Tangley Place, Worplesdon, Guildford, Surrey GU3 3LQ, England*

WALSH, J. F., *Department of Biology, University of Salford, Salford M5 4WT, England*

WILESMITH, J. W., *Ministry of Agriculture, Fisheries and Food, Central Veterinary Laboratory, Epidemiology Unit, New Haw, Weybridge, Surrey KT 15 3NB, England*

YOUNG, A., *9 Spence Street, Glasgow G20, Scotland*

LORD ZUCKERMAN, FRS, *The Zoological Society of London, Regent's Park, London NW1 4RY, England*

Chairman of Symposium and Chairmen of Sessions

CHAIRMAN OF SYMPOSIUM

SIR ANDREW HUXLEY, PRS, on behalf of the Zoological Society of London

CHAIRMEN OF SESSIONS

SIR CYRIL CLARKE, KBE, FRS, *Nuffield Unit of Medical Genetics, The University, PO Box 147, Liverpool L69 3BX, England*

SIR ANDREW HUXLEY, PRS, *Department of Physiology, University College London, Gower Street, London WC1 6BT, England*

W. H. G. REES, *Chief Veterinary Officer, Ministry of Agriculture, Fisheries and Food, Hook Rise South, Tolworth, Surbiton, Surrey KT6 7NF, England*

D. A. J. TYRRELL, CBE, FRS, *Medical Research Council Common Cold Unit, Harvard Hospital, Coombe Road, Salisbury, Wilts SP2 8BW, England*

Introduction by Sir Andrew Huxley

I was both surprised and pleased when Lord Zuckerman invited me to be Chairman of this Symposium — surprised because I have no special knowledge of animal disease, but pleased because I am devoted to the cause of animal conservation. Animal conservation gets discussed, of course, at enormous length in the press, but almost entirely in relation to side effects of modern technology. One can divide these effects into two classes. One is the directly damaging effects of pollution, in which the technical problems may be difficult and the solutions may be expensive, but in which there is seldom any moral conflict, at least of a kind which would be likely to worry any of those who come to a meeting such as this. But in contrast to that, there is another class of effects in which there is a direct conflict between wildlife conservation on the one hand and legitimate human needs on the other. Here technology comes in again, principally through medical and other advances which have led to the rapid increase of human population, with a corresponding increase in demand for land and for natural products of one sort and another. This in turn has caused the loss of habitat — notably in tropical forests, but there are many other examples — which is the principal threat to wildlife all over the world. And it is this second aspect that is probably the more serious, and at the same time the more difficult to deal with because of the real conflict of interests. For example, a friend of ours has worked on the rhinoceros in Nepal, and he told us it was almost exterminated by the use of DDT. Of course DDT itself does no harm to a rhinoceros, but by eliminating malaria it made the swampy bottom of the Indus valley habitable to humans who drove the animal out. Fortunately a small area was kept as a nature reserve for the animals.

I think it will become clear during this meeting, if it is not already clear, that there is a similar distinction in the connections between animal disease and conservation. On the one hand there is the straightforward destruction of wildlife by disease, and on the other, the deliberate human destruction of wildlife in order to protect man or his domestic animals from disease: foxes in relation to rabies, big game in Africa in relation to tsetse fly and trypanosomiasis, and, of course, badgers in relation to bovine tuberculosis. This second category includes also the converse case where humans deliberately introduce a disease in order to control a pest — for example, myxomatosis to keep down rabbits. We all look forward to being made clearer in our own minds about ways in which we can avoid disastrous effects on wild populations from these conflicts between directly human interests and the conservation of wildlife.

London, November 1981

Contents

SPECIFIC DISEASES

The Effects of Rinderpest and Rinderpest Control on Wildlife in Africa

WALTER PLOWRIGHT

Trypanosomes, Trypanosomiasis and Tsetse Control: Impact on Wildlife and its Conservation

D. H. MOLYNEUX

Rabies in Wildlife

F. STECK

Myxomatosis: The Natural Evolution of the Disease

JOHN ROSS

Botulism in Waterfowl

G. R. SMITH

ANIMALS AS RESERVOIRS OF DISEASE

Influenza in Nature

MARTIN M. KAPLAN

Leishmanial Parasites of Mammals in Relation to Human Disease

RALPH LAINSON

Carrion-feeding Cannibalistic Carnivores and Human Disease in
Africa with Special Reference to Trichinosis and Hydatid Disease in Kenya

G. S. NELSON

GENERAL EPIDEMIOLOGICAL PRINCIPLES AND POLICIES

Major Factors in the Spread of Infections

C. E. GORDON SMITH

Surveillance and Control of Bubonic Plague in the United States

ALLAN M. BARNES

CONSERVATION

Conservation in Relation to Animal Disease in Africa and Asia

D. M. JONES

The Control of Disease in Wildlife when a Threat to Man and Farm Livestock

W. M. HENDERSON

The Effects of Rinderpest and Rinderpest Control on Wildlife in Africa

WALTER PLOWRIGHT

Department of Microbiology, Institute for Research on Animal Diseases, Compton, Newbury, Berkshire, England   .

SYNOPSIS

Rinderpest caused a "virgin-ground" panzootic in Africa during the period 1889—1898, spreading from the north-east to the Cape and west coast and causing catastrophic losses in many game animals, as well as cattle. It was eliminated from southern Africa but became established in sub-Saharan regions, where it caused periodic waves of infection affecting both cattle and wild Artiodactyla. A focus of permanent, often mild, infection was established by the early 1930s in the Serengeti region of northern Tanganyika and adjacent parts of southern Kenya. Its existence was commonly attributed to virus maintenance in the very large herds of wildlife, as mass vaccination of cattle from the late 1940s failed to eliminate it by the early 1960s. Where smaller game communities were involved, as in southern Tanganyika, rinderpest was repeatedly eradicated by vaccination of cattle and movement controls.

Clinically apparent disease last occurred in Serengeti buffaloes and wildebeest in 1960—1961; restricted sub-clinical infection occurred in 1962—1963 and 1965—1966 but no serological evidence of virus activity has been detected since. It was concluded that, by themselves, the large numbers of susceptible game animals in the Serengeti region in 1962—1963 were not capable of maintaining the strains of virus current at that time. Monitoring of the buffalo and wildebeest populations in the Serengeti area showed that they increased greatly during the decade from 1961, buffalo from *c.* 30 000 to >60 000; wildebeest from 260 000 to 770 000. This was probably largely due to the reduction of juvenile mortality attributable to rinderpest.

The threat to the present susceptible herds of wild ungulates in east Africa, posed by rinderpest travelling from the north, is emphasized; it requires contingency planning to combat it. The removal of the threat depends ultimately on rinderpest elimination from the cattle of north-east Africa.

INTRODUCTION

Rinderpest or cattle plague is without doubt the most lethal and potentially dangerous infectious disease which affects wild Artiodactyla, although its terrors for domestic animals are now much reduced by effective vaccines. It is unique in its almost indiscriminate attack on many wild and domestic species, a factor which for decades in Africa created a community of interest between those concerned for wildlife conservation and those responsible for the health and productivity of domestic cattle. Sometimes, it is true, there has been an apparent conflict of attitudes; the veterinarians could claim that the game were reservoirs and disseminators of infection for their charges and the wardens could assert that without cattle the rinderpest would disappear from game. The truth seems to be somewhere in between and this paper will present some of the facts which should make possible a balanced judgement.

The paper will start with an historical introduction, an outline of the great panzootic and its sequelae which led, probably for the first time, to the establishment of persistent rinderpest in sub-Saharan Africa.

THE EARLIER HISTORY OF RINDERPEST IN AFRICA

According to Curasson (1932) rinderpest was introduced to the Nile Valley (Egypt) in 1841–1842 and again in 1863 when it spread westwards to Senegal, etc. It was also apparently present in that country as early as 1828 but whether or not it disappeared in the intervening periods is not known. The first great epizootic to be observed by Europeans was that which began in Somalia or Abyssinia in 1889 (Mack, 1970 and Fig. 1), probably following the importation of zebu cattle from India for the Italian armies. It spread along trade routes to reach Uganda and northern Kenya by 1889–1890 and the Masailands of southern Kenya and northern Tanganyika by 1890. By July 1892 it was present at the northern end of Lake Nyasa and, progressing down the Lake, reached the Zambesi River, which impeded its progress until March 1896, when it was confirmed in Bulawayo. By the end of the year it had reached the Cape — about 3000 miles from its point of introduction (Mack, 1970). The virus not only spread southwards but also west along the southern border of the Sahara to reach the west coast and equatorial Africa in 1890–1892. The losses there in cattle indicated that the population was completely susceptible (Curasson, 1932). Apart from enormous losses in domestic cattle — estimated at 5.3 million in southern Africa alone — the mortality in wild ruminants was devastating.

Simon (1962) quoted an early warden, F. J. Jackson, as estimating that at least 90% of Kenya's buffalo succumbed and referred to a widely held belief that bongo "were almost exterminated by rinderpest in the 1890s". An old Masai man, alive at the time of the pan-

FIG. 1. The course of the great rinderpest panzootic in Africa, 1889–1897. From Mack (1970), with acknowledgements to the author and *Tropical Animal Health and Production.*

zootic, told Branagan & Hammond (1965), with macabre clarity, how the corpses of cattle and people were "so many and so close together that the vultures had forgotten how to fly". The catastrophic effect of the rinderpest mortality in cattle and game on the well-being of pastoral peoples is a factor not to be forgotten — especially when it was combined with smallpox, jigger fleas and bovine pleuropneumonia, as in east Africa (Ford, 1971)

There is evidence that the mortality in wild ruminants was so high that the tsetse fly, *Glossina morsitans* in particular, was severely reduced or died out in some areas, as for example the Kruger National Park, for lack of suitable hosts on which to feed (Stevenson-Hamilton, 1911; Duke, 1919; Carmichael, 1933), though Stevenson-Hamilton (1957) later queried the proposed association. The implications of a possible reduction by rinderpest in the number of *G. morsitans* were important for wildlife conservation. First, the association was later seized upon as a justification for game eradication in the control of trypanosomiasis. Secondly, Buxton (1955) considered that in parts of east Africa the reduction of grazing and browsing animals allowed the encroachment of bush which extended the range of the fly; there is little doubt that the presence of large numbers of tsetse flies has protected game in some areas from the incursion of cattle-owning or agricultural peoples.

In the panzootic there were obvious differences between species in the time at which they were infected and the severity of the disease. Generally, however, buffalo, eland, warthog and bushpig suffered badly and at an early time, followed in east Africa by giraffe, greater and lesser kudu, roan antelope, bushbuck and finally wildebeest (Simon, 1962). In South Africa blesbok and bontebok, gemsbok, duiker, reedbuck, waterbuck and springbok also succumbed but hartebeest, impala and wildebeest were said to have escaped (see Scott, 1970). Nevertheless, all these and many other species have been found to suffer severely in subsequent outbreaks. It may be postulated that these anomalies were due to differences between the strains of virus or the intensity of exposure but there are no facts to support these suggestions. The reduction in numbers of the most susceptible species was so great that doubts were expressed as to whether they would ever recover and yet within ten years or so buffalo were so numerous that in some places measures were necessary to reduce their numbers. Other species, such as greater kudu, roan antelope and bongo, were so reduced in some localities that they appeared never to have recovered (Simon, 1962).

The course of the panzootic showed clearly that the cattle and wild ungulate populations were fully susceptible and had no previous exposure to rinderpest virus. In fact, the disease literally burned itself out in southern Africa, aided by vaccination and movement controls, so that by 1903 it was confined to Zululand and south-west Africa. It remained periodically epizootic or enzootic, however, in Egypt, the Sudan, east and north-east Africa generally and in parts of west Africa. Thus a wave of infection spread south from Abyssinia in 1897–1901 (Simon, 1962) and major extensions from the Sudan, spreading to Tchad (which had been free for 20 years) and Equatoria took place in 1913 and 1917 (Pecaud, 1924); this wave also engulfed west Africa and east Africa, including southern Tanganyika, during and after the First World War. It was eventually halted by establishing a zone of immunized cattle and elimination of game along the Rhodesia and Nyasaland borders. By 1922 the disease was restricted to the area north of the Central Railway (Tabora to Dar-es-Salaam) and by 1928 it was again confined to the Northern and Lake Provinces (Branagan & Hammond, 1965).

Subsequent epidemics spread from Kenya to Northern Tanganyika in 1927 and from Uganda to the Congo, affecting the Albert National Park in 1929–1931. The former outbreak was still smouldering away in 1930–1931, about which time it is possible that a permanent focus of infection had established itself in game animals of the Serengeti region. A dangerous surge southwards, in which infected game movements were a decisive factor, gained momentum in 1936–1937 when rinderpest again reached the Southern Highlands of Tanganyika. In this area, around Lake Rukwa and the Kilombero Valley, rinderpest caused considerable losses, not only in buffalo and eland but also in greater kudu. The threat of further extension to central Africa was averted by the creation of belts of immune cattle and by construction of a game-proof fence, eventually 167 miles long, from the southern end of Lake Tanganyika and towards the northern end of Lake Nyasa; the cost, incidentally, at that time was about £60 per mile (Vaughan-Jones, 1953). A 25-mile game-free strip was to be created on either side of the fence (Lowe, 1942). By the end of 1941 the disease was once more virtually eliminated south of the Central Railway but the fence and cordon were continued until 1952.

Further west, in Uganda, infection spread in game from the Sudan to the Lakes Region in 1942 and from there to the Congo, while similar excursions from the Sudan to the Congo also occurred in 1956 and 1960. A major extension south from the Ethiopian border into Kenya, which severely affected many species including buffalo, eland, warthog, bushbuck and giraffe, as well as impala, oryx and bongo more moderately, was recorded in 1960–1962 by Stewart (1964). A virus (RGK/1) was isolated in this case from a sick reticulated giraffe and caused a mortality of about 60% in experimental cattle (Liess & Plowright, 1964).

RINDERPEST AS A DISEASE OF WILD UNGULATES

Whilst all members of the Order Artiodactyla are probably susceptible (Scott, 1964), the manifestations of rinderpest virus infection in susceptible wild species vary considerably, from acute lethal disease, comparable to that in cattle, to very mild or undetectable in species such as Thomson's gazelle. Tables Ia and Ib give a grading of susceptibility for many important African species, being based largely on field observations of sick or dead animals, sometimes supplemented by experimental data or the records of outbreaks in zoos. The Tables also tend to reflect the frequency with which these forms are affected when exposed to natural infection.

It must be emphasized, however, that there are differences in the

TABLE Ia

Wild ungulates of high susceptibility to rinderpest virus infection

Level of susceptibility	Common name	Scientific name
Very high	Buffalo	*Syncerus caffer*
	Warthog	*Phacochoerus aethiopicus*
	Eland	*Taurotragus oryx*
	Kudu	*Tragelaphus strepsiceros* and *T. imberbis*
High	Giraffe	*Giraffa camelopardalis* and *G. reticulata*
	Bushbuck	*Tragelaphus scriptus*
	Bushpig	*Potamochoerus porcus*
	Sitatunga	*Tragelaphus spekei*
	Uganda kob	*Adenota kob*
	Giant forest hog	*Hylochoerus meinertzhageni*
	Bongo	*Boocercus euryceros*
	Wildebeest	*Connochaetes* spp.

TABLE Ib

Wild ungulates of lower susceptibility to rinderpest virus infection

Level of susceptibility	Common name	Scientific name
Moderate	Reedbuck	*Redunca* spp.
	Topi	*Damaliscus korrigum*
	Blesbok and bontbok	*Damaliscus albifrons* and *D. pygargus*
	Gemsbok	*Oryx gazella*
	Roan and sable antelopes	*Hippotragus equinus* and *H. niger*
	Oribi	*Ourebia ourebi*[a]
	Impala	*Aepyceros melampus*[a]
	Springbok	*Antidorcas marsupialis*
Low	Waterbuck	*Kobus ellipsiprymnus* and *K. defassa*
	Duiker	*Cephalophus* spp.[a]
	Oryx	*Oryx beisa*
	Grant's gazelle	*Gazella granti*[a]
	Dikdik	*Rhynchotragus kirkii*
	Hartebeest	*Alcelaphus* spp.
Very low	Thomson's gazelle	*Gazella thomsoni*
	Hippopotamus	*Hippopotamus amphibus*
	Gerenuk	*Litocranius walleri*

[a] Species whose natural susceptibility is reportedly very variable.

virulence of different virus strains for cattle and this may well be true also for wild species. Thus, for example, hartebeest are generally unaffected but in Somalia in 1897 (Simon, 1962) and in west Africa in 1913—1917 (Pecaud, 1924) they suffered severely. Similarly, Grant's gazelles are not usually observed to be sick but they were said to show diarrhoea and blindness in the Tanga Province of Tanganyika in 1945. It is in fact remarkable that corneal opacity or blindness, which does not occur in domestic animals with rinderpest, has been seen several times in wild species, including giraffe, as in the 1960—1961 outbreak in northern Kenya (Simon, 1962: Plowright, unpublished).

OBSERVATIONS ON RINDERPEST OUTBREAKS IN GAME ANIMALS IN AFRICA

Until around 1960, methods for assessing the prevalence of rinderpest infection in wild or domestic animal populations were essentially the same as those which had been available in the previous century. Populations could be observed for sickness and deaths; dead animals could be counted and subjected to autopsy; occasionally more detailed pathological observations (e.g. Thomas & Reid, 1944) or attempts to transmit the disease to cattle or goats, which were presumed susceptible, were undertaken (e.g. Cornell, 1934; Carmichael, 1938). Sometimes the diagnoses were undoubtedly mistaken but, when typical clinical and pathological signs were exhibited and the disease was known to be present in the local cattle population, then identification of the disease in game could be regarded as reasonably secure.

In the late 1950s techniques fortunately became available for the rapid, more economical, detection of rinderpest virus and its antigens or antibodies in animal tissues. Agar-gel double-diffusion tests for the antigens (White, 1958) and cell culture methods for the virus and its antibodies (Plowright & Ferris, 1961, 1962a) made possible diagnosis and epidemiological surveys on a scale hitherto impossible. Some species which have been shown by serological means to have been infected frequently, e.g. the hippopotamus in western Uganda (Plowright, Laws & Rampton, 1964) and Thomson's gazelle in the Serengeti area (Plowright & McCulloch, 1967), have never been observed to be affected clinically.

The magnitude of the territories and populations involved precluded, and still severely limits, quantitative observations on the effect of rinderpest epizootics (or any other disease); Tanganyika alone covered about 370 000 square miles and in 1964 had 8.8 million cattle (Branagan & Hammond, 1965). Another complicating factor is the speed with which sick animals, and the carcases of those which die, are removed by predators and scavengers. Sinclair (1977) estimated that 10 000 buffalo calves died yearly in the Serengeti area, but few were ever found moribund or in a state suitable for veterinary examination. Furthermore, it is only over the last 20 years (approximately) that reliable figures have become available, through aerial census, for the numbers of major species in the most spectacular accumulations of wild ungulates, such as those in the Serengeti and contiguous regions. These comprise about 5000 square miles of country (Grzimek & Grzimek, 1960; Stewart & Talbot, 1962). The size of ungulate populations in more broken bush country, with smaller and more fragmented communities, was much worse documented. Hence our primary efforts in trying to elucidate the epidemiology of rinderpest were concentrated in the Serengeti area.

ENZOOTIC RINDERPEST IN EAST AFRICA

By the late 1920s, a pattern emerged in Tanganyika which was quite different from that observed in earlier epizootics. Infection in the north was often limited to young animals, those which had not been present in the previous wave and had lost maternal immunity. In them the disease was frequently so mild by 1931 as to be barely recognizable by experienced observers but at other times it was severe and accompanied by a high mortality. Both cattle and wild species, wildebeest particularly, showed these fluctuations — sometimes simultaneously, at other times independently. It was frequently suggested that strains of virus recovered from game animals were of reduced virulence for cattle (Cornell, 1934; Wilde, 1953). Carmichael (1938) reported similar suspicions in Uganda but said that passage in cattle restored the virulence. However, the first demonstration that a strain from game was of stable and low pathogenicity was made by Robson *et al.* (1959), who passaged an eland isolate nine times serially in cattle without increasing its virulence. This strain was, nevertheless, similar to others recovered from cattle in Tanganyika and Kenya (Lowe *et al.*, 1947; Plowright, 1963a). One buffalo strain (Essimongor) was passaged 11 times serially in cattle with only a slight increase in its low pathogenicity (Plowright, 1963a). On the other hand, a strain spreading south from the Kenya/Tanganyika border in 1945 was extremely virulent for game and cattle; it killed waterbuck, oryx, Grant's gazelle and impala, species which were normally resistant (Branagan & Hammond, 1965).

It was also increasingly realized that sick game animals did contribute to the spread of disease. Buffalo and, perhaps, eland were especially dangerous, as affected herds tended to fragment and move more quickly and further than usual in search of water with which to slake their thirst, enhanced by diarrhoea and dry seasons. Buffalo also became unusually aggressive (Carmichael, 1938; Branagan, 1966). Nevertheless it was stated that a spread by game over a distance greater than 50 miles had not been encountered (Hornby, 1939). Wide dissemination by game was less likely in dry seasons when supplies of water and available grazing were limited to river courses, etc. On the other hand it could be encouraged by shooting (Lowe, 1942).

A second important characteristic of rinderpest was noted, particularly in Tanganyika; this was that the virus tended to persist only in areas with large aggregations of wild ungulates. Lowe (1942) voiced a widespread opinion when he said that "but for the game, rinderpest would be eradicated from Tanganyika in the near future". Reid (1949) emphasized that rinderpest died out in smaller game communities, that "large concentrations" of susceptible species (especially in the Serengeti region) were important for virus maintenance, and that a thinning-out process might be necessary to control what had "come to be known as the uncontrollable problem of game transmission". He continued

> more and more I am convinced that the only way is to separate off the massive game concentrations of the National Park from the cattle populations of the territory by fencing, ditching, organised shooting and driving of game from those areas linking up the Park with the rest of the territory. This is likely to be an expensive policy to introduce or enforce but I am convinced that, in the long run, it is the only method of eradicating cattle plague from Tanganyika and E. Africa (Reid, 1949).

Two years later (Reid, 1951) he noted that the eradication policy pursued in Tanganyika over 30 years had been unsuccessful, the disease was still as great a menace as ever and "the game factor" had been responsible for the failure. Similar opinions were expressed by several relevant authorities as late as 1959 (see Libeau & Scott, 1960).

It is not difficult to understand the note of despair which permeates the Tanganyikan Annual Reports during the later 1940s and to the end of the 1950s. Cattle were being vaccinated at the rate of about 1.5 million a year in 1948, rising to 2.5 million in 1949 and yet there was in game an apparently independent and persistent focus of infection which was beyond control. In spite of this, Thomas & Reid wrote in 1944

> there is ample room in Africa for game in plenty as well as for livestock but their interests need not and must not be allowed to clash. In other words it is becoming more and more evident that wild animals will have to be segregated in sanctuaries as much for their own protection as for that of livestock and agriculture.

As we shall see shortly, for reasons that are not properly understood, rinderpest eradication did not, in fact, demand the measures which for so long seemed necessary to those best informed and able to judge the situation.

VACCINATION OF CATTLE AGAINST RINDERPEST

The potential for a conflict of interests between game and cattle was greatest in the "Serengeti region" of east Africa, where the Masai, a semi-nomadic tribe of pastoralists, occupied vast tracts of country in which rinderpest had remained enzootic for about 30 years by the late 1950s. Tanganyika Masailand alone extended over 23 000 square miles and their cattle numbered "upwards of 800 000" when Branagan & Hammond were writing in 1965. The Lake Province of Tanganyika into which the Serengeti wildebeest migrated was also a potential flashpoint. In addition the Masai utilized large areas of the Amboseli National Park and what was later to become the Masai-Mara National Park in Kenya; furthermore their grazings in the Kajiado and Ngong Districts were contiguous with the Nairobi National Park.

It was particularly important, therefore, that as many cattle as possible in the Masailands and Sukumuland (Lake Province) should be kept immunized against rinderpest to prevent the spread of disease from sick game and thence to the herds of neighbouring peoples. Somewhat similar problems existed in the Karamoja District of Uganda and the Northern Frontier Province of Kenya, which were intermittently threatened by the disease moving in trade cattle or game from the north-west part of Somalia, the southern Sudan and Ethiopia. This northern enzootic area, however, was distinguished from the southern by the virulence of many of the reported outbreaks, especially in game (see, for example, Stewart, 1964) and the lack of a wild population which was apparently large enough to maintain the virus independently of susceptible cattle.

The scale of the immunization campaigns, even before the advent of international programmes with multilateral finance, was considerable. Thus the total annual issues of vaccine for Kenya, Tanganyika, Uganda, Somaliland, Sudan and Ethiopia exceeded 8.5 million doses in 1959 and by that time cattle south of the Central Railway in Tanzania were no longer vaccinated (Libeau & Scott, 1960). In many areas vaccination was compulsory and free; the product generally in use from 1940 up to that time was the Kenya goat-attenuated virus (KAG) which conferred a long-lasting, probably life-time immunity on animals which were devoid of colostral antibody (Brown & Rashid, 1965).

In Tanganyika vaccination campaigns were carried out annually in June/July immediately after the long rains and all animals were inoculated twice — once as calves and again the following year as yearlings. From 1950, all susceptible cattle north of the Central Railway, especially where game contacts were possible, were covered by annual campaigns of mass vaccination. Ear punches and branding were used to identify vaccinated animals. In the years between 1940 and 1954 over 24 million cattle in Tanganyika were inoculated with KAG (Dawe, 1957), the number in the following year being virtually three million. These vaccinations were not without risk. Whilst normally, in healthy animals, associated with small losses ($<2\%$ within three weeks of inoculation), the vaccination mortality could attain about 20% in calves suffering from intercurrent protozoal infections, such as coccidiosis, theileriasis and trypanosomiasis (Branagan, 1965). This relatively high pathogenicity of the goat-attenuated vaccine for young animals — incidentally very similar to that encountered in some west African zebus (Plowright, 1957) — had the effect of inhibiting owners from bringing forward their calves for vaccination in the first year of life. Such animals, which would accordingly be susceptible to rinderpest for a period of some months, between the disappearance of passive colostral immunity (at six to 12 months) and the succeeding vaccination campaign, did of course help to maintain the virus in conjunction with game animal species and hence to vitiate attempts to eradicate the disease. Such outbreaks as were reported in cattle commonly affected calves and yearlings almost exclusively.

The answer to this problem of vaccination lay in the development of more attenuated vaccines by passage of virulent virus in cell cultures (Plowright & Ferris, 1959a, b, 1962b; Johnson, 1962). Culture-attenuated virus was innocuous for cattle of all ages and still retained the capacity to produce prolonged immunity, lasting more than 10 years (Plowright & Taylor, 1967; Plowright, 1972). It also facilitated the international campaign (JP-15) for the control of rinderpest, Phase I of which began in west Africa in 1962 and covered the countries of the Lake Tchad Basin (Nigeria, Niger, Chad and Cameroun). Phases II and III gradually incorporated other countries in west and central Africa with cattle populations totalling 32.75 millions and involved over 81 million vaccinations. Phase IV began in east Africa (Kenya, Uganda, Tanzania, Sudan and Somalia) in 1968 and later incorporated Ethiopia, which alone had an estimated 25 million cattle (DeTray, 1970).

It was, unfortunately, only after the initiation of the JP-15 campaign in east Africa that statistically adequate surveys of the effect of a vaccination campaign on the immune status of cattle populations were made possible by serological testing. Some data were obtained for animals which were believed to have received only KAG vaccine as calves or yearlings in the years up to and including 1961 and tested in 1970.

Thus in Table II it will be seen that, whether or not brands indicating past vaccination were present, 80—95% of cattle which had

TABLE II

The effect of mass rinderpest vaccination on the immune status of adult cattle[a] in Tanzania, April to October, 1970 (Plowright, 1972)

Presence of vaccination brand	No. with antibody/No. tested in age group			
	3—4 years	5—6 years	7—8 years	≥9 years
+	605/643 (87.5%)	314/351 (89.5%)	231/262 (88.2%)	111/138 (80.4%)
−	51/69 (73.9%)	38/51 (74.5%)	18/24 (75%)	20/21 (95.2%)
Years of vaccination	1966—1967	1964—1965	1962—1963	Before 1962
Vaccine employed	Cell culture	Cell culture	Caprinized or cell culture	Caprinized

[a] 135 herds in six districts of Lake and Northern Provinces were tested. Vaccine was administered only to calves and yearlings in successive years.

been vaccinated with KAG before 1962 still had rinderpest antibody, indicating that they had been effectively immunized. A minority may have suffered natural exposure as "wild" virus was still circulating at that time (see below). Table II also shows the effectiveness of later campaigns, using culture-attenuated virus, which conferred an immunity on almost 90% (87.5—89.5%) of branded cattle. If we compare the proportion of immune cattle (those possessing virus neutralizing antibody) in various age groups before and after the annual campaigns (Table III) it is evident that, in areas such as Ngorongoro, where vaccination was carried out with great care, the proportion of young animals (up to 12 months old) which became susceptible between annual campaigns increased from about 5—19% to 40—50%. We know that in other areas of Masailand the figures were decidedly worse; i.e. as many as 80—90% of animals up to two years old became susceptible (Table IV). Assuming that these age groups constituted 35—40% of the total cattle population, it can be seen that

TABLE III

The effect of rinderpest vaccination on the immune status of cattle[a] in the Ngorongoro District of Tanzania, 1969–1970

Date	No. of localities	No. of owners	No. positive/No. tested in age group				
			≤6 months	7–12 months	13–24 months	≥25 months	All
August 1969 (after vaccination)	4	23	64/68 (94.1%)	114/119 (95.8%)	68/73 (93.1%)	351/383 (91.6%)	597/643 (92.8%)
25 May 1970 to 2 June 1970 (before vaccination)	4	21	55/117 (47%)	50/82 (60.9%)	108/128 (84.4%)	235/273 (86%)	448/600 (74.6%)
1 July 1970 to 4 July 1970 (after vaccination)	4	16	61/75 (81.3%)	56/60 (93.3%)	137/142 (96.5%)	145/165 (87.9%)	399/442 (90.3%)

[a] As indicated by the possession of virus-neutralizing antibody in their serum.

TABLE IV

The effect of rinderpest vaccination on the immune status of cattle in the Kajiado District, Masailand, 1969–1970

Date	No. of localities	No. of owners	No. positive/No. tested in age group				
			≤6 months	7–12 months	13–24 months	≥25 months	Total
17 January 1970	5	17	5/62 (8.1%)	5/86 (5.8%)	16/68 (23.5%)	64/80 (80%)	90/296 (30.4%)
17 January 1970 to 13 February 1970	3	18	1/46 (2.2%)	8/72 (11.1%)	4/42 (9.5%)	23/43 (58.1%)	38/203 (18.7%)
12 January 1970 to 6 February 1970	3	17	13/42 (31%)	12/61 (19.7%)	19/84 (22.6%)	80/133 (60.1%)	124/320 (38.7%)

the problem of maintenance of immunity in cattle herds was both difficult and persistent.

THE DISAPPEARANCE OF RINDERPEST FROM THE SERENGETI ENZOOTIC AREA

To return to Tanganyika in the early 1960s, cell culture vaccine was introduced gradually from 1962 and by 1964 was the only product employed (Plowright, 1972). It is perhaps coincidental that these were also years when a remarkable change took place in the areas with enzootic rinderpest in cattle and game. The infection disappeared and has not returned except for a very restricted focus which was identified in the Northern Region in 1965 (Taylor & Watson, 1967). The efforts of at least 30 years were crowned with success and Tanganyika was at last free of cattle plague. While it is understandable how this was brought about for cattle, the reasons for its disappearance from the game population are more difficult to find.

An account of the earlier virological investigations on game animals in northern Tanganyika and southern Kenya from 1960 to 1963 was published in 1967 (Plowright & McCulloch, 1967) (Fig. 2) and the important facts established, with sera from a few hundred animals, were as follows.

In October 1960 the last severe outbreak of clinical rinderpest occurred in the main wildebeest herds on the Serengeti Plains; it proved at first very difficult to obtain permission from the Park authorities to shoot any sick animals from which to obtain material for confirmation of diagnosis by virus recovery. Eventually, in November, five convalescent animals (seven to nine months old) were shot for this purpose and all had virus-neutralizing antibody apparently induced by recent infection. A sixth animal, found dead, also yielded virus-specific antigen and this was the only occasion on which such proof of aetiology could be obtained. Collections of wildebeest sera, predominantly from the main migratory herds, did not take place again until the period January 1962 to June 1963. By ageing the animals as accurately as possible, from the known calving season, body size and dentition, it was possible to correlate age with the possession of rinderpest antibody as shown in Table V (Plowright & McCulloch, 1967).

Table V also gives the results of examination of sera from the same migrant population, collected three years later (1965–1966) by Taylor & Watson (1967), who, in addition, employed scars left by the corpora lutea of pregnancy to assist ageing. The two sets of

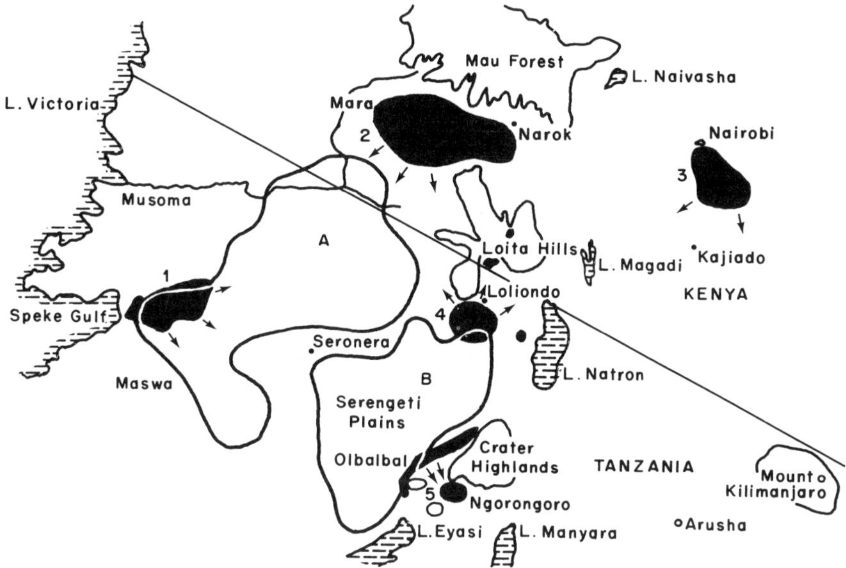

FIG. 2. The main wildebeest populations of east Africa (with acknowledgements to Watson, 1967). A, dry season dispersal area of the main Serengeti migrant population (July—November). B, wet season concentration area of Serengeti migrant wildebeest: (1) Kirawira resident group; (2) Mara National Park: Loita Plains group; (3) Kajiado: Nairobi National Park group; (4) Loliondo resident group; (5) Ngorongoro resident group. Arrows indicate the direction of dry-season dispersal.
Reprinted from Plowright & McCulloch (1967) with acknowledgements to *Journal of Hygiene, Cambridge*.

TABLE V

Rinderpest antibody in Serengeti migratory wildebeest (Ol Balbal) — sampled 1962—1966

Year of birth	No. positive/ No. tested 1962—1963	% Positive	No. positive/ No. tested 1965—1966	% Positive
⩽1958	38/44	86	—	—
1959	33/37	89	39/47	83
1960	25/33	76	13/15	87
1961	54/80	67	7/11	64
1962	3/15	20	1/16	6
	Plowright & McCulloch (1967)		Taylor & Watson (1967)	

figures were in close agreement, showing a decline in the proportion of animals with evidence of past infection from 1959—1960, when positives frequently exceeded 80%, to an overall 13% in animals born in 1962. As expected from studies in other species, rinderpest neutralizing antibody was remarkably persistent in recovered animals. It does not, like some antibodies, disappear quickly from a significant proportion of animals. The most remarkable example of its enduring character is probably provided by the hippopotamus (*Hippopotamus amphibius*) in western Uganda, which probably retained antibody for more than 30 years after natural epizootics (Plowright *et al.*, 1964). It was shown by titration that the antibody titres in Serengeti migrant wildebeest probably declined rapidly at first following severe infection but then stabilized by two years or so (Plowright & McCulloch, 1967) at levels which were maintained for at least three years.

The evidence for infection in 1961 of more than 60% of animals born early that year was not accompanied by any reports of disease or mortality in wildebeest, although both were observed at this time in buffalo around the Ngorongoro Crater (Kinloch, 1963). Similarly, no reports of disease were received in 1962 but the ususal annual mortality was seen in the adjacent Mara District of Kenya. Serological results from the latter area, which has a partially resident population, are given in Table VI. Again it was evident that infection was very frequent in 1960—1961, declining perhaps in 1962 and absent in 1963.

TABLE VI

Rinderpest antibody in wildebeest of the Mara District, Kenya — sampled 1960—1966

Year of birth	No. positive/ No. tested 1960	% Positive	No. positive/ No. tested 1966	% Positive
≤1957	12/13	92	—	—
1958	3/3	—	—	—
1959	5/6	83	—	—
1960	0/6[a]	—	10/12	83
1961	—	—	3/4	75
1962	—	—	2/13	15
1963	—	—	0/11	—
	Plowright & McCulloch (1967)		Taylor & Watson (1967)	

[a] Calves 5—8 months old.

Other non-migrant groups, including one (10—15 000) of which some (35% possibly) commute seasonally between the Ngorongoro Crater (July—December) and the Ol Balbal plains immediately to the west (January—June), were also investigated and details are given in Table VII. It was confirmed that widespread infection occurred in 1960, again in1961 but spreading slowly, and finally late in 1962 (August onwards). The frequency of infection in the Ngorongoro

TABLE VII

Rinderpest antibody in wildebeest of the Mara District, Kenya — sampled 1960— area — 1961—1966

Year of birth	No. positive/ No. tested 1961—1963	% Positive	No. positive/ No. tested 1966	% Positive
⩾1958	19/19[a]	100	—	—
1959	6/8	75	—	—
1960	2/3	66	9/10	90
1961	6/52	12	1/1	—
1962				
(Ol Balbal)	3/19	—	4/6	66
(Crater)	0/24			
1963	0/26	—	0/1	—
1964	—	—	0/6	—
1965	—	—	0/5	—
	Plowright & McCulloch (1967)		Taylor & Watson (1967)	

[a] All figures refer to animals beyond the age when passively acquired antibody may still have been present, i.e. 5—6 months.

resident population born in 1962 and sampled in 1966 (66%) contrasted with the very much lower figures for the comparable Serengeti migrant and Mara animals (6—20%).

The Loliondo District of northern Tanzania also has a resident wildebeest population, of about 5000 head, which was investigated in 1965—1966 following localized outbreaks of rinderpest in cattle, after a period of three years without detection of the disease in the country (Taylor & Watson, 1967). The results are shown in Table VIII together with those for another small resident group (Kirawira) immediately to the north of the Serengeti National Park.

TABLE VIII

Rinderpest antibody in resident wildebeest of the Loliondo and Kirawira groups, 1965—1966 (from Taylor & Watson, 1967)

Locality	Year of birth	No. positive/ No. tested	Locality	Year of birth	No. positive/ No. tested
Kirawira	1960	4/10	Loliondo	1960	4/6
	1961	1/4		1961	—
	1962	0/4		1962	0/4
	1963	0/3		1963	1/2
	1964	0/8		1964	1/16
	1965	0/4		1965	0/14
	1966	0/19		—	—

Whilst infection was again widespread in calves born in 1960 and possibly 1961, it was surprising to find single positives at Loliondo in animals born in 1963 and 1964 but none in the 1965 calf crop. The mild virus recovered from cattle in these investigations (Taylor & Watson, 1967) did not, therefore, establish itself in wildebeest, although it had probably remained enzootic in the Loliondo area since the 1960–1961 epizootics. No rinderpest has been reported in Tanzania after 1966.

THE OCCURRENCE OF RINDERPEST INFECTION IN OTHER GAME ANIMAL SPECIES IN EAST AFRICA

The wildebeest has a particular importance in rinderpest epizootiology because it occurs in such large numbers, in the Serengeti region particularly. Other species have, as we noted earlier, considerable significance because like buffalo, warthog, eland and giraffe they are highly susceptible to many strains of virus, exhibit severe disease and frequently die.

It was not, unfortunately, possible to sample these species for serological investigations on an adequate scale, if at all, during the time when rinderpest was still enzootic. The first collections of buffalo sera were in fact obtained in 1967, as a result of a planned cropping operation, but following an error in preservation only about 30 of them were suitable for employment in the tests then available for rinderpest antibody. The results for these and others collected in the following years are given in Table IX. Rinderpest antibody was

TABLE IX

Rinderpest neutralizing antibody in the sera of buffaloes in the Serengeti National Park

Year of sampling	Age at collection of serum						
	$\leqslant 5$ months	6–12 months	1–2 years	3 years	4 years	5 years	$\geqslant 6$ years
1967	1/1	5/5	0/5	0/1	1/2		15/19
1968/69	0/1	0/7	0/6	0/3	0/3	1/4	11/19

present in 27 out of 42 buffaloes which were five years or more old; one four-year-old animal was also positive in 1967, showing that infection had occurred in 1963. No infections were detected in the years 1964 to 1969, although passively acquired antibody was present in some calves (up to 12 months old). The statistical validity of epidemiological conclusions based on such small numbers is of course questionable but the data are at least consistent with what had been observed earlier in wildebeest. If larger collections had been made available they would have been well received!

Only the figures for one other species will be presented here. These are for warthogs collected in the period 1967—1970 and a comparison is shown in Table X between animals from the Serengeti

TABLE X

Rinderpest neutralizing antibody in the sera of warthogs in the Serengeti and Queen Elizabeth National Parks

Park	Years	Rinderpest antibody in warthogs of age group					
		$\leqslant 5$ months	6—12 months	13—24 months	25—36 months	37—48 months	$\geqslant 48$ months
Serengeti (Tanganyika)	1967— 1969	$1/5^a$ (1967)	$1/15^b$ (1969)	$1/15^b$ (1969)	$1/13^b$ (1969)	1/12	3/10 (all 1969)
Queen Elizabeth (Uganda)	1968— 1970	0/13	0/23	0/12	0/19	0/32	0/3

[a] Positive was a two-months-old animal, antibody possibly colostral.
[b] Titres were very low, $\leqslant 0.4$ ($\log_{10} VN_{50}$), probably non-specific.

and others from the Ruwenzori (Queen Elizabeth) National Parks. In Tanganyika, three out of 10 animals which were four or more years old in 1969 had antibody to rinderpest virus but it was surprising to find some animals with low-titre antibody as late as 1969 at a time when they were at most two to three years old. It is impossible to say whether or not these indicated very low level infection but no similar activity was detected in sera from Uganda where the virus was last reliably reported in 1944—1945 (Plowright *et al.*, 1964).

THE EFFECT OF RINDERPEST ON THE POPULATION DYNAMICS OF GAME ANIMALS IN THE SERENGETI REGION

The absence of data and limitation of space preclude consideration of many species, but two, the wildebeest and buffalo, have been well enough studied to make possible an assessment of the effects of rinderpest. In 1961, Talbot & Talbot noted in the Narok (Mara) District of Kenya that predation and disease accounted during 1959—1961 for the loss of about 85% of the annual wildebeest calf crop, predators accounting for 45% and disease for 40%. The latter was an annual event, beginning in October or November and continuing through to January; Talbot & Talbot (1961) postulated that prolonged drought, extended migration, predator activity and the flush of short grass induced by rain at this season, could have contributed a "stress factor" which increased mortality due to the disease. Indeed, it is possible that inapparent rinderpest infections, such as were proved to have occurred in the Serengeti and Ngorongoro areas in 1961 and 1962, may have been associated with a reduction in these contributory causes. The annual "wildebeest disease" was proved, as already shown, to be primarily associated with rinderpest virus infection (see Plowright, 1963b; Plowright & McCulloch, 1967).

The wildebeest and buffalo populations of the Serengeti region have been monitored periodically since the late 1950s and the results were summarized recently by Sinclair (1977) with adjustments for differences in counting techniques (Figs 3, 4). Between 1961, when

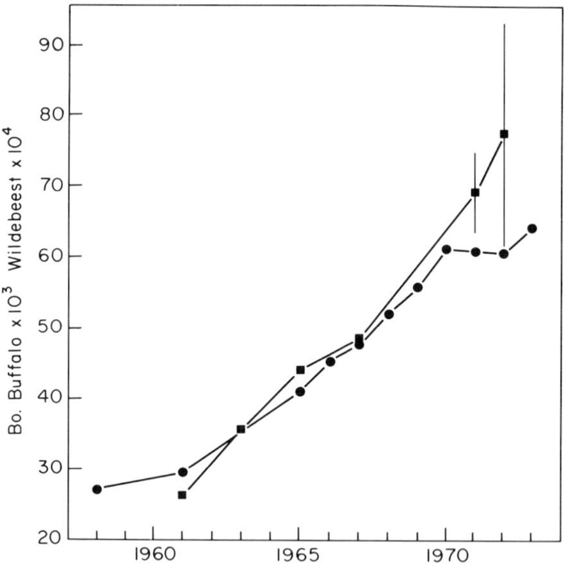

FIG. 3. The observed increases in buffalo and wildebeest populations (circles and squares) in the Serengeti National Park, 1958–1973. Vertical lines are 95% confidence limits. From Sinclair (1977), University of Chicago Press.

widespread rinderpest last occurred in the main Serengeti migrant herds, and 1965, the numbers of wildebeest increased from 263 000 to 439 000; by 1972 the estimate had reached 773 000. Sinclair

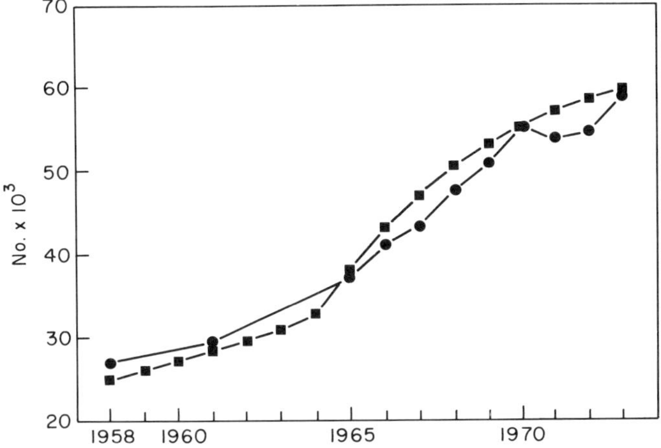

FIG. 4. The buffalo population of the Serengeti National Park predicted from a model with constant rinderpest mortality prior to 1964: comparison with observed figures. ■ — ■, population with rinderpest mortality before 1964, thereafter regulated by adult mortality alone. ● — ●, observed number of adults and yearlings. From Sinclair (1977), University of Chicago Press.

(1977) calculated that, when rinderpest was acting, 1.73 times more yearling wildebeest died than when it was absent. So far as the buffalo is concerned, he estimated that the Serengeti population increased from about 29 500 in 1961 to 61 600 by 1970. In virulent rinderpest outbreaks such as that in 1959 in the Lake Manyara area (outside the Serengeti Park) he was of the opinion that 50% of the total buffalo population could have been killed whereas in other areas, such as the Tarangire and Mount Meru reserves, rinderpest had disappeared by the early 1950s and populations had already stabilized at relatively high levels.

Going back even further, Sinclair considered that the Serengeti and neighbouring buffalo populations did not recover from the 1889 holocaust and subsequent rinderpest epizootics until the 1960s and that rinderpest was the major "external regulator" of buffalo numbers. His projections backwards of trend lines from 1965 led to the conclusion that there was a major shift in juvenile mortality at about that time and this was a key factor in the subsequent population expansion. Rinderpest or, more accurately, its removal by veterinarians, was the artificial perturbation which caused the increase in population from the early 1960s and facilitated analysis of its regulation by factors, such as rainfall and food supply, which exert their effects on older animals.

EPIDEMIOLOGICAL CONSIDERATIONS OF THE DISAPPEARANCE OF RINDERPEST FROM THE SERENGETI AREA

In 1963, the overwhelming evidence of a focus of rinderpest infection in game animals of the Serengeti region which had persisted for over 30 years led me to write

> it is impossible to say whether rinderpest would disappear if cattle-game contacts were broken, but I think it is perfectly feasible to suppose that the Serengeti game population, taken as a whole, is capable of continuous maintenance of the virus, even in the absence of cattle.

Events have proved otherwise and it may be instructive to look for the reasons.

The relevant population, taken as a whole, included at that time not only wildebeest (330 000) and buffalo (20 000) but also about 600 000 Thomson's gazelle, 20 000 topi, 10 000 giraffe and impala, and 5 000 hartebeest (Watson, 1965). The young are protected from infection in the first few weeks or months of life by antibody absorbed from the colostrum of the dam. When this disappears, at a time dependent on the variable initial level and rate of catabolism (which is, broadly speaking, inversely proportional to the body size) the offspring become susceptible to infection.

The number of susceptibles at any one moment therefore depends on the time of year (e.g. wildebeest calves usually begin to become susceptible, having lost antibody, about April and virtually all are vulnerable by September of the year of birth) and the rate of recruitment to the population. According to Watson (1966) this latter may approach 50% of the adult population at birth, reducing to 30% at one year old; assuming the latter figure and a total Serengeti migrant population of 330 000, then there could have been *c*. 100 000 susceptible yearling wildebeest in the area by August/September 1963. In addition there was an added proportion of two-year-old or more mature animals which had escaped in previous outbreaks.

The density of the population also varies greatly between a low point in the dry season dispersal area, increasing on the wet season concentration and breeding ground. There is also a remarkable concentration during the twice-yearly migration (June and November). Transfer of an infection such as rinderpest can obviously take place more easily as population density increases. Finally, the calving or farrowing season varies for different species; wildebeest, for example, in the Serengeti area calve mainly in January (December to February) and warthogs farrow predominantly in September and October. In spite of these variables it seems reasonable to suppose that six to

12 months after the last outbreak of rinderpest a total susceptible population of at least 120—150 000 would have been present in the Serengeti region.

Bearing these factors in mind, it is remarkable that rinderpest which last affected buffalo in 1963 (Sinclair, 1977) did not spread at that time to other susceptible species, as it had done virtually every year from the early 1930s. If we look at the critical size required for the maintenance of a closely-related virus, i.e. measles of man, in island communities, we find that it is now well established that breaks in endemicity occur when total populations fall below 500 000 (Black, 1966). The transmission chain is broken in smaller communities because the virus is very labile outside the body, does not generally persist in the convalescent or recovered individual, confers long-lasting immunity and in this case does not have alternative natural hosts (Matumoto, 1969). However, man is much longer-lived and reproduces far more slowly than wild ungulates and Bartlett (1957, 1960) found, in *urban* communities in Britain and the USA, that fade-out occurred when there were fewer than 4000—5000 cases per year, the minimal size for a city to produce these being 300 000 inhabitants. It is therefore even more surprising that accumulations of more than 100 000 susceptible game animals in relatively small areas could not maintain rinderpest.

In the Serengeti region, at the periphery of which cattle—game contacts were frequent, it is possible that for the decades up to 1962—1963 a feed-back mechanism, of cattle-to-game and game-to-cattle transmission, was necessary to maintain the virus continuously. Such transfers as may have occurred in wildebeest in later years, as from cattle in 1965—1966, or from buffalo in 1963, failed to establish a chain of infection and outbreaks remained confined to small foci, from which the virus rapidly died out. A similar situation presumably obtains still in small communities of game animals where rinderpest is still enzootic in cattle.

RINDERPEST AND THE FUTURE OF LARGE BOVID POPULATIONS IN EAST AFRICA

The present situation in east Africa is a dangerous one in that there are now very few rinderpest-immune animals left in the vast herds of the Serengeti region. Species of the highest susceptibility (buffalo, eland, warthog, wildebeest) are probably present in greater numbers than at any time since the great panzootic. Rinderpest virus periodically invades Kenya and Uganda from the north and it is not difficult to imagine how infection could spread across the Tana and Galana Rivers down the coast and into the Tsavo, Amboseli or Kajiado Districts. This could be effected by illicit or undetected movements of infected cattle or by wild animals, particularly buffalo and eland which, as already noted, migrate more actively when the herds are infected. Once established in the Kenya Parks it would be difficult to prevent spread into the Loliondo and Mara areas and finally into the Ngorongoro and Serengeti Parks. In such a scenario it would be futile to question the "regulatory" potential of disease, as has sometimes been the tendency of ecologists in rinderpest-free environments (Eltringham, 1979). Vigorous and rapid attempts would be required to limit game and cattle movements, similar to those successfully adopted on previous occasions in Tanganyika, to prevent the spread of rinderpest to southern Africa (1913—1921, 1937—1941). Should an invasive and virulent strain of virus become established again in the Serengeti area the losses could be enormous — of the order of tens of thousands of buffalo alone.

It is now possible to safely vaccinate buffalo, eland and oryx for example with the standard culture vaccine used for cattle (Plowright & King, unpublished) but this needs to be administered parenterally and there could be no reasonable hope of applying such a measure on an adequate scale in the foreseeable circumstances of an outbreak. Harthoorn & Lock (1960) estimated that a team of four people could immobilize and vaccinate 20—30 buffalo daily!

The only way to protect wildlife reliably and over the longer term in Africa is to eliminate rinderpest from cattle. This, as we have seen, has been accomplished in Tanzania and southern Kenya but north-east Africa continues to present a source of virulent strains of virus which are not yet adequately controlled. Until this and other reservoirs of the virus, in combined cattle—game communities, are finally eliminated from the continent it would be unwise to assume that fences many miles long and the elimination of game from belts on either side (see Scott, 1970) will never be required in the future.

Finally, if a rinderpest-like disease should afflict again the animals in major National Parks in Africa, it is essential that access should be given immediately to qualified scientific staff to investigate its cause and to collect pathological and epidemiological data on an adequate scale. Sometimes, unfortunately, in the past cooperation has been given grudgingly, after a delay, and with mutual suspicion; the dangers of such attitudes in the present situation in east Africa should be obvious to all.

REFERENCES

Bartlett, M. S. (1957). Measles periodicity and community size. *Jl R. statist. Soc.* (A) **120**: 48–60.

Bartlett, M. S. (1960). The critical community size for measles in the United States. *Jl R. statist. Soc.* (A) **123**: 37–44.

Black, F. F. (1966). Measles endemicity in insular populations. *J. theoret. Biol.* **11**: 207–211.

Branagan, D. (1965). Observations on post-vaccinal sequelae to rinderpest vaccination using caprinised vaccine on cattle in Tanganyika Masailand. *Bull. epiz. Dis. Afr.* **13**: 5–10.

Branagan, D. (1966). Behaviour of buffalo infected with rinderpest. *Bull. epiz. Dis. Afr.* **14**: 341–342.

Branagan, D. & Hammond, J. (1965). Rinderpest in Tanganyika: a review. *Bull. epiz. Dis. Afr.* **13**: 225–246.

Brown, R. D. & Rashid, A. (1965). Duration of rinderpest immunity in cattle following vaccination with caprinised rinderpest virus. *Bull. epiz. Dis. Afr.* **13**: 311–315.

Buxton, P. A. (1955). *The natural history of tsetse flies*. London: H. K. Lewis.

Carmichael, J. (1933). The virus of rinderpest and its relation to *Glossina morsitans* (Weston). *Bull. ent. Res.* **24**: 337–342.

Carmichael, J. (1938). Rinderpest in African game. *J. comp. Path.* **51**: 264–268.

Cornell, R. L. (1934). Rinderpest in wildebeest. In *Rep. Dept. vet. Sci. anim. Husb., Tanganyika* 1933: 37–42.

Curasson, G. (1932). *La peste bovine*. Paris: Vigot Frères.

Dawe, E. C. S. (1957). In *Rep. vet. Dept., Tanganyika* 1956.

DeTray, D. E. (1970). Joint campaign against rinderpest in Africa. *Proc. A. Meet. U.S. anim. Hlth Assoc.*, **74**: 225–229.

Duke, H. L. (1919). An enquiry into the relations of *Glossina morsitans* and ungulate game, with special reference to rinderpest. *Bull. ent. Res.* **10**: 7–20.

Eltringham, S. K. (1979). *The ecology and conservation of large African mammals*. London and Basingstoke: Macmillan.

Ford, J. (1971). *Role of the trypanosomiases in African ecology*. Oxford: Clarendon Press.

Grzimek, M. & Grzimek, B. (1960). Census of plains animals in the Serengeti National Park, Tanganyika. *J. Wildl. Mgmt* **24**: 27–37.

Harthoorn, A. M. & Lock, J. A. (1960). A note on the prophylactic vaccination of wild animals. *Br. vet. J.* **116**: 252–254.

Hornby, R. E. (Ed.) (1939). *Rep. Dept. vet. Sci. anim. Husb., Tanganyika* 1938.

Johnson, R. H. (1962). Rinderpest in tissue culture. Use of the attenuated strain as a vaccine for cattle. *Br. vet. J.* **118**: 141–150.

Kinloch, B. G. (1963). In *Rep. Game Dept., Tanganyika* 1961.

Libeau, J. & Scott, G. R. (1960). Rinderpest in eastern Africa to-day. *Bull. epiz. Dis. Afr.* **8**: 23–26.

Liess, B. & Plowright, W. (1964). Studies on the pathogenesis of rinderpest in experimental cattle. I. Correlation of clinical signs. viraemia and virus excretion by various routes. *J. Hyg., Camb.* **62**: 81–100.

Lowe, H. J. (1942). Rinderpest in Tanganyika Territory. *Emp. J. exp. Agric.* **10**: 189–202.

Lowe, H. J., Wilde, J. K. H., Lee, R. P. & Stuchbery, H. M. (1947). An outbreak of an aberrant type of rinderpest in Tanganyika Territory. *J. comp. Path.* 57: 175—183.

Mack, R. (1970). The great African cattle plague epidemic of the 1890s. *Trop. anim. Hlth Prod.* 2: 210—219.

Matumoto, M. (1969). Mechanisms of perpetuation of animal viruses in nature. *Bacteriol. Rev.* 33: 408—418.

Pecaud, G. (1924). Contribution à l'étude de la pathologie vétérinaire de la Colonie du Tchad. *Bull. Soc. Path. exot.* 17: 196—207.

Plowright, W. (1957). Recent observations on rinderpest immunisation and vaccines in Northern Nigeria. *Br. vet. J.* 113: 385—399.

Plowright, W. (1963a). Some properties of strains of rinderpest virus recently isolated in E. Africa. *Res. vet. Sci.* 4: 96—108.

Plowright, W. (1963b). The role of game animals in the epizootiology of rinderpest and malignant catarrhal fever in East Africa. *Bull. epiz. Dis. Afr.* 11: 149—162.

Plowright, W. (1972). The standardisation of procedures for the production of rinderpest vaccines and for the laboratory confirmation of rinderpest diagnosis. *Cento Seminar on Viral Diseases, Istanbul, Turkey:* 48—54.

Plowright, W. & Ferris, R. D. (1959a). Studies with rinderpest virus in tissue culture I. *J. comp. Path.* 69: 152—172.

Plowright, W. & Ferris, R. D. (1959b). Studies with rinderpest virus in tissue culture II. *J. comp. Path.* 69: 173—184.

Plowright, W. & Ferris, R. D. (1961). Studies with rinderpest virus in neutralisation tests. *Arch. ges. Virusforsch.* 11: 516—533.

Plowright, W. & Ferris, R. D. (1962a). Studies with rinderpest virus in tissue culture IV. A technique for the detection and titration of virulent virus in cattle tissues. *Res. vet. Sci.* 3: 94—103.

Plowright, W. & Ferris, R. D. (1962b). Studies with rinderpest virus in tissue culture. The use of attenuated culture virus as a vaccine for cattle. *Res. vet. Sci.* 3: 172—182.

Plowright, W., Laws, R. M. & Rampton, C. S. (1964). Serological evidence for the susceptibility of the hippopotamus (*Hippopotamus amphibius* Linnaeus) to natural infection with rinderpest virus. *J. Hyg., Camb.* 62: 329—336.

Plowright, W. & McCulloch, B. (1967). Investigations on the incidence of rinderpest virus infection in game animals of N. Tanganyika and S. Kenya 1960/63. *J. Hyg., Camb.* 65: 343—358.

Plowright, W. & Taylor, W. P. (1967). Long-term studies of the immunity in East African cattle following inoculation with rinderpest culture vaccine. *Res. vet. Sci.* 8: 118—128.

Reid, N. R. (1949). In *Rep. Dept. vet. Sci. anim. Husb., Tanganyika* 1948.

Reid, N. R. (1951). In *Rep. Dept. vet. Sci. anim. Husb., Tanganyika* 1950.

Robson, J., Arnold, R. M., Plowright, W. & Scott, G. R. (1959). The isolation from an eland of a strain of rinderpest virus attenuated for cattle. *Bull. epiz. Dis. Afr.* 7: 97—102.

Scott, G. R. (1964). Rinderpest. *Adv. vet. Sci.* 9: 113—224.

Scott, G. R. (1970). Rinderpest. In *Infectious diseases of wild mammals:* 20—35. Davis, J. W., Karstad, L. H. & Trainer, D. O. (Eds). Iowa: Iowa State University Press.

Simon, N. (1962). *Between the sunlight and the thunder.* The wild life of Kenya. London: Collins.

Sinclair, A. R. E. (1977). *The African buffalo: A study of resource limitation of populations*. Chicago and London: University of Chicago Press.

Stevenson-Hamilton, J. (1911). The relation between game and tsetse flies. *Bull. ent. Res.* 2: 113.

Stevenson-Hamilton, J. (1957). Tsetse fly and the rinderpest epidemic of 1896. *S. Afr. J. Sci.* 58: 216–218.

Stewart, D. R. M. (1964). Rinderpest among wild animals in Kenya 1960–62. *Bull. epiz. Dis. Afr.* 12: 39–42.

Stewart, D. R. M. & Talbot, L. M. (1962). Census of wildlife on the Serengeti, Mara and Loita Plains. *E. Afr. agric. for. J.* 28: 58–60.

Talbot, L. M. & Talbot, M. H. (1961). Preliminary observations on the population dynamics of the wildebeest in Narok District, Kenya. *E. Afr. agric. for. J.* 27: 108–116.

Taylor, W. P. & Watson, R. M. (1967). Studies on the epizootiology of rinderpest in blue wildebeest and other game species of Northern Tanzania and Southern Kenya, 1965–67. *J. Hyg., Camb.,* 65: 537–545.

Thomas, A. D. & Reid, N. R. (1944). Rinderpest in game: a description of an outbreak and an attempt at limiting its spread by means of a bush fence. *Onderst. J. vet. Sci. anim. Ind.* 20: 7–23.

Vaughan-Jones, T. (1953). Notes on the rinderpest fence and cordon — Northern Rhodesia-Tanganyika Border. *Bull. epiz. Dis. Afr.* 1: 286–290.

Watson, R. M. (1965). Game utilisation in the Serengeti: preliminary investigations. Part I. *Br. vet. J.* 121: 540–546.

Watson, R. M. (1966). Game utilisation in the Serengeti: preliminary investigations. Part II. Wildebeeste. *Br. vet. J.* 122: 18–27.

Watson, R. M. (1967). *The population ecology of the wildebeest* (Connochaetes taurinus albojubatus *Thomas) in the Serengeti*. PhD Dissertation: University of Cambridge.

White, G. (1958). A specific diffusible antigen of rinderpest virus demonstrated by the agar double-diffusion precipitation reaction. *Nature, Lond.* 181: 1409.

Wilde, J. K. H. (1953). The game animal factor in the control of rinderpest in tropical Africa. *Int. vet. Congr.* 15 (1): 283–287.

DISCUSSION

Steck — I have two questions. First, do you have strong evidence that the detection of antibody equates with immunity? Secondly, the live vaccine which was used, may it perhaps have spread into the game animal population and have been producing immunity by contact?

Plowright — None of the vaccine strains spreads experimentally from cattle to cattle by natural routes; so I think you can eliminate that. The answer to the other question is that neutralizing antibody is as stable as that in measles virus infection of man and equally well indicates, if it is actively produced, the immune status of the individual. Rinderpest-neutralizing antibody is extraordinarily long-lived, and for example in the hippopotamus was found to persist for as much as 30 years after probable infection; furthermore, it was still present to a reasonably high titre. A similar stability is also found for measles antibody in man. We think that actively acquired antibody indicates invariably a state of immunity to rinderpest.

Nelson — I witnessed a very devastating epidemic of rinderpest in northern Uganda, in the West Nile District, in 1955—1956, which eliminated the buffalo, and I can confirm that it also eliminated the tsetse flies. I was Medical Officer of Health during that period, and I was obliged to set aside one ward for patients who were injured by buffaloes which were suffering from rinderpest. But the interesting point is that it was on the west bank, where the white rhinoceros occurred, and rinderpest didn't seem to affect this species or the elephant. There may even have been a suggestion that the elephant population suddenly exploded across the river in the Murchison National Park because rinderpest had eliminated most of the buffalo population from the area at that time.

Plowright — An interesting point. There is no evidence, of course, that rhinoceros or elephant are susceptible to rinderpest. We've looked at some sera and have never found specific antibody, but perhaps we did not look at enough sera. This business was always bedevilled by the small numbers of sera available, except in some of those cropping programmes which were carried out by NUTAE, for example, in the lakes in western Uganda and up in the Murchison Park. This was also true in the Serengeti area later on, where we did get collaboration which was really effective. In the early years it was conspicuous by its absence in the Serengeti National Park.

Jewell — Dr Plowright emphasized the need to be prepared for possible new outbreaks of rinderpest, particularly amongst the huge populations of hoofed animals in the Serengeti. I would like to emphasize how important it is that more background information on the ecology of these populations should be obtained. It is predicted that the wildebeest may rise to three million if good rainy seasons continue. This density has a profound effect on the grazings and, in a complicated way, on the carrying capacity. Vegetational litter is reduced, fires are less frequent and more forage appears to be available for many species that are also increasing in numbers. Obviously these population events could profoundly affect the manner in which rinderpest may spread. With the importance of this ecosystem in mind, biologists at Cambridge University are promoting a cooperative scheme of research with Tanzania.

Plowright — I have in fact recommended that the Serengeti Research Institute should try to get staff who are capable of doing the kind of collaborative studies with local institutes that we were able to do earlier, in the 1960s particularly. They do not have them now, to my knowledge at least.

Kaplan — If another plague occurred, is there any possibility that your tissue culture vaccine could be used to immunize a significant fraction of the wildlife population ahead of the plague?

Plowright — In my written paper I quoted some figures of Harthoorn & Lock (1960).[1] In a day's work they could immunize 30 to 40 buffalo, and I just wonder how many operators would be killed.

Kaplan — I meant by darts.

Plowright — I suspect that even using darting teams they could never cover the numbers required. I believe it would be essential to do what was done before, i.e. to clear a belt at least tens of miles wide across a wide stretch of country and maintain it free of game species, especially the susceptible ones. During the last war they built a game-proof fence across the south of Tanzania from Lake Rukwa to the northern end of Lake Nyasa, and it was expensive enough at that time: it would obviously cost a great deal more now. But something of this character would, I am sure, be necessary if a virulent strain of rinderpest were really to get on the move down from the north of Kenya towards the Serengeti.

---

[1] See list of references, pp. 25—27.

Symp. zool. Soc. Lond. (1982) No. 50, 29–55

Trypanosomes, Trypanosomiasis and Tsetse Control: Impact on Wildlife and its Conservation

D. H. MOLYNEUX

Department of Biology and Centre for Development Studies, University of Salford, Salford, M5 4WT, England

SYNOPSIS

Trypanosomes of the genus *Trypanosoma* (Protozoa: Kinetoplastida) are parasites found in the blood and tissues of many vertebrate hosts. They are, however, important pathogens of man and domestic livestock in various parts of the world. The African pathogenic trypanosomes are transmitted by the bite of an infective tsetse fly of the genus *Glossina*. There are 22 species of *Glossina* and all are capable of transmitting trypanosomes. Many species of African game mammal act as reservoir hosts of pathogenic trypanosomes but show few if any signs of these infections. Similarly, other vertebrate hosts (fishes, birds, reptiles, amphibians and various mammalian orders) which are naturally infected suffer no serious ill effects.

Glossina control, in association with rural development in tsetse-infested areas, has and will continue to have profound effects on wildlife. This is exemplified by the game destruction policies for tsetse control carried out in Zimbabwe during the 1940s and 1950s or directly through deliberate modification of the fly habitat by vegetation removal which has been shown to be an effective tsetse control measure but which will have a deleterious effect on wildlife. The use of pesticides for *Glossina* control, although it has not been shown to seriously endanger large mammals, causes mortality in lower vertebrates and non-target invertebrates, particularly insects.

Appreciation of the complexity of the problem of tsetse and trypanosomiasis has resulted in the evaluation of wider ecological, geographical, agricultural and other factors in the planning of trypanosomiasis control (of which tsetse control is a component part) particularly the definition of the most effective land-use strategy. Man and his activities will make a considerable impact on wildlife as well as on tsetse populations. Solutions to particular trypanosomiasis problems require detailed planning on the basis of economic and scientific judgements; such planning must account for environmental and conservation considerations.

INTRODUCTION

Trypanosomes of the genus *Trypanosoma* are the causative agents of a variety of diseases – trypanosomiases found in man and domestic livestock. They are flagellated parasitic Protozoa of the Order Kinetoplastida. All vertebrate classes are parasitized by trypanosomes but the infections in most of these vertebrate hosts are not associated with pathogenicity or disease conditions. Non-mammalian trypanosomes have been studied less often than their mammalian congeners but even in the mammals many infections have little or no pathogenic effect and it is only when hosts are subject to 'stress' or have been, in evolutionary terms, only recently associated with trypanosomes that pathogenicity and disease result.

In this Symposium attention is focussed on animal diseases which are relevant to animal conservation, the effects of such diseases on fluctuations of natural populations of animals and their pathogenic effects when transmitted to other species. The trypanosomal diseases and the vectors which transmit them are particularly relevant in this context in view of widespread natural infections of trypanosomes, the role of wildlife as reservoirs of animal and human trypanosomiasis and the devastating consequences which these diseases may have on man and his livestock. However, the control of the trypanosomiases through vector control is itself highly relevant to this meeting in view of the methods applied to control the insect. Such methods as selective game destruction, habitat destruction through clearing the vegetation ("bush clearing"), or more recently the use of pesticides applied from the ground or from the air all have implications for the wildlife species in the variety of habitats where such activities have been or will be carried out. In addition, man's own activities influence these diseases by reducing game populations through hunting, removing wood for fuel and generally changing land use patterns during development.

BIOLOGY OF TRYPANOSOMES IN GAME ANIMALS

The trypanosomiases which will form the bulk of the subject matter of this chapter are those organisms transmitted cyclically by species of *Glossina*, the tsetse flies (Diptera: Glossinidae), between game animals of Africa. When these organisms infect domestic animals and man they cause the diseases nagana and sleeping sickness respectively. There is a considerable volume of information on the infections of wild game animals with the various sub-genera and species of the genus *Trypanosoma* (Ashcroft, 1959; Wells & Lumsden, 1968; Baker, 1968; Keymer, 1969; Geigy & Kauffmann, 1973; Allsopp, 1972a; Carmichael & Hobday, 1975; Drager & Mehlitz, 1978; Dillman & Townsend, 1979; Mehlitz *et al.*, in press). There are five sub-genera of trypanosomes which have been found to infect what are known as "game" animals. These are *Trypanozoon* (e.g. *Trypanosoma brucei*), *Nannomonas* (e.g. *T. congolense*), *Duttonella* (e.g. *T. vivax*), *Pycnomonas* (e.g. *T. suis*) and *Megatrypanum* (e.g. *T. ingens*). The last two sub-genera are relatively unimportant in terms of either the numbers of animals infected and restricted distribution (*T. suis*) or of limited, if any, pathogenicity in the case of *Megatrypanum* species (*T. theileri*-like organisms) which are cyclically transmitted by other Diptera such as *Tabanus* by contamination and not by bite (Keymer, 1969; Hoare, 1972).

The sub-genera *Trypanozoon*, *Nannomonas* and *Duttonella* are salivarian trypanosomes transmitted by species of *Glossina* as a result of the bite of infective flies. The cycle of development of trypanosomes in *Glossina* involves morphological and physiological changes. The factors that influence the establishment of infection, the mechanisms of transmission and infection rates in flies can be found in Buxton (1955), Mulligan (1970), Hoare (1972), Jordan (1974) and Molyneux (1977, 1980). Less common but sometimes of epizootiological importance is mechanical (or non-cyclical) transmission of infections which occurs as a result of contamination of mouthparts and interrupted feeding by biting flies such as *Stomoxys* and *Tabanus*; this tends to occur when infected animal populations are living together in high density; *T. (D.) vivax* and *T. (T.) evansi* (a close relative of *T. (T.) brucei*) are transmitted in this way outside the tsetse belt of Africa.

Although there have been extensive studies of trypanosome infections of game animals these are almost exclusively in eastern and central Africa. These studies have concentrated on the rates of infection with the different trypanosome sub-genera, the species of game infected and the zoonotic potential of different game species as reservoirs of human sleeping sickness. There has been no doubt for many years as to the importance of game animals as hosts of trypanosomes pathogenic to livestock. The publications and results of studies which describe trypanosome infection in game animals are referred to above. A cautionary note regarding such data should be given at this stage. Earlier studies in Mozambique and Zimbabwe relied exclusively on thick blood film examination for the detection of infections; there is no doubt (see Dillman & Townsend, 1979) that this technique is less sensitive when compared with present methods of parasitological diagnosis such as microhaematocrit centrifugation (Bennett, 1962; Woo, 1970), miniature anion exchange centrifugation techniques (MAECT) (Lumsden *et al.*, 1981) or animal inoculation techniques into mice, nursling rats or *Mastomys* which have been employed in more recent surveys (Geigy & Kauffmann, 1973;

Dillman & Townsend, 1979; Mehlitz *et al.*, in press). There is still no really suitable laboratory animal which can reliably be infected with *T. (D.) vivax* and thus a high proportion of infections with the subgenus *Duttonella* may escape detection; similarly it is known that there are stocks of *Trypanozoon* which do not readily grow in nursling rats or *Mastomys natalensis* (Drager & Mehlitz, 1978). Isolates obtained from subinoculation from wild or domestic animals to laboratory animals will also not necessarily be a representative sample of the trypanosome population circulating in the host from which the sample was taken.

There is strong evidence (Willett, 1972) that some trypanosome infections predispose cattle to infection with other species of trypanosome. Willett (1972) found an unexpectedly high frequency of multiple infections of *Trypanozoon*, *Nannomonas* and *Duttonella* in the Lambwe Valley, Kenya. Such multiple infections were much more frequent than would have been expected on the basis of chance infection but this was only observed if *T. (T.) brucei* was one of the infective agents. Animals with *T. (T.) brucei* were three times more prone to *T. (N.) congolense* infection and six times more prone to *T. (D.) vivax* than healthy cattle. A similar pattern of enhanced susceptibility, possibly due to the immunosuppressive effects of *T. (T.) brucei* group parasites (Goodwin, Green, Guy & Voller, 1972), is discerned in game surveys (Geigy, Mwambu & Kauffmann, 1971; Allsopp, 1972a). In the area around Lugala in Uganda, Wilson, Dar & Paris (1972) also found that despite low *Trypanozoon* infection rates in *Glossina* and relatively high *Duttonella* and *Nannomonas* infections *Trypanozoon* parasites always appeared first in naive cattle brought into the area.

There are a great many different trypanosome populations circulating in game animals, although overall infection rates as determined by parasitological methods seem to vary from area to area. There is evidence from serological studies (Drager & Mehlitz, 1978) on game animals of more frequent trypanosome infections by comparison with the results of Carmichael & Hobday (1975) who used less sensitive techniques on the same animal species in a similar area. Drager & Mehlitz (1978) believe that "serological examinations appear to give a more realistic picture of trypanosome infections in wildlife than do parasitological findings". The Drager & Mehlitz (1978) study is also one of the first to relate parasitological and serological findings to the age groups of a population of game. Six hundred and five buffalo, *Syncerus caffer*, were examined and divided into four age groups (one year, one—three years, four—seven years and over seven years). Buffalo calves under one year had low antibody titres and only one of 74 calves examined was parasitologically positive; in the one—three year group antibody titres rose and the percentage parasitologically positive was 34.4% of 90 animals; the four—seven years age group had still higher antibody titres which persisted in the older animals, but parasitologically positive animals dropped to 22.4% (four—seven years) and 4.3% in older animals. The suggestion was made that the infection was suppressed in one-year-old calves by innate resistance or maternal antibodies. This study is one of the few where a large number of animals were sampled and infection rates in age groups as well as serological studies were carried out giving a pattern of infection. Drager & Mehlitz (1978) were also able to correlate the relationship between parasitological and serological results and "high" and "low" fly density.

Techniques of game immobilization have improved over recent years and more information on trypanosome infections has been forthcoming from these advances (Allsopp, 1972b; Drager, Patterson & Breton, 1976; Kupper, Drager, Mehlitz & Zillman, 1981) because earlier studies had depended on game culling or game control exercises. It has been shown that compounds currently in use as immobilization agents have no effect on trypanosome viability (Rickman & Rottcher, 1980).

Although most infections of game animals can be attributed to the acquisition of infection by bites from *Glossina* it has been found that carnivores have an unexpectedly high infection rate with trypanosomes (Baker, Sachs & Laufer, 1967; Baker, 1968; Geigy & Kauffmann, 1973; Dillman & Townsend, 1979). Lion, *Panthera leo*, and hyaena, *Crocuta crocuta*, were the carnivores examined. Studies of *Glossina* blood meals from areas where carnivores are found reveal that very low percentages of blood meals are taken from them (Moloo, 1973). The suggestion that carnivores are infected when feeding on infected prey is supported by laboratory studies (Heisch, 1963; Moloo, Losos & Katuza, 1973) which have demonstrated the ease with which dogs and cats can be infected by eating meat from killed infected animals (goats). The relevance of these observations to large mammal biology in the Serengeti is discussed by Bertram (1973).

The paper by Bertram (1973) is important because it highlights the view of the role of trypanosome infections from the viewpoint of a game biologist rather than a parasitologist or epidemiologist. It points out the problems of representative sampling and the problems of interpretation of infection rate in relation to changes in parasitaemia. The likelihood of a high percentage of lions being infected with trypanosomes is to be expected as they live in closely knit permanent social groups; trypanosomes will enter through small wounds in buccal mucosa or through wounds on the nose or muzzle which become covered in blood when lions are feeding. If 7.5% (Geigy *et al.*, 1971) of prey is infected in the area surveyed lions will be frequently susceptible to becoming infected which will be enhanced because older or sick animals are more frequently preyed on and are likely to be more parasitaemic (sickness being either a direct effect of infection by trypanosomes or parasitaemia with trypanosomes being exacerbated by other infections). Lions could also infect each other by social grooming (licking wounds or sores) which occurs extensively within the pride. Although sick ungulates will be eliminated by predators rapidly, a sick lion (Sachs, Schaller & Baker, 1967; Mortelmans & Kageruka, 1971) with a comparatively high parasitaemia could be maintained within the pride, which would increase the likelihood of infection for other members of the group, through grooming or mechanical transmission by biting flies.

PATHOLOGY IN GAME AND DOMESTIC ANIMALS

Game Animals

There is very limited information on the effects of trypanosome infection on game animals. Only recently have controlled experiments been undertaken using game animals bred in captivity or captured in tsetse-free areas and infected with defined trypanosome populations. Earlier studies (Ashcroft, 1959) show that some species of game when experimentally infected succumb to infections of *Trypanozoon*, but limited store should be put on such experiments in view of the stressed condition of the animals. The scarcity of natural records of pathogenicity of trypanosome infections to game animals could be due to the tolerance of game to these parasites or to predation on any unfit animal (Bertram, 1973; Carmichael & Hobday, 1975). Reports of pathogenicity in two zebra with central nervous system involvement of *T. (T.) brucei* suggest that the disease was due to the trypanosome (McCulloch, 1967) and pathogenicity could be related to the infrequent exposure to trypanosomes through limited feeding by tsetse on zebra (Weitz, 1963). Sachs *et al.* (1967) found a sick lion cub and suggested a *Trypanozoon* infection was the agent responsible. Mortelmans & Kageruka (1971) infected two lion cubs with *T. (T.) brucei* in Europe and found small changes in haematological parameters associated with low grade parasitaemia.

However, Losos & Gwamaka (1973) have found significant histological lesions in two impala (despite low frequency of tsetse bites on this host), one Thomson's gazelle, three Coke's hartebeests and two lions, which they attributed to trypanosome infections because the myocarditis and meningoencephalitis found were similar to lesions of domestic stock caused by pathogenic trypanosomes. Recent work (Olubayo, 1979 and Murray, Grootenhuis *et al.*, 1980) has shown that captive-bred eland, though they become parasitaemic with *Trypanozoon* and *Nannomonas* parasites, do not show significant changes in blood parameters (packed cell volume, red blood cells, haemoglobin, white blood cells). Similarly in captive-bred bushbuck (*Tragelaphus scriptus*) no clinical signs have been observed after infection with *T. (T.) b. rhodesiense* (Etat 10) blood stream forms. This confirms the observations of Knottenbelt (1974) who found no pathogenic effects of trypanosomiasis in bushbuck and kudu shot in Zimbabwe. However, there is an early skin reaction (chancre) to *T. (N.) congolense* and *T. (T.) brucei* in eland and waterbuck (*Kobus defassa*); this reaction, however, is significantly reduced compared with that in domestic animals; only half the successful feeds by tsetse on eland and waterbuck resulted in chancres whereas in domestic livestock almost all tsetse bites produced chancres. Calculations of the direct effects *Glossina* have on individual animals in terms of the volume of blood removed per day is reported by Lamprey *et al.* (1962), who estimated a warthog would lose 13—27 g of blood per day to *G. swynnertoni* and bushbuck 7.5 g per day in *G. pallidipes* infested areas.

The use of animals which have been maintained for some time or bred in captivity will at least reduce the problem of stress in captivity such as immobilization, restraint and subsequent artificial confinement (Olubayo, 1979). Ferguson, Herbert & McNeillage (1970) have shown that stressed mice survived for longer periods when infected with *T. (T.) brucei* than single mice kept in isolation, thus suggesting social groupings can influence parasitaemia and survival (see also Jackson & Farmer, 1970). In view of the innate tolerance of game animals generally to trypanosome infections and their higher productivity in more arid environments, arguments for large-scale game meat production have been advanced. The undoubted higher productivity is countered by the problems and expense of harvesting such meat, the unsuitability of most species (except eland) as a source of dairy products, the marketing and acceptability problems, the public health aspects of game meat utilization and the need therefore for its supply to local rather than export markets. Field (1979) has extensively reviewed the current information on ranching, farming and cropping of game in Africa;

evaluation of productive potential of game in Africa is given by Reull (1979).

Pathology in Livestock

Despite the comparative lack of information on the pathogenic effects on game animals many studies on the pathology of salivarian trypanosomes in laboratory animals, cattle and small ungulates have been carried out. There are several pertinent reviews which summarize recent work (Ormerod, 1970; Goodwin, 1970, 1974; Losos & Ikede, 1972; Murray, 1974; IDRC-132e, 1979). Anaemia is the "cardinal sign" in trypanosomiasis in susceptible bovines from which other sequelae result (Murray, 1978). Clinical and pathological findings indicate three phases of anaemia in trypanosomiasis in cattle; phase one is characterized by an anaemia as a result of haemolytic anaemia, erythrophagocytosis and haemodilution; there is a large spleen and parasites are detectable in the blood; fever through parasite lysis accompanies this phase. In phases two and three trypanosomes are difficult to detect and erythrocyte destruction occurs probably due to an expanded mononuclear phagocytic system. In phase three of the infection there is an inactive bone marrow and a small spleen. Death occurs through circulatory disorder, damage to the heart and the anaemia. There is, however, wide variation in the response of bovines, as well as sheep and goats, to trypanosomes; some factors involved in the response to parasites are parasite dose (or "challenge"), strain of parasite, breed of host (and therefore degree of tolerance), nutritional and immunological state. There are also well-defined effects on the reproductive system resulting in the female in abortion storms which can be mistakenly diagnosed as brucellosis but in fact are trypanosomiasis. Recent studies also indicate the importance of a variety of biologically active substances in the pathology and pathogenesis of the disease. Such substances are produced from dead or dying trypanosomes or are released by the host in response to infection (Murray, Davis & Morrison, 1981).

The first phase of trypanosome infection usually lasts some weeks and if cattle survive, a second phase of several months occurs, during which parasites are difficult to detect but the animal retains a low grade parasitaemia. The third disease phase is one where parasites are difficult to find but the disease condition persists.

Although different trypanosome species cause disease in bovines the same generalized disease manifestations are produced as a result of infection except where a haemorrhagic syndrome with some *T. (D.) vivax* organisms occurs. The outcome of infection in the bovine is variable, ranging from death to recovery; however, a syndrome of lethargy, stunting and wasting associated with an anaemia produces what Murray (1978) regards as the trypanosomiasis syndrome. Recent studies in Kenya show that *T. b. rhodesiense* infections in cattle cause meningoencephalitis with histological lesions similar to those in man with sleeping sickness. Thus lethargy in the bovine may not be due only to anaemia.

EFFECTS OF TRYPANOSOMES ON *GLOSSINA*

Recent studies of *T. (T.) brucei* in *Glossina* have demonstrated that trypanosomes can affect the feeding behaviour of *Glossina*, increasing the likelihood of transmission (Jenni, Molyneux, Livesey & Galun, 1980; Molyneux & Jenni, 1981). These differences were associated with the attachment of parasites to mechanoreceptive sensilla in the labrum, the entanglement of sensilla with parasite rosettes and the change of flow over sensilla as a result of the colonization of the labrum of *Glossina* (Thevenaz & Hecker, 1980; Livesey, Molyneux & Jenni, 1980). These findings have interesting epidemiological implications for transmission of salivarian trypanosomes amongst game populations and can explain some anomalies in data on fly and game infection rates. The many factors affecting the acquisition and transmission of infections by *Glossina* are reviewed by Molyneux (1977, 1980).

TRYPANOSOME INFECTIONS IN OTHER VERTEBRATES

Lower vertebrates

It has long been known that trypanosomes are found frequently in fishes, amphibians, reptiles and birds as well as mammals. Although extensive studies have been undertaken on the life-cycles and vectors of these organisms (see reviews of Lom, 1979 on fish trypanosomes; Bardsley & Harmsen, 1973, anuran trypanosomes; Baker, 1976, bird trypanosomes; Hoare, 1972, mammalian trypanosomes) relatively little work has been carried out on the possible pathogenic effects of these infections. It is generally believed that these widespread infections cause little pathology in their natural hosts and only rarely are associated with disease states. High trypanosome parasitaemias may be associated with suppression of the immune state by intercurrent infections but as these observations have been carried out on wild animals that have been caught, the influence of other factors is difficult to determine. Recent studies on experimentally infected marine and freshwater fish have indicated reversible changes in haematological parameters and serum changes associated with trypanosome infection (e.g. decreased haemoglobin, increased erythrocyte sedimentation rate, reduced packed cell volume) (Tandon & Joshi, 1973; Cottrell, 1977; Khan, 1977; Khan, Barrett & Campbell, 1980). Histopathological changes associated with piscine and avian trypanosome infections have been demonstrated by Dykova & Lom (1979) and Molyneux (in preparation) respectively. There are two examples of pathogenicity associated with amphibian trypanosomes (see Bardsley & Harmsen, 1973) but, in general, infection with lower vertebrate trypanosomes is benign. It can therefore be concluded that in view of the widespread occurrence of these organisms in non-mammalian vertebrates, the low parasitaemias and their tendency to be restricted in their host preferences they do not play a significant role as agents of disease.

Mammalian Trypanosomes

The comparative ease with which some mammalian species of the sub-genus *Herpetosoma* can be studied has resulted in comparatively more observations under more controlled conditons. *T. (H.) lewisi* of rats and *T. (H.) musculi* of mice have been incriminated with changes in the spleen, liver and kidney (glomerulonephritis) and anaemia. In pregnant female rats infected with *T. (H.) lewisi* resorption of foetuses and death if the foetuses were not resorbed (see Molyneux, 1976 and Wiger, 1977, 1979a,b for references), has been observed. The vast aggregations of *T. (H.) musculi* observed in the placentae of female mice infected in early pregnancy are a possible cause of these effects. However, in *T. (H.) musculi* we have to date observed no histopathological effects on mouse placenta (Molyneux, unpublished observations).

The effects of fighting on trypanosome (*T. (H.) musculi*) parasitaemias investigated in mice by Jackson & Farmer (1970) demonstrated that subordinated mice develop slightly but not significantly higher parasitaemia than dominant animals. Mice allowed to fight at the time of infection developed significantly lower parasitaemias than isolated non-fighting controls.

Wiger (1971) has found a very high infection rate of the trypanosomes of lemming, *Lemmus lemmus (T. (H.) lemmi)*, when the species is most abundant. This ensures that after the population crash a high proportion of survivors are also infected; despite this interesting finding Wiger (1977, 1979a,b) concludes on the basis of a study of blood parasites of *Clethrionomys glareolus* and *Microtus agrestis* (including the trypanosomes *T. (H.) evotomys* and *T. (H.) microti*) that these parasites are of little importance as mortality factors in the population ecology of cyclic populations of sub-Arctic rodents.

The parasite *T. (Schizotrypanum) cruzi* which causes American Trypanosomiasis (Chagas' Disease) has been found in many species of wild mammals in South and Central America. Several mammalian orders are infected — Rodentia, Primates, Marsupialia, Carnivora, Edentata, Chiroptera. However, despite the vast assemblage of mammals from which *T. (S.) cruzi* and related infections have been recorded and the knowledge that in man as well as in experimental animals infections can be, and often are, fatal, most naturally infected reservoir hosts show few if any symptoms.

A check-list of *T. (S.) cruzi*-like infections which have been found in mammals has been given by Hoare (1972). Brener (1973), Barretto (1976), Miles (1979) and Zeledón & Rabinowitch (1981) review the role of animals in the epidemiology and ecology of Chagas' Disease. Although this disease is a well defined zoonosis the control of disease by reducing the wild animal reservoir is not feasible nor does the organism appear to play a role in the regulation of natural mammal populations in the Neotropics.

VECTOR CONTROL AND ITS IMPACT ON CONSERVATION

Habitat Modification

Those aspects of trypanosomiasis which directly affect wild vertebrates have already been discussed but in the control of trypanosomiasis the conservation of wild animal species is of great relevance. African game animals represent a vast reservoir of trypanosomes infective to man and his livestock. The observations of the effect of the rinderpest panzootic in southern Africa on *G. morsitans*, perhaps the most important vector of trypanosomiasis during the early years of this century, and the resulting retraction of the fly belts due to the absence of game indicated that the removal of game animals would serve as a means of allowing livestock into previously infested areas (see Ford, 1971). Thus large-scale game destruction was practised in southern Africa in the early part of this century up to the 1950s and 1960s to free land from *G. morsitans* and render it safe for livestock development. Organized game destruction in Zimbabwe drove back *Glossina morsitans* and permitted the introduction of livestock. These activities are described in detail by Buxton (1955) and Ford (1971). Such techniques are fortunately no longer practised, because not only are they regarded as uneconomic but they emphasize the incompatibility of maintenance of game animals in areas which authorities consider desirable for livestock development. The development of an appreciation of game animals for their own sake and recognition of their value as a source of foreign currency and thus a national resource has been instrumental in changing attitudes. However, the evidence that other species of *Glossina* such as *G. pallidipes* and particularly the riverine species of the *G. palpalis* group can be controlled in this way is not persuasive (Buxton, 1955). Cockbill (1967) described more recent activities in Zimbabwe where preferred hosts only were removed — all Suidae, elephant, buffalo, kudu and bushbuck. Duiker, reedbuck, waterbuck, zebra and large herds of impala were not removed; as a result of this policy *G. morsitans* populations were "virtually exterminated". Some game species such as zebra, wildebeest and small antelope play little or no part in supporting tsetse populations and some herbivore species (impala, waterbuck, hartebeest), although numerous and infected with trypanosomes, are only infrequently fed on by *Glossina*. Further north in Uganda, 8000 square miles were reclaimed by game population destruction between 1946 and 1966 (Wooff, 1966). The desirability of keeping livestock away from areas where game is kept has been a recognized technique for trypanosomiasis control by the use of game and stock fences together with tsetse pickets. This technique has been used in Botswana, Zimbabwe and Zambia together with authorized hunting of game between game and stock fences. However, the efficiency and cost-effectiveness of this method must be questioned, for fences are difficult to maintain, do not completely keep out various game animals and tsetse pickets or deflying techniques are of limited effectiveness for preventing flies moving to areas where livestock are kept.

Impact of Insecticides on Non-target Fauna

However, the use of insecticides in tsetse control operations has posed a further threat to exposed animal species. The environmental consequences on non-target animal species during tsetse control activities have been intensively studied (Koeman, Balk & Takken, 1980). It must however be said that if tsetse control using pesticides as a means of disease control is embarked upon, the future use of land for livestock will have a more long-lasting and widespread effect on resident animal species than acute effects of any insecticide. It is also important to emphasize that the proportion of insecticide used in tsetse control compared with other pest control activities in Africa is extremely small, and that the effects of these other non-monitored operations is likely to be proportionally greater. Any tsetse control operation will undoubtedly result in dramatic habitat changes which will result from an increase in livestock numbers, establishment of subsistence agriculture, removal of game by local hunting and habitat destruction primarily for providing land for subsistence crops but also because of the extensive demands for firewood.

The reports of environmental studies during control operations indicate that the insecticides currently in use have a variety of effects on non-target organisms (Koeman, 1977). Of particular importance in this context, however, are the effects of residual applications which have been applied from helicopters in west Africa. This is largely because ground spray application of residual insecticide has become both selective and discriminative and although some mortality has been observed (Koeman, Rijksen, Smies, Na'Isa & Mac-Lennan, 1971), the method of aerial application by helicopter has more serious consequences for non-target species, the precise effects being dependent on the insecticide used.

The organochlorine compounds, DDT and dieldrin, after residual application have been recorded as producing acute mortality in mammals (rodents and primates), birds and cold blooded vertebrates. Insectivorous birds have been identified as especially vulnerable, particularly flycatchers (Muscicapidae) and robin chats. Many non-target insects are destroyed but owing to their high productivity rates will recover more quickly than *Glossina*. Endosulfan has been extensively used for *Glossina* control but has, however, been associated with serious effects on cold blooded vertebrate populations (particularly fishes) when applied as a residual application adjacent to watercourses; this is a highly undesirable effect particularly in areas where fish provide a high proportion of the protein needs of the population. Recent reports state that endosulfan, even when applied as a non-residual insecticide as doses of 6–12 g active ingredient/hectare (ai/ha) for five to six cycles, can cause changes in physiological parameters, produce pathological lesions and affect the behaviour of fishes in the Okavango Delta in Botswana. Clearly endosulfan or any other insecticide applied as a non-residual will have some side-effects on non-target insects but the recovery capacity will be greater in species other than *Glossina* which is also particularly susceptible to endosulfan. Mattheissen (1981) has described reversible changes in haematological parameters of fishes after application of non-residual endosulfan (6—12 g ai/ha — six cycles) in the Okavango Delta in Botswana. The blood of the affected species, however, returned to normal values within six months after the spraying ceased. The haematological parameters measured were red blood cell count, leucocyte count, haemoglobin and plasma protein concentrations. Blood cell counts were significantly elevated during spraying and plasma protein levels often disturbed. There was a significant decline in haemoglobin/erythrocyte in *Tilapia* from 59 to 39 pg (Mattheissen, 1981). These studies indicate that endosulfan has metabolic effects and P. Mattheissen & R. Roberts (cited by Mattheissen, 1981) have described pathological lesions in the liver and brain of several fish species which occurred simultaneously with the onset of spraying. In sprayed areas *Tilapia rendalli* nest behaviour was abnormal and nesting density as determined by aerial photography was reduced 75% compared with control areas (Fox, in preparation).

These studies indicate that significant sublethal effects occur when endosulfan is applied at the lowest doses effective against *Glossina* as a non-residual insecticide. Application as a residual application near water is known to have catastrophic consequences for fishes (Park, 1979; Everts, 1979).

Although residue analyses of pesticide levels in animal tissues have been carried out after tsetse control operations, prespray base line data are often not available, or if they are, it is difficult to ascertain what residues present are due to tsetse control or previous agricultural insecticide applications; frequently such activities are not known to have been undertaken by those investigating environmental effects of tsetse control. However, Allsopp (1978) working in the Lambwe Valley, Kenya, studied prespray and postspray levels of dieldrin (HEOD) and its more acutely toxic photo-isomer (PIPD)[*] (the photo-isomer however has a shorter biological half-life) in buck (*Redunca redunca*) and oribi (*Ourebia ourebi*), hyaena (*Crocuta crocuta*) and civet (*Viverra civetta*).

In general the residues in the kidney and brain of the antelope and in the brain and liver of the carnivores sampled after spraying 144 g dieldrin per hectare remained below those considered hazardous to the animals' health. Within two years of spraying 409 g dieldrin per

---

[*] Dieldrin = HEOD (1,2,3,4,10,10 hexachloro-6,7 epoxy-1,4,4a,5,6,7,8,8a-octahydro-exo 1,4, endo-5,8, dimethanonaphthalene) to the photo-isomer PIPD (3,exo-4,5,6,6,7-hexachloro-11,12-exo-epoxy-pentacyclo $(6.4.0.0^{2,10}.0^{3,7}.0^{5,9}$. dodecane). This isomer is more acutely toxic to mammals than HEOD (Henderson & Crosby, 1967) but has a shorter biological half life (Brown, Robinson & Richardson, 1967).

hectare residues in game tissues and exposed vegetation had returned to prespray levels. The photoisomerization of dieldrin did not increase the hazard of this insecticide to wildlife. No observation of increased mortality of any of the larger fauna was associated with the air spray trials.

Although the synthetic pyrethroids have only relatively recently been used in antitsetse operations, side effect studies on both residual and non-residual applications have been carried out (Takken, Balk, Jansen & Koeman, 1978; Koeman, Balk & Takken, 1980; Roman, 1979 in Everts, 1979).

There was no acute toxic effect of any application of 50 g ai/ha of permethrin to fish or crustaceans but decamethrin applied as 12.5 g ai/ha caused a depletion in the populations of the crustaceans *Caridinia* and *Macrobrachium*. *Macrobrachium* is a large, commercially caught decapod which re-establishes more slowly in contaminated rivers than *Caridinia*. Pyrethroids, therefore, should be applied with caution in areas where crustaceans may have an economic importance either themselves or in food chains.

Proposals to minimize the impact of insecticides applied in tsetse control operations to fauna have been set out by Koeman, Balk & Takken (1980). These include (1) a prespray survey to ascertain ecological value of habitats to be sprayed and identification of vulnerable areas; (2) the use of the more discriminative technique of ground application in vulnerable areas, and the use of helicopter and fixed wing aircraft where ecological value is "low"; (3) the use of non-residual application at low dose rates wherever possible; (4) provision for adjacent unsprayed areas from which species could reinvade should spraying destroy the fauna. Koeman, Balk & Takken (1980) have identified "indicator species" of animals which can be used as susceptible monitors of the effects of insecticides applied for *Glossina* control.

Tsetse control is frequently regarded as an end in itself but should always be related to the control of trypanosomiasis; in addition any tsetse control measures should be undertaken only after a consideration of the way the land resource is to be utilized (Jordan, 1979). Clearly some areas would not be suitable for livestock development and other methods of land utilization would be more efficient for geographical and other reasons — such as the presence of easily degraded slopes, rocky or mountainous areas, areas required for wildlife conservation, woodland/forest areas which will provide firewood, forestry plantations as well as land for a balanced arable economy.

LAND USE AND TSETSE CONTROL: EFFECTS ON CONSERVATION

Human occupation and the subsequent development of areas previously infested with *Glossina* particularly of the *morsitans* group lead to regression of this species, associated with habitat destruction, or with removal of game either directly by the hunting activities of the local population or as a result of the degradation of specific habitats. It is normally believed that where human population densities of around $40 \, \text{km}^{-2}$ are exceeded then *G. morsitans* populations in the drier savannas are not maintained (Nash, 1948). In contrast, however, the *G. palpalis* group of flies can exist more easily in small, isolated areas of forest, in peridomestic habitats in the more humid areas (Baldry, 1980) and in areas of intensive cultivation such as mango, sugarcane and banana plantations (Finelle, 1980).

However, occasionally *G. morsitans* exists in areas of more intense human habitation as in Gambia where the warthog (*Phacochoerus aethiopicus*) provides almost 90% of the diet of *G.m. submorsitans* (Snow & Boreham, 1979). Because of the absence of any systematic attempt by man to eliminate or reduce warthog (and where the animal cannot be eaten for religious reasons), the animals have achieved pest status by damaging crops. This results in the maintenance of *G.m. submorsitans* populations which cause a serious trypanosomiasis problem in local cattle in areas of high human population density.

The activities of man in relation to the destruction of both a *Glossina* population and the wild game on which that population depended have been described by Onyiah (1978). Onyiah (1978) describes the results of "flyrounds" in the Anara Forest Reserve near Kaduna, Nigeria, which were used by Nash & Page (1953) for the classical studies on the ecology of *G. palpalis*. In the 1970s the population of fly collapsed through man's continued expansion of cultivation, the destruction of game for meat and the vegetation for firewood. The same effects on *G. morsitans* belts around Kaduna have also been observed (Jordan, 1979). Thus human activities which play such an important part in wildlife ecology interact with the fly—trypanosome—game cycle by destroying the natural habitat of the insect and its food host, thereby providing land for development by subsistence farmers and pastoralists often without the need for any specific anti-tsetse measures. Population pressure itself is the force behind the destruction of the natural flora and fauna. However, in these circumstances residual pockets of riverine flies remain.

The deliberate destruction of *Glossina* habitat by removal of vegetational elements on which *Glossina* depends for resting sites has been used for the control of the riverine vectors of human sleeping sickness, as has savanna woodland clearing to destroy *G. morsitans* group flies. Thus large areas of Kenya and Uganda have been rendered tsetse-free by ruthless vegetation clearing using caterpillar-tractors dragging heavy anchor chains between them, or bulldozers. Such operations are undertaken in open *Acacia* savanna, *Acacia-Combretum—Albizzia* woodland, or in secondary forest, More restricted "partial" clearing of vegetation in riverine habitats has been used for the control of human sleeping sickness (Buxton, 1955). The impact of such procedures on wildlife seems not to have been assessed but there can be little doubt that such extensive habitat removal has a major impact; this impact would, however, also be achieved even when livestock development had occurred, and other methods not initially as destructive to vegetation have been used to reduce or eradicate *Glossina*. The knowledge that removal of the vegetation will reduce *Glossina* populations has been exploited by Tarimo & Pallotti (1979) who have reported on the use of the defoliant DNOC at a dosage of 5.9 kg/ha. It is to be hoped that this practice is not pursued further!

However, some of the most interesting data relevant to this are derived from forestry science. Roche (1975) provides data on firewood consumption in Africa. The extent of the degradation of natural forests in Africa and the failure to replace them will certainly have a major impact on *Glossina* and wildlife populations, as already described by Onyiah (1978). This natural forest is not being replaced at an appropriate rate to satisfy domestic firewood demand by the year 2000. The present annual requirements for firewood are 38 million m^3 in Tanzania, 26 million m^3 in Sudan, 24 million m^3 in Ethiopia, 14 million m^3 in Kenya, 5 million m^3 in Zambia. This is harvested from natural forests which are declining rapidly and are not being replaced; what replacement does occur is usually by exotics. The shortfall of planting compared with estimated annual requirements in the year 2000 is 90 000 ha in Nigeria, and 49 000 ha in Ethiopia.

The reliance of *Glossina* species on specific habitats and the evidence that in Africa forest is being massively and indiscriminately destroyed and not replaced, suggest that a reduction of savanna populations will occur; however, it seems likely that persistent pockets of riverine fly or man-made habitats will always present problems. The conclusion is that in Africa deforestation is leading to destruction of the natural ecosystems and to occupation of such lands by cattle to the exclusion of the natural fauna (inclusive of *Glossina*).

Ormerod (1976) has proposed that tsetse control is responsible, at least in part, for large-scale climatic changes such as those associated with the Sahelian drought. The complex relationships between cattle and rainfall, overgrazing and carrying capacity of land is discussed in relation to *Glossina* by Ormerod (1978) and Bourn (1978). The ideas of Ormerod, however, have been strongly criticized (see Haskell, 1977 and discussion). Ormerod believes that *Glossina* eradication campaigns are widespread in west Africa; this is not correct as only in a small proportion of west Africa are extensive long-term campaigns in operation despite Ormerod's claim that his theory applied throughout the Sahel.

Bourn (1978) has used the regression equation of Coe, Cumming & Phillipson (1976) to analyse the effects of tsetse on the carrying capacity of different areas of Africa particularly Ethiopia. Coe *et al.* (1976) examined data from 20 large wildlife ecosystems in Africa and found a high correlation between $\log_{10}$ mean annual rainfall and $\log_{10}$ large herbivore biomass. Using this equation they predicted that in areas with a mean annual rainfall of less than 700 mm the herbivore biomass could be calculated. These data on optimum carrying capacity were used in various regions in relation to known cattle biomass to evaluate levels of stocking in Ethiopia. This approach will permit prediction of areas where over-stocking and thus the creation of desert are most likely to occur. The regression line in tsetse infested zones falls below that line of Coe *et al.*. (1976), indicating understocking and suggesting that should tsetse be controlled or eradicated the carrying capacity could be increased or, through redistribution, permit reduced stocking in tsetse-free overgraded areas. This leads back to implications of the effects of other factors on land use and tsetse populations on which there is so much debate (see Ormerod, 1976; De Vos, 1975; Cloudsley Thompson, 1977).

SUMMARY

This chapter has attempted to identify the relationships between trypanosomes, the diseases they cause and the wildlife and domestic animals so affected, either directly or indirectly. The subject has been traditionally controversial and clearly will remain so; if any progress is to be made it will only be achieved by a realization that each situation or problem merits its own and perhaps unique solution. The time has fortunately passed when trypanosomiasis control was adopted without rational consideration of other land use requirements. The web of interrelationships described in this chapter and the implication for conservation in Africa certainly will provide a rich field for study and remain a source of controversy as relevant land use strategies are developed.

ACKNOWLEDGEMENTS

I am grateful to Drs R. W. Ashford, S. L. Croft, L. Ryan, D. Mehlitz, and particularly Dr L. S. Goodwin, FRS, for their comments on this manuscript.

REFERENCES

Allsopp, R. (1972a). The role of game animals in the maintenance of endemic and enzootic trypanosomiases in the Lambwe Valley, South Nyanza District, Kenya. *Bull. W.H.O.* **47**: 735–746.

Allsopp, R. (1972b). Immobilization of game animals in trypanosomiasis research. *Bull. W.H.O.* **47**: 815–819.

Allsopp, R. (1978). The effect of Dieldrin, sprayed by aerial application for tsetse control, on game animals. *J. appl. Ecol.* **15**: 117–127.

Ashcroft, M. T. (1959). The importance of African wild animals as reservoirs of trypanosomiasis. *E. Afr. med. J.* **36**: 3–11.

Baker, J. R. (1968). Trypanosomes of wild mammals in the neighbourhood of the Serengeti National Park. *Symp. zool. Soc. Lond.* No. 24: 147–158.

Baker, J. R. (1976). Biology of the trypanosomes of birds. In *Biology of the Kinetoplastida* 1: 131–174. Lumsden, W. H. R. & Evans, D. A. (Eds). New York and London: Academic Press.

Baker, J. R., Sachs, R. & Laufer, I. (1967). Trypanosomes of wild mammals in an area northwest of the Serengeti National Park, Tanzania. *Z. Tropenmed. Parasit.* **18**: 280–284.

Baldry, D. A. T. (1980). Local distribution and ecology of *Glossina palpalis* and *G. tachinoides* in forest foci of West African human trypanosomiasis, with special reference to associations between peridomestic tsetse and their hosts. *Insect Sci. Appl.* **1**: 85–93.

Bardsley, J. E. & Harmsen, R. (1973). The trypanosomes of Anura. *Adv. Parasit.* **11**: 1–63.

Barretto, M. P. (1976). Possible role of wild mammals and triatomines in the transmission of *Trypanosoma cruzi* to man. In *New approaches in American trypanosomiasis research*. Proceedings of an International Symposium, Belo Horizonte, Minas Gerais, Brazil, 18–21 March 1975. *Sci. Publs Pan Am. Hlth Org. Wash.* No. 318: 307–318.

Bennett, G. F. (1962). The haematocrit centrifuge for laboratory diagnosis of haematozoa. *Can. J. Zool.* **40**: 124–125.

Bertram, B. C. R. (1973). Sleeping sickness survey in the Serengeti area (Tanzania) 1971. III. Discussion of the relevance of the trypanosome survey to the biology of large mammals in the Serengeti. *Acta trop.* **30**: 36–48.

Bourn, D. (1978). Cattle, rainfall and tsetse in Africa. *J. arid Environ.* **1**: 49–61.

Brener, Z. (1973). Biology of *Trypanosoma cruzi*. *A. Rev. Microbiol.* **27**: 347–383.

Brown, V. K. H., Robinson, J. & Richardson, A. (1967). Preliminary studies on the acute and sub-acute toxicities of a photoisomerisation product of HEOD. *Food Cosm. Toxicol.* **5**: 771–779.

Buxton, P. A. (1955). *The natural history of tsetse flies* (London School of Tropical Medicine & Hygiene Mem. **10**). London: H. K. Lewis.

Carmichael, I. H. & Hobday, E. (1975). Blood parasites of some wild Bovidae in Botswana. *Onderstepoort J. Vet. Res.* **42**: 55–62.

Cloudsley-Thompson, J. L. (1977). *Man and the biology of arid zones.* London: Arnold.

Cockbill, G. F. (1967). The history and significance of Trypanosomiasis problems in Rhodesia. *Proc. Trans. Rhod. Sci. Ass.* **52**: 7.

Coe, M. J., Cumming, D. & Phillipson, J. (1976). Biomass and production of large African herbivores in relation to rainfall and primary production. *Oecologia* **22**: 341–354.

Cottrell, B. J. (1977). A trypanosome from the plaice, *Pleuronectes platessa* (L). *J. Fish Biol.* **11**: 35–47.

De Vos, A. (1975). *Africa, the devastated continent?* Man's impact on the ecology of Africa. The Hague: Junk.

Dillman, J. S. S. & Townsend, A. J. (1979). A trypanosomiasis survey of wild animals in the Luangwa Valley, Zambia. *Acta trop.* **36**: 359–356.

Drager, N. & Mehlitz, D. (1978). Investigations on the prevalence of trypanosome carriers and the antibody response in wildlife in Northern Botswana. *Tropenmed. Parasit.* **29**: 223–233.

Drager, N., Patterson, L. & Breton, D. (1976). Immobilisation of African buffaloes (*Syncerus caffer caffer*) in large numbers for veterinary research. *E. Afr. Wildl. J.* **14**: 113–120.

Dykova, I. & Lom, J. (1979). Histopathological changes in *Trypanosoma danilewskyi* Laveran & Mesnil, 1904 and *Trypanoplasma borreli* Laveran & Mesnil, 1902 infections of goldfish *Carassius auratus* (L). *J. Fish Dis.* **2**: 381–390.

Everts, J. W. (Ed). (1979). *Side effects of aerial insecticide applications against tsetse flies near Bouaflé, Ivory Coast.* Wageningen: Department of Toxicology, Agricultural University, Report.

Ferguson, W., Herbert, W. J. & McNeillage, G. J. C. (1970). Infectivity and virulence of *Trypanosoma (Trypanozoon) brucei* to mice. 3. Effects of social stress. *Trop. Anim. Hlth Prod.* **2**: 59–64.

Field, C. R. (1979). Game ranching in Africa. *Appl. Biol.* **4**: 63–101.

Finelle, P. (1980). Repercussions des programmes d'aménagement hydraulique et rural sur l'epidemiologie et l'epizootiologie des Trypanosomiases. *Insect Sci. Appl.* **1**: 95–98.

Ford, J. (1971). *The role of the trypanosomiases in African ecology.* Oxford: Clarendon Press.

Geigy, R. & Kauffmann, M. (1973). Sleeping sickness survey in the Serengeti Area (Tanzania) 1971. I. Examination of large mammals for trypanosomiasis. *Acta trop.* **30**: 12–23.

Geigy, R., Mwambu, P. M. & Kauffmann, M. (1971). Sleeping sickness survey in Musoma District, Tanzania. IV. Examination of wild mammals as a potential reservoir for *T. rhodesiense* infections. *Acta trop.* **28**: 211–220.

Goodwin, L. G. (1970). The pathology of African trypanosomiasis. *Trans. R. Soc. trop. Med. Hyg.* **64**: 797–812.

Goodwin, L. G. (1974). The African scene: mechanisms of pathogenesis in trypanosomiasis. In *Trypanosomiasis and Leishmaniasis with special reference to Chagas Disease:* 107–119. (Ciba Foundation Symp. No. 20 (new series)). Elliott, K., O'Connor, M. & Wolstenholme, G. E. W. (Eds). Amsterdam.

Goodwin, L. G., Green, D. G., Guy, M. W. & Voller, A. (1972). Immunosuppression during trypanosomiasis. *Br. J. exp. Path.* **53**: 40–43.

Haskell, P. T. (1977). Chairman. Problem of land use and tsetse control. Contributors, Ford, J. & Ormerod, W. E. and discussion. — 17th Seminar on Trypanosomiasis. *Trans. R. Soc. trop. Med. Hyg.* **71**: 12–15.

Heisch, R. B. (1963). Presence of trypanosomes in bush babies after eating infected rats. *Nature, Lond.* **169**: 118.

Henderson, G. L. & Crosby, D. G. (1967). Photodecomposition of dieldrin and aldrin. *J. Agric. Food Chem.* **15**: 888–893.

Hoare, C. A. (1972). *The trypanosomes of mammals.* A zoological monograph. Oxford: Blackwell.

IRDC-132e. (1979). Pathogenicity of trypanosomes. *Proceedings of workshops held at Nairobi, Kenya, 20–23 Nov. 1978.* Losos, G. & Chouinard, A. (Eds). Ottawa: IRDC.

Jackson, L. A. & Farmer, J. N. (1970). Effects of host fighting behaviour on the course of infection of *Trypanosoma duttoni* in mice. *Ecology* **51**: 672–679.

Jenni, L., Molyneux, D. H., Livesey, J. L. & Galun, R. (1980). Feeding behaviour of tsetse flies infected with salivarian trypanosomes. *Nature, Lond.* **283**: 383–385.

Jordan, A. M. (1974). Recent developments in the ecology and methods of control of tsetse flies (a review). *Bull. ent. Res.* **63**: 361–399.

Jordan, A. M. (1979). Trypanosomiasis control and land use in Africa. *Outlook on Agric.* **10**: 123–129.

Keymer, I. F. (1969). A survey of trypanosome infections in wild ungulates in the Luangwa Valley, Zambia. *Ann. trop. Med. Parasit.* **63**: 195–200.

Khan, R. A. (1977). Blood changes in the Atlanta cod (*Gadus morhua*) infected with *Trypanosoma murmanensis*. *J. Fish. Res. Bd Can.* **34**: 2193–2196.

Khan, R. A., Barrett, M. & Campbell, J. (1980). *Trypanosoma murmanensis*: its effects on the Longhorn Sculpin, *Myxoxocephalus octodecemsponsus*. *J. Wildl. Dis.* **16**: 359–362.

Knottenbelt, D. C. (1974). An investigation into the incidence and pathology of natural trypanosomiasis of Bushbuck (*Tragelaphus scriptus*) and Kudu (*T. strepsiceros*). *Trop. Anim. Hlth Prod.* **6**: 131–143.

Koeman, J. H. (1977). Effects of tsetse fly control measures on non-target organisms. *Meded. Rijks. fac. Landbouwwet. Gent* **42**: 889–896.

Koeman, J. H., Balk. F. & Takken, W. (1980). The environmental impact of tsetse control operations. *FAO Anim. Prod. Hlth Pap.* **7**: (Rev. 1) 1–71.

Koeman, J. H., Rijksen, M., Smies, B., Na'Isa, B. & MacLennan, K. J. R. (1971). Faunal changes in a swamp habitat in Nigeria sprayed with insecticide to exterminate *Glossina. Neth. J. Zool.* **21**: 443–463.

Kupper, W., Drager, N., Mehlitz, D. & Zillman, U. (1981). On the immobilisation of Hartebeest and Kob in Upper Volta. *Tropenmed. Parasit.* **32**: 58–60.

Lamprey, H. F., Glasgow, J. P., Lee-Jones, F. & Weitz, B. (1962). A simultaneous census of the potential and actual food sources of the tsetse fly, *Glossina swynnertoni* Austen. *J. Anim. Ecol.* **31**: 151.

Livesey, J. L., Molyneux, D. H. & Jenni, L. (1980). Mechano-receptor-trypanosome interactions in the labrum of *Glossina*: fluid mechanics. *Acta trop.* 37: 151—161.

Lom, J. (1979). Biology of the trypanosomes and trypanoplasms of fish. In *Biology of the Kinetoplastida* 2: 269—337. Lumsden, W. H. R. & Evans, D. A. (Eds.). New York and London: Academic Press.

Losos, G. J. & Gwamaka, G. (1973). Histological examination of wild animals naturally infected with pathogenic African trypanosomiasis. *Acta trop.* 30: 57—63.

Losos, G. J. & Ikede, B. O. (1972). Review of pathology in domestic and laboratory animals caused by *Trypanosoma congolense, T. vivax, T. brucei, T. rhodesiense* and *T. gambiense. Vet. Path.* (suppl.) 9: 1—71.

Lumsden, W. H. R., Kimber, C. D., Dukes, P., Haller, L., Stangellini, A. & Duvallet, G. (1981). Field diagnosis of sleeping sickness in the Ivory Coast. I. Comparison of the miniature anion-exchange centrifugation technique with other protozoological methods. *Trans. R. Soc. trop. Med. Hyg.* 75: 242—250.

Mattheissen, P. (1981). Haematological changes in fish following aerial spraying with endosulfan insecticide for tsetse fly control in Botswana. *J. Fish Biol.* 18: 461—469.

McCulloch, B. (1967). Trypanosomes of the *brucei* subgroup as a probable cause of disease in wild Zebra (*Equus burchelli*). *Ann. trop. Med. Parasit.* 61: 261—264.

Mehlitz, D., Zillman, U., Scott, C. M. & Godfrey, D. G. (In press). Epidemiological studies on the animal reservoir of Gambian sleeping sickness. Part IV. Characterisation of *Trypanozoon* stocks by isoenzymes and sensitivity to human serum. *Tropenmed. Parasit.*

Miles, M. A. (1979). Transmission cycles and the heterogeneity of *Trypanosoma cruzi*. In *Biology of the Kinetoplastida* 2: 117—196. Lumsden, W. H. R. & Evans, D. A. (Eds). New York and London: Academic Press.

Moloo, S. K. (1973). Relationships between hosts and trypanosome infection rates of *Glossina swynnertoni* Aust. in the Serengeti National Park, Tanzania. *Ann. trop. Med. Parasit.* 67: 205—211.

Moloo, S. K., Losos, G. J. & Katuza, S. S. (1973). Transmission of *Trypanosoma brucei* to cats and dogs by feeding on infected goats. *Ann. trop. Med. Parasit.* 67: 331—334.

Molyneux, D. H. (1976). Biology of the trypanosomes of the subgenus *Herpetosoma*. In *Biology of the Kinetoplastida* 1: 285—325. Lumsden, W. H. R. & Evans, D. A. (Eds). London and New York: Academic Press.

Molyneux, D. H. (1977). Vector relationships in the Trypanosomatidae. *Adv. Parasit.* 15: 1—82.

Molyneux, D. H. (1980). Host-trypanosome interactions in *Glossina. Insect Sci. Appl.* 1: 39—46.

Molyneux, D. H. & Jenni, L. (1981). Mechanoreceptors, feeding behaviour and trypanosome transmission in *Glossina. Trans. R. Soc. trop. Med. Hyg.* 75: 160—162.

Mortelmans, J. & Kageruka, P. (1971). Experimental *Trypanosoma brucei* infections in lions. *Acta trop.* 28: 229—333.

Mulligan, H. W. (Ed.) (1970). *The African trypanosomiases*. London: George Allen & Unwin.

Murray, M. (1974). The pathology of African trypanosomiasis. *Prog. in Immun.* 4: 181—192.

Murray, M. (1978). Anaemia of bovine African trypanosomiasis: an overview. In *IRDC-132e. Pathogencity of trypanosomes*. Proceedings of a workshop held at Nairobi, Kenya, 20—23 Nov. 1978: 121—127. Losos, G. J. & Chouinard, A. (Eds.). Ottawa: IRDC.

Murray, M., Davis, C. E. & Morrison, W. I. (1981). Important aspects of the pathology of African animal trypanosomiasis. *Working paper for FAO Panel of Experts on Ecological and Technical Aspects of the Programme for the Control of African Animal Trypanosomiasis (ASA/TRYP/EA/81/ 20). 1—5 June 1981 Rome Italy.*

Murray, M., Grootenhuis, J. G., Akol, G. W. O., Emery, D. L., Shapiro, S. Z., Moloo, S. K., Dar, F., Bovell, D. L. & Paris, J. (1980). Potential application of research on African trypanosomiasis in wildlife and preliminary studies on animals exposed to tsetse infected with *Trypanosoma congolense*. In *Wildlife disease research and economic development:* 40-45. Karstad, L., Nestel, B. & Grahan, M. (Eds). Ottawa, Canada: IRDC-179e.

Nash, T. A. M. (1948). *Tsetse flies in British West Africa. IV. The Gambia.* London: H.M.S.O.

Nash, T. A. M. & Page, W. A. (1953). The ecology of *Glossina palpalis* in Northern Nigeria. *Trans. ent. Soc. Lond.* **104**: 71—169.

Olubayo, R. (1979). Trypanosomiasis of game animals. In *IRDC-132e. Pathogenicity of trypanosomes*. Proceedings of a Workshop held at Nairobi, Kenya, 20—23 Nov. 1978: 87-88. Losos, G. & Chouinard, A. (Eds). Ottawa: IRDC.

Onyiah, J. A. (1978). Fluctuations in numbers and eventual collapse of a *Glossina palpalis* (R-D) population in Anara Forest Reserve of Nigeria. *Acta trop.* **35**: 253—261.

Ormerod, W. E. (1970). Pathogenesis and pathology of trypanosomiasis in man. In: *The African trypanosomiases:* 587—601. Mulligan, H. W. (Ed.). New York: George Allen & Unwin.

Ormerod, W. E. (1976). Ecological effect of control of African trypanosomiasis. *Science, Wash.* **191**: 815—821.

Ormerod, W. E. (1978). The relationship between economic development and ecological degradation: How degradation has occurred in West Africa and how its progress might be halted. *J. arid Environ.* **1**: 357—379.

Park, P. O. (1979). Notes on the eradication of tsetse in Niger, February 1977. In (ISCTRC) International Scientific Council for Trypanosomiasis Research and Control. Fifteenth meeting, Banjul, 1977. *Publ. OAU/STRC* No. 110: 537—540.

Reull, R. H. (1979). Production potential of wild animals in the tropics. *World Anim. Rev.* **32**: 18—24.

Rickman, L. R. & Rottcher, D. (1980). A study of the effects of some game animal tranquillising drugs on *Trypanosoma brucei rhodesiense* infections in white rats. *East Afr. Med. J.* **57**: 327—331.

Roche, L. (1975). Forestry and the conservation of plants and animals in the tropics. *Forest Ecol. Mgmt* **2**: 103—122.

Roman, B. (1979). Ichthyological studies in the Comoe River near Folonzo (Upper Volta) one year after experimental insecticide treatments for the control of tsetse flies by WHO. In *Side effects of aerial insecticide application against tsetse flies near Bouaflé, Ivory Coast*: 73—75. Everts, J. W. (Ed.). Wageningen: Department of Toxicology, Agricultural University Report.

Sachs, R., Schaller, G. B. & Baker, J. R. (1967). Isolation of trypanosomes of the *T. brucei* group from lion. *Acta trop.* **24**: 109–112.

Snow, W. F. & Boreham, P. F. L. (1979). The feeding habits and ecology of the tsetse fly *Glossina morsitans submorsitans* Newstead in relation to nagana transmission in the Gambia. *Acta trop.* **36**: 47–51.

Takken, W., Balk, F., Jansen, R. C. & Koeman, J. H. (1978). The experimental application of insecticides from a helicopter for the control of riverine populations of *Glossina tachinoides* in West Africa. VI. Observation on side effects. *P.A.N.S.* **24**: 455–466.

Tandon, K. S. & Joshi, B. D. (1973). Studies on the physiopathology of blood of freshwater fishes infected with two new forms of trypanosomes. *Z. wiss. Zool.* **185**: 207–221.

Tarimo, C. S. & Pallotti, E. (1979). Residues of aerially sprayed DNOC in biological materials. In (ISCTRC) International Scientific Council for Trypanosomiasis Research and Control, Fifteenth meeting, Banjul, 1977. *Publ. OAU/STRC* No. 110: 572–578.

Thevenaz, P. H. & Hecker, H. (1980). Distribution and attachment of *Trypanosoma (Nannomonas) congolense* in the proximal part of the proboscis of *Glossina morsitans morsitans*. *Acta trop.* **37**: 163–175.

Weitz, B. (1963). The feeding habits of *Glossina*. *Bull. W.H.O.* **28**: 711–729.

Wells, E. A. & Lumsden, W. H. R. (1968). Trypanosome infections of wild mammals in relation to trypanosome diseases of man and his domestic stock. *Symp. zool. Soc. Lond.* No. 24: 133–145.

Wiger, R. (1971). Trypanosomiasis in the Norwegian Lemming, *Lemmus lemmus* (L). *Norw. J. Zool.* **19**: 83–87.

Wiger, R. (1977). Some pathological effects of endoparasites on rodents with special reference to the population ecology of microtines. *Oikos* **29**: 598–606.

Wiger, R. (1979a). Seasonal and annual variations in the prevalence of blood parasites in cyclic species of small rodents in Norway with special reference to *Clethrionomys glareolus*. *Holarctic Ecol.* **2**: 169–175.

Wiger, R. (1979b) Demography of a cyclic population of the bank vole *Clethrionomys glareolus*. *Oikos* **33**: 373–385.

Willett, K. C. (1972). An observation on the unexpected frequencey of some multiple infections. *Bull. W.H.O.* **47**: 757–750.

Wilson, A. J., Dar, F. K. & Paris, J. (1972). A study on the transmission of salivarian trypanosomes isolated from wild tsetse flies. *Trop. Anim. Hlth Prod.* **4**: 14–22.

Woo, P. T. K. (1970). The haemocrit centrifuge technique for the diagnosis of African trypanosomiasis. *Acta trop.* **27**: 384–386.

Wooff, W. R. (1966). The eradication of the tsetse *Glossina morsitans* Westw. and *Glossina pallidipes* Aust. by hunting. In (ISCTRC) International Scientific Council for Trypanosomiasis Research, Twelfth meeting, Bangui, 1968. *Publ. OAU/STRC* No. 102: 267–286.

Zeledón. R. & Rabinovitch, J. E. (1981). Chagas disease: An ecological appraisal with special emphasis on its insect vectors. *A. Rev. Ent.* **26**: 101–133.

DISCUSSION

Huxley (Chairman) — In what way is the forest environment necessary for the *Glossina?*

Molyneux — The pupae are usually found and the larvae are usually deposited in moist soil. There are 22 species of tsetse, some of which have very well-defined ecological needs. The species which survive the dry habitats are the *morsitans* group; they tend not to penetrate into the forest zones; the *fusca* group are restricted to forest habitats; the *palpalis* group to riverine habitats. The species which exist in areas of the lowest humidity do require shade, particularly for the deposition of larvae, and at the end of the dry season even the *morsitans* group tend to behave like riverine flies.

Fielding — It seems from the way you have been talking that we have a choice to make: either the conservation of wildlife in its natural environment in these areas, or the increase in population and the development of mankind. Have you formed your own opinion on which choice, and if the choice is that we should both share the environment, how do you suggest that trypanosomiasis should be controlled?

Molyneux — I certainly was not trying to come out on one side or the other. I believe, and FAO believes, that the problem of trypanosomiasis, in animals anyway, is one that should be solved through a balanced land use policy and that involves the preservation of those natural habitats which are applicable for the conservation of wild species of game animals. I don't think there is any question that one has to adopt the right approach and the right balance, and this has to be done by the involvement and the recognition by national authorities that planning of any tsetse control activity has to be done in association with real development planning. I think there is no doubt about that. As far as the human disease is concerned, I think it's totally right to say that human disease control should take place wherever it's necessary. Now, human trypanosomiasis is at present a very serious problem in many parts of Africa, particularly in Uganda, and it's true to say that human trypanosomiasis in epidemic form results frequently from civil unrest and disorder. Where epidemic sleeping sickness occurs there is an obligation on national governments and health authorities to intervene by whatever method is necessary. But, generally speaking, in this situation wildlife is not directly threatened. There have been suggestions that wildlife is preserved in large areas of Africa through the presence of tsetse fly, as well as the opposite view that tsetse is preventing economic development of large parts of Africa. I don't subscribe to either of these extremes. I believe one has to seek a balance, and there is a balance which is obtainable. I think it important to get the problem into context, for the relationship between tsetse and wildlife is not so important, as that if there is wildlife in an area and there is a shortage of protein, the population will exploit it. The only way you can protect wildlife is by laws — anti-poaching laws — and punishment. The amount of game meat which is still consumed in Africa is very substantial indeed. It will be difficult to prevent exploitation of available game if sufficient protein is not available from other sources.

Jewell — Bearing in mind all that Professor Molyneux said about the problems of maintaining exotic domestic animals, particularly cattle, in regions where diseases are strongly adverse, why is it that game animals are not ranched? Indigenous wild animals are resistant not only to trypanosomiasis but to other African diseases that affect cattle. Pilot schemes, as at Galana in Kenya and Dr Hopcraft's ranch, have demonstrated the productive potential of cropping a community of large wild herbivores. It would seem sensible to provide research effort and adequate financial support to mount full-scale trials of this kind of husbandry.

Molyneux — I can't answer precisely why. I think it's a political question, and the answer is a political one. I believe there is a lobby from the veterinarians and the livestock industry against the idea that game meat can be used and farmed effectively. The arguments against game farming are, first, the administrative problems; then there is the logistic problem of culling animals, the problem of acceptability and marketability of meat, and there are veterinary public health arguments in relation to the diseases which game animals could transmit to man. On the other hand, I don't think there is any doubt that the large herbivores of Africa use the grassland far more effectively and convert far more efficiently than cattle. I would be very happy to see more work — and indeed, there was a symposium recently in Nairobi on this very topic — on the utilization of game animals for protein production. Myers has recently produced a popular review indicating that you could take off 10% of the protein of the Serengeti per annum and still have a healthy animal population. It's a problem, I think, of organization and logistics and perhaps convincing the authorities that it would be an economical and acceptable form of protein.

Ormerod — One point that you have not considered in your discussion of the control of trypanosomiasis involving the eradication of tsetse, is the possibility of transmission by other means. *Trypanosoma vivax*, for example, is transmitted in South and Central America on a scale that is of increasing economic importance. Would you like to discuss this point?

Molyneux — I think if you asked that question of a trypanosomiasis control man he would say that where tsetse control and eradication has taken place, say in Nigeria, trypanosomiasis ceases to be a problem after the tsetse have gone. I would accept that there is evidence of mechanical transmission — we know that — because of transmission of *T. evansi* and *T. vivax* outside the range of *Glossina*. I think it is probably also true to say that there has not been sufficient research on the vectorial capacity of the biting flies which are supposed to transmit the infection. I don't have any doubt that it occurs but I think an assessment of the species which are involved and their capacity as mechanical vectors is required. I would also reiterate that where tsetse control has taken place, trypanosomiasis as a problem tends not to recur.

Khalil — Might I just come back to the question of ranching of game animals. I can't answer this from the trypanosome point of view, but from that of gastro-intestinal nematodes. We have some evidence that once you keep game animals in ranching conditions similar to cattle and sheep, worm burdens and infection rates increase. I think we should tread very carefully in the question of ranching game animals on a large scale until thorough studies are undertaken.

Cox — In cattle, two species of trypanosomes are highly pathogenic, but *Trypanosoma brucei brucei* is not really important. Game may be infected with one, two or three species and there have been relatively few studies on the pathogenicity caused by different trypanosomes. To what extent are particular trypanosomes pathogenic to game animals?

Molyneux — There are several records of trypanosomes being incriminated as causing pathology in game animals which have been shot in the wild. However, as you will appreciate, because you shoot a game animal and you see pathology which is similar to that which is caused by a trypanosome in an experimentally infected cow, you really do not know what is the cause of that pathology. I think recent work shows that when you take a captive-reared animal and inoculate a defined infection, or give a fly-transmitted infection, the pathogenicity is remarkably reduced, and nothing comparable with that seen in the bovine or the goat (IRDC-MDe, 1980).[1] I only showed *T. congolense*, but in fact in eland and waterbuck infected with fly-transmissible strains of *T. vivax* there is no parasitaemia, whereas of course — and this is an east African haemorraghic *T. vivax* — cattle die within a matter of weeks. Thus, using captive-reared animals, which have never been under challenge, with defined isolates of *congolense, vivax* and *brucei*, evidence is that game animals show little evidence of any pathogenic effects.

---

[1] IRDC-MDe (1980). *Wildlife disease research and economic development.* Ottawa, Canada.

Symp. zool. Soc. Lond. (1982) No. 50, 57–75

Rabies in Wildlife

F. STECK

Vet.-Bakteriologisches Institut, Universität Bern, Postfach 2735,
Länggass-Strasse 122, CH-3001 Bern, Switzerland

SYNOPSIS

Rabies is distributed throughout the world, except in Oceania and many other islands elsewhere. The epidemiological pattern is, however, characterized for a given region by the dominant role played by usually one main vector species. Other animals are only secondarily involved. Rabies virus strains appear to be adapted to the principal hosts, but the mechanisms of adaptation are still ill-understood. Antigenically all rabies virus strains show a high degree of cross-reactivity. With the use of monoclonal antibodies it is, however, possible to detect characteristic differences between strains originating from foxes, bats or laboratory strains. This new refined tool may help to clarify epidemiological problems. Other rhabdo-viruses such as Mokola, Duvenhagé and Lagos bat virus, which share mainly internal antigens with rabies virus, do not have the tendency to spread worldwide.

The spread and incidence of rabies is greatly influenced by the population density of the main vector species. For large parts of Europe this is the red fox (*Vulpes vulpes*). Its distribution and density are influenced by the carrying capacity (food and shelter) of a given landscape. Foxes have adapted well to human activity, so that most of rural Europe has a high population density.

Rabies together with control measures leads to a dramatic population reduction and the temporary disappearance of rabies. Fox populations recover, however, within three to five years, after which rabies makes a cyclic reappearance.

The prophylactic reduction of fox population as a control measure has been successfully applied in certain areas such as Denmark. In many regions, however, a sufficient effort was not maintained over longer periods. Rabies does not naturally lead to an immune population. Attempts to control rabies by oral immunization have yielded encouraging results but are still under experimental scrutiny.

Rabies has persisted in all larger areas reached by the present epizootic. The mechanisms which terminated a similar outbreak in the last century are unknown.

Rabies has a worldwide distribution. With the exception of Australia and many larger and smaller islands, the disease occurs on all continents. Dog rabies was known in Babylon as early as 2300 B.C. as a disease transmittable from dog to man. The Spanish Conquistadores met vampire bat rabies when they invaded Central and South Amer-

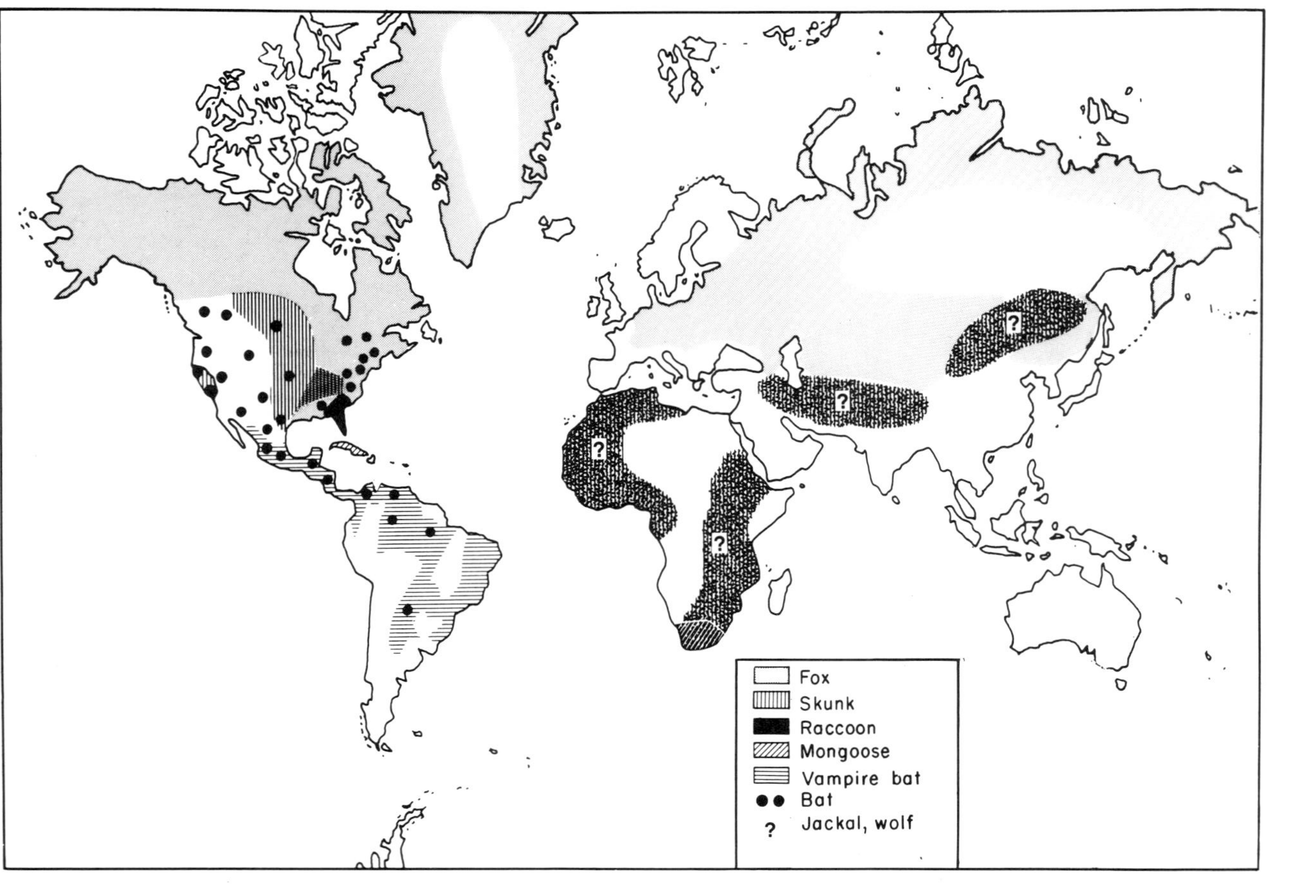

FIG. 1. Major world distribution of wildlife rabies.

ica during the sixteenth century. Some diseases such as dikowanje or polar madness in Arctic foxes, derriengue and mal de caderas in cattle in Latin America and oulo fato in dogs in Africa were shown actually to be rabies only in the last 60 years. With the classical method of virus neutralization and even immunofluorescence, all the virus strains involved turned out to be closely related or indistinguishable (Eichwald & Pitzschke, 1967; Baer, 1975a). (Other rhabdo-virus isolates, such as Lagos bat, Mokola and Duvenhagé virus, which are only partially related to rabies virus, are discussed on p. 61).

Despite the antigenic similarity of the rabies virus strains, epidemiological patterns show great differences in different geographical regions. The most striking observation is that for a particular area, which is usually very large, one particular species of vector animal predominates — to the extent that if rabies disappears from this main vector species, it also tends to disappear in all other species, which are, therefore, considered to be only secondary hosts. This has in several instances been convincingly demonstrated by the effect of control of rabies in dogs, foxes and also in vampire bats.

The species acting as principal vectors are the following (Fig. 1):

Polar foxes (*Alopex lagopus*) in the Arctic zones (Kantorovich, 1964; Rausch, 1958; Crandell, 1966).

Foxes (*Vulpes vulpes, V. fulva*) throughout the temperate zone of Eurasia and parts of North America (Winkler, 1975; Wandeler *et al.*, 1974; Kauker, 1966; Johnston & Beauregard, 1969).

Skunks (*Mephites mephites, Spilogale putorius*) (Parker, 1975).

Racoon (*Procyon lotor*) in Florida, Georgia, Alabama, South Carolina (Bigler, McLean & Revino, 1973; McLean, 1975).

Mongooses (various species) in South Africa and in some Caribbean Islands (Everard, Baer & James, 1974).

Vampire bat (*Desmodus rotundus*) in Latin America.

Insectivorous bats (several species) in Latin America and spreading out through the USA and Canada.

Jackals in Africa and south-east Asia.

Dogs (strays and community dogs) are the most important reservoir and vector in tropical and subtropical zones of Central and South America, Africa andAsia (WHO).

It is interesting to note that rabies does not easily switch from one principal host into another, even under ecological conditions which would favour this event. Probably the most striking recent example is the spread of racoon rabies from Florida through Georgia into areas where previously only fox or skunk rabies was present (Bigler *et al.*,

1973). Two factors play a key role: one is the ill-understood adaptation of rabies virus strains to certain animal species; the other, high population densities of the main vector species.

The characterization of rabies viruses has for a long time been limited to biological properties, which could only be measured by comparing the behaviour of virus strains in various host systems:

Virus: *Host* (animal, tissue culture, embryonated eggs):
Infectivity   Susceptibility
Virulence   Clinical picture
Dose   Incubation period
              Virus excretion
              Immunogenicity

Sikes (1962), by determining susceptibility and virus excretion, has made interesting speculations about mechanisms which may hamper the transfer of infection from foxes to skunks and vice versa:

Foxes are susceptible even to very low doses of rabies virus. If infected with high doses of skunk virus they tend to die after a short incubation without virus excretion in their saliva, which of course prevents further transmission. If infected with low doses they will usually shed virus in their saliva in sufficient quantities to infect foxes, but insufficient to infect skunks.

Skunks are infected only by relatively high doses of virus, and excrete large quantities of virus in their saliva usually leading to infection in other skunks; in foxes, however, these high doses lead again to rapid death without excretion as mentioned above.

This certainly does not explain the whole epidemiological pattern but may represent an expression of adaptation. Dog rabies, prevalent through large areas of Yugoslavia over many decades, has not spread to foxes. Only in recent years have the northern parts of the country been invaded by fox rabies. The conditions were apparently suitable, but the dog rabies virus did not have a chance to cross the species barrier (M. Petrovitch, personal communication). Similarly, Hungary has eradicated dog rabies by systematic dog vaccination and dog control. It was, however, quite independently in recent years invaded by fox rabies (Kauker, 1975).

The virus properties responsible for adaptation to certain species are so far unknown, too subtle to be measured by classical laboratory methods. A new tool, the study of the fine antigenic structure of the virus by the use of monoclonal antibodies, has, however, opened new perspectives (Wiktor & Koprowski, 1978). By the use of a whole battery of such monoclonal antibodies, each of which is directed by definition against only one specific antigenic determinant, it becomes possible to classify different virus strains by their antigenic pattern. It turns out that there are measurable differences between strains isolated from bats or foxes or between certain laboratory strains (Schneider, in press; Wiktor, in press). Even if this does not in itself explain the mechanism of adaptation it will certainly be useful in clarifying complex epidemiological situations.

There are some rhabdo-virus strains which share with rabies virus certain properties, in particular structure and inner nucleoprotein antigens, but whose surface antigens are so different that they are considered to be separate virus species. They have, however, great similarities to rabies virus in their pathogenic potential (Murphy *et al.*, 1973a,b; Schneider *et al.*, 1973). Duvenhagé virus was isolated from a man with a rabies-like encephalitis (Meredith, Rossouw & Van Praag Koch, 1971), who had been bitten a few weeks earlier by a bat. Lagos bat virus has been isolated from several species of African bats (Crick, Tignor & Moreno, 1981). Mokola virus was isolated from shrews and from two human patients with central nervous system disease in Nigeria (Shope, 1975). These rabies-related strains have so far not shown the tendency of rabies virus to spread in an epizootic dimension.

This situation, however, strongly suggests that refined diagnostic methods should be included in routine diagnosis and in particular in clarification of vaccination failures in order to detect such antigenically aberrant rhabdo-viruses (Wiktor, Flamand & Koprowski, 1980). The fact that rabies virus strains from such divergent hosts as bats, dogs or Arctic foxes show strong cross-reactions in serum neutralization nevertheless demonstrates that the antigenic variability of rabies virus is, in comparison to many other viruses, rather limited.

If we turn now to rabies in Europe, we realize that dog rabies is prevalent or recurrent in some countries of southern or south-eastern Europe, as it probably has been for centuries. In recent years great progress has been made in the control or eradication of dog rabies from Mediterranean countries (Kauker, 1966, 1975), as it was several decades earlier in most other regions, by stray dog control and vaccination. In the temperate zone of Europe, great concern has arisen about the steady progression of fox rabies, starting out from Poland (WHO, 1979) during the Second World War and reaching gradually most of the central and western European states, but spreading also eastwards into the USSR (Selimov *et al.*, 1980; Toma & Andral, 1977; Wachendörfer, 1977) (Fig. 2). The epizootic has developed completely independently of dog rabies. If we look at the

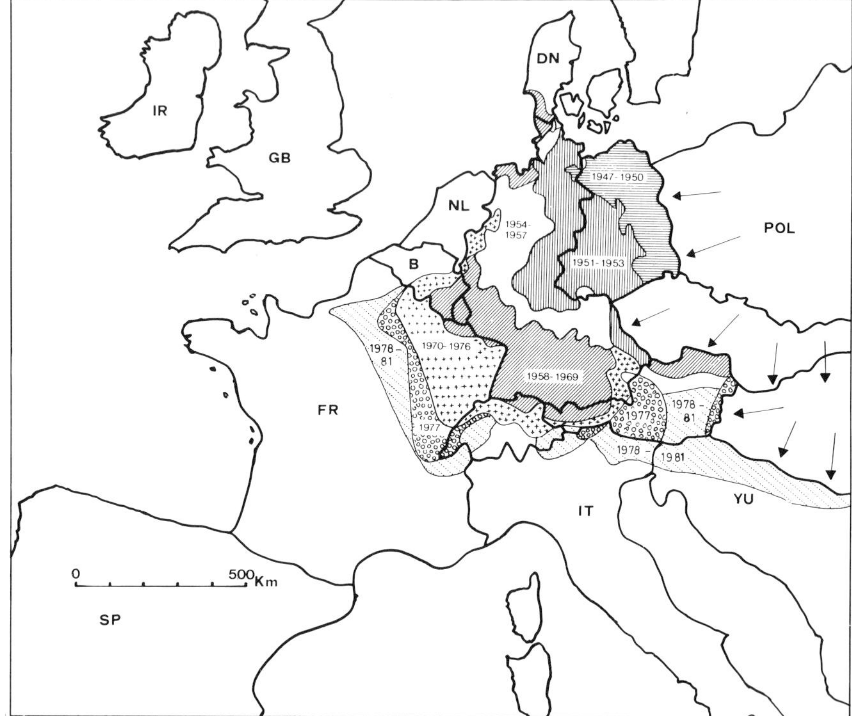

FIG. 2. Spread of fox rabies in central and western Europe, 1947–1981.

animal species involved we see that foxes are the principal vectors. There has been no convincing evidence that rabies is maintained over any length of time without the dominant involvement of foxes. Of course, this has important implications for all attempts to control the disease, a question which will be discussed later (Table I).

Rabies is much less frequently encountered in any other animal species. Badgers and stone martens could be considered as potential vectors. But even if we allow for a certain number of undetected cases, they are about 20 times less frequently involved than foxes. In badgers it appears that short chains of infection can be maintained. Similar observations are made in racoon dogs (*Nyctereutes procyonoides*) particularly in the USSR (V. L. Cherkasskij, unpublished).

The fairly large number of rabid stone martens seems to reflect sporadic cases within a relatively large population. In none of these three species (fox, badger, stone marten) were we able to find the presence of virus in the salivary gland without concurrent infection of the central nervous system. Virus excretion in saliva is, however, frequent and at high titres in rabid foxes and badgers. Of rabid stone

TABLE I

Animal species affected by rabies in Switzerland (laboratory results 1967–1978)[a]

Species	Total no. examined	No. rabies positive	% of all rabies cases
Wild animals			
fox (*Vulpes vulpes*)	19 054	7252	76.6
badger (*Meles meles*)	1387	397	4.2
marten (mostly stone marten) (*Martes foina*)	2847	243	2.6
polecat (*Putorius putorius*)	120	6	0.06
stoat (*Mustela erminea*) and weasel (*M. nivalis*)	202	0	0
wildcat (*Felis sylvestris*)	1	1	0.01
lynx (*Felis lynx*)	1	0	0
roe deer (*Capreolus capreolus*)	3476	449	4.7
red deer (*Cervus elaphus*)	115	6	0.06
chamois (*Rupicapra rupicapra*)	133	13	0.14
wild boar (*Sus scrofa*)	25	0	0
hare (*Lepus europaeus*)	800	0	0
squirrel (*Sciurus vulgaris*)	884	0	0
marmot (*Marmota marmota*)	13	1	0.01
rat (*Rattus norvegicus, R. rattus*)	109	0	0
mouse and vole (*Apodemus, Mus, Microtus, Arvicola, Clethrionomys*)	529	(1)[b]	(0.01)
dormouse (*Glis glis*)	24	0	0
hedgehog (*Erinaceus europaeus*)	140	0	0
mole (*Talpa europea*)	23	0	0
shrew (*Sorex, Crocidura*)	22	0	0
bat (*Chiroptera*)	21	0	0
bird	195	0	0
Total wild animals	30 121	8368	88
Domestic animals			
dog	1207	45	0.5
cat	7851	397	4.2
cattle	1443	327	3.5
sheep	1267	256	2.7
goat	191	32	0.3
swine	35	6	0.06
horse	77	23	0.2
donkey	10	2	0.02
rabbit	143	3	0.03
antelope	2	2	0.02
others	134	0	0
Domestic and zoo animals	12 360	1093	12
Total examined	42 481	9461	100

[a] Source: Steck, Wandeler, Nydegger *et al.* (1980).
[b] Parentheses = Virus isolation from salivary gland of one *Apodemus sylvaticus* after the third blind intracerebral passage in white mice.

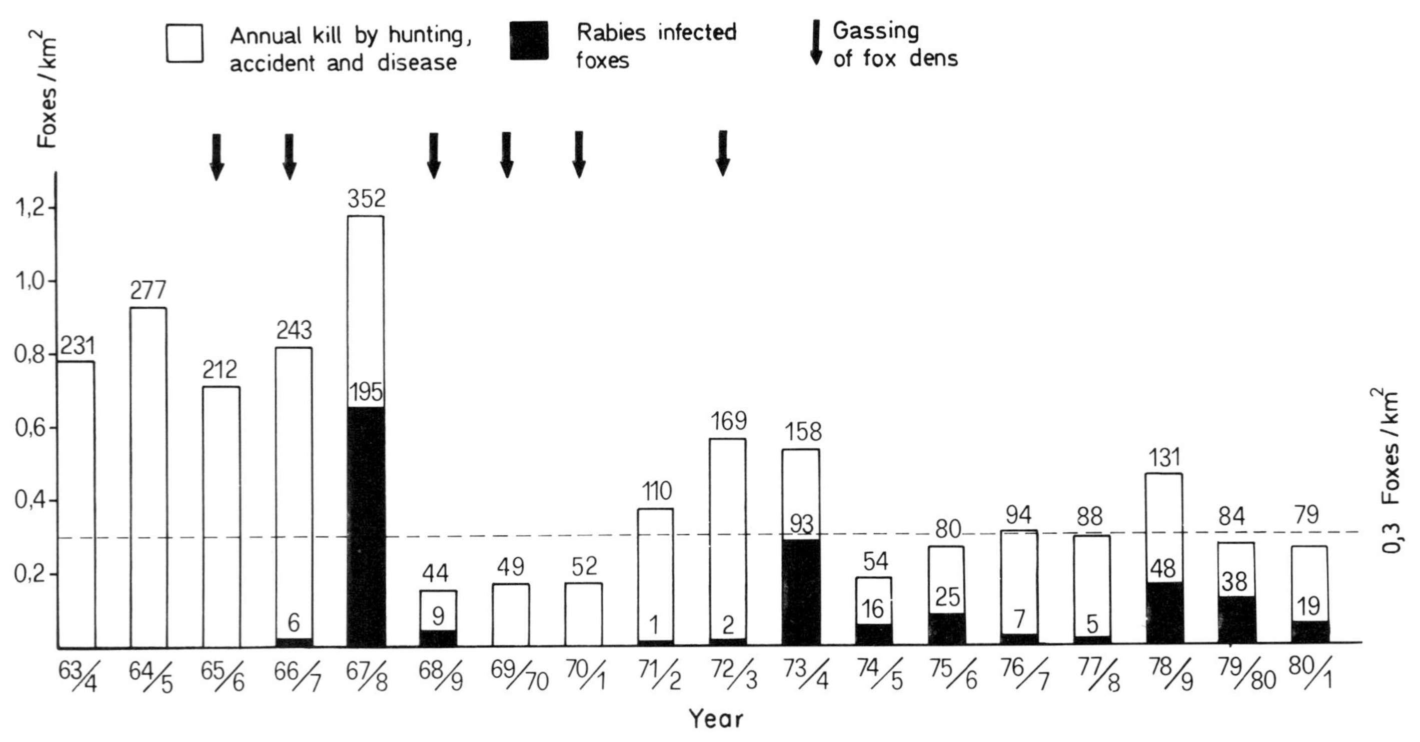

FIG. 3. Fox density-dependent cyclic reappearance of rabies in Canton Schaffhausen, Switzerland.

martens only half are salivary gland positive at the time of death and the amount of virus present is considerably less (Wandeler *et al.*, 1974). This may explain in part their less frequent involvement in the epizootic despite a susceptibility to the virus similar to that of foxes. Other mustelids (weasels, stoats, polecats) are only sporadically found infected with rabies.

Domestic cats with their uncontrolled way of life quite often become infected through their frequent contacts with foxes. Again there seems to be no spread from cat to cat. They are, however, often aggressive and pose serious public health hazards. Compulsory dog vaccination has reduced the occurrence of rabies in Switzerland, as it has in the USA, by a factor of about ten, in comparison to countries where dog vaccination is either prohibited or facultative.

It is interesting to note that domestic and wild ungulates (cattle, sheep and roe deer) are found to be rabies-infected as often as badgers or stone martens; even so they certainly have to be considered as dead-end hosts in most instances. They are of course a great hazard to humans such as farmers and their families, hunters, etc. (Steck, Addy *et al.*, 1968; Steck & Wandeler, 1980; Steck, Wandeler, Nydegger *et al.*, 1980).

Rabies in foxes appears to be strongly density-dependent (Fig. 3). The dependence on density of the main vector species is a feature which in rabies is common to all epidemiological situations. Below a certain threshold rabies tends to die out. In the European situation the limiting population size is reflected by a figure of about 0.3 foxes killed/km^2 during regular hunting (H. Moegle & K. Dietz, pers. comm.; Steck, Wandeler, Nydegger *et al.*, 1980). Similar information was gained much earlier in the United States (Dean, 1955). This figure does of course reflect not only fox population density but also the intensity and type of hunting and should, therefore, be interpreted with care. Nevertheless, it has proven to be quite useful in predicting the potentials of an outbreak. The carrying capacity of certain types of landscapes such as high alpine regions (Steck & Wandeler, 1980; Steck, Wandeler, Nydegger *et al.*, 1980), low marshlands (Fischer, Boegel & Hoppe, 1981) and to some extent also large uninterrupted forest complexes, appears to support only small fox populations which are below the critical threshold. These landscapes, therefore, constitute natural barriers to the spread of the disease (WHO consultations, 1979, 1981). The same is true for areas of small game hunting where fox populations are kept low by their continual destruction (Gessler, 1979; Gessler & Spittler, 1981). Foxes in continental Europe have not adapted to urban environments to the same extent as they have in the Greater London area (Macdonald, 1977, in press), which has helped to keep rabies out of the major cities. Nevertheless, the carrying capacity of landscapes is greatly influenced by human activity. A large proportion of fox food consists of garbage and slaughter offal, and also of synanthropic rodents and rodents in cultivated land (Wandeler & Hörning, 1972). The main factors controlling population density are hunting, diseases such as sarcoptic mange and in endemic areas, of course, rabies itself, accidents and probably a few minor causes. In areas not affected by rabies a fairly stable equilibrium has been reached between reproduction and factors influencing mortality. Sarcoptic mange, which is also occurring at high densities, appears to stop spreading at a higher population density than rabies (Wandeler *et al.*, 1974).

Rabies itself can have a detrimental effect on fox populations. This is particularly true when the front wave of an epizootic hits an area densely populated by foxes. Together with gassing of dens, this has resulted in a more than ten-fold reduction, from one to six foxes killed annually/km^2 down to $0.1/km^2$. Between 30 and 60% of all annual losses (hunting, disease, accidents) are caused by rabies (Steck & Wandeler, 1980). Among foxes with clinically abnormal behaviour, the proportion of rabid foxes is as high as 86%, among those found dead 60%, 9% among foxes caught in traps and 2% of "normal" foxes killed during hunting. At lower fox densities, this effect is less pronounced (Wandeler *et al.*, 1974). The epizootic wave, together with intensified fox destruction, usually leads to a temporary interruption of the infective cycle until after two or three years the fox population has regained a level sufficient for transmission to occur. The reintroduction of the disease comes from neighbouring areas (Wittmann & Kokles, 1967; Steck & Wandeler, 1980). The only other species suffering similar losses in central and western Europe is the badger. Certainly den gassing has a very strong effect, but Moegle & Knorpp (1978) came to the conclusion that rabies itself can be devastating for badgers. All other species, being only secondarily involved, remain unaffected in their population size.

In Switzerland, with fairly intensive fox hunting during autumn and winter, the fox population appears to have a rapid turnover; this is certainly true for the fraction of foxes up to one year old. They constitute around 70% of the total annual losses, which again are compensated by the annual reproduction (Wandeler, 1968; Lloyd *et al.*, 1976).

It is interesting to note that young foxes become involved in the rabies transmission cycle only gradually towards the end of their first year of life. Even though they represent up to 70% of the actual population in May and June, they constitute only between 15 and

40% of those found to be rabies-infected. This feature is certainly caused by the different social behaviour of juveniles and adults, and not by differences in susceptibility: adults within their territories attack intruders whereas juveniles avoid outsiders (Wandeler *et al.*, 1974; Steck & Wandeler, 1980).

Attempts have been made to bring fox populations below the critical threshold by intensified hunting, poisoning and gassing of fox dens. In suitable geographical areas such as in Denmark, this has successfully stopped the spread of rabies (Müller, 1971). Difficulties were, however, met in maintaining this effort over longer periods of time, unless there was a traditional interest in doing so through game wardens in reserves for small game. Fox dens have to be searched for and located in topographically suitable areas and this effort adapted to changing habits of foxes by a well conceived strategy, otherwise population reduction will not be achieved. It has been very difficult to motivate people to pursue gassing operations over many years, which is from a certain point of view quite understandable. Nevertheless, it has been shown that the spread of rabies can be stopped by fox population reduction. The incidence of rabies in other wild and domestic animal species and, in parallel, the exposure of humans, are under these circumstances much reduced or may even become zero at low fox population density. It is, on the other hand, in no way justified to apply reduction measures to wild animal species which are only secondarily involved in the epidemic cycle, such as weasels, stoats, polecats, stone- and pine martens.

Several groups involved in rabies research in North America and Europe have been investigating for several years the feasibility and potential of oral vaccination of foxes for rabies control (Baer, 1975b; Wachendörfer, 1976). The results obtained so far are encouraging. A major drawback of this method is, however, the fact that one may have to use a live attenuated virus, which still possesses a certain degree of virulence. This certainly precludes the prophylactic use of such a vaccine without immediate threat of a rabies outbreak.

Field experiments with chicken head baits and liquid vaccine carried out in Switzerland have shown that it may be possible to stop the advance of rabies by creating a zone in which about 60% of the fox population are immunized against rabies. Part of the success was due to a particularly favourable topographical situation, because the entrance to a large alpine valley was chosen for the experiment. Further studies will show if this method is also applicable in more open country without strong natural barriers (Steck, Wandeler, Bichsel & Capt, in press).

The exposure to virulent virus during an epizootic wave does not lead to substantial immunity in the surviving fox population (Wittmann & Kokles, 1967; Wandeler *et al.*, 1974). The pathogenesis of rabies, where the virus follows a neural pathway within the organism, only occasionally induces the formation of circulating antibodies and the course of the disease is with rare exceptions fatal (Murphy, 1977). Less than 2% of 239 foxes survived and became immune after experimental infections (cumulative data from several sources cited by Steck & Wandeler, 1980). The evaluation of the immune status of the fox population in endemic areas in the field was hampered by the poor quality of the sera obtained from killed foxes. The estimate, however, varies between 0 and 8%. Similar conclusions were also reached by D. H. Johnston & R. O. Ramsden (pers. comm.) through age structure analysis. This immune fraction of the population would, however, soon be replaced through the rapid population turnover by young susceptible foxes.

It is, therefore, justified to assume that naturally acquired immunity is no major factor in the epidemiology of fox rabies. This is of course in marked contrast to most other epidemic infectious diseases. It also appears that predisposing factors which could have an influence on non-specific resistance such as malnutrition, parasitism, etc. play no role, or only a minor one, in fox rabies.

The major factors which control the spread and incidence of rabies are largely linked with population densities and with the seasonal migratory and territorial behaviour of the main vector species. The maximum spread is naturally limited by their distribution range. It is, therefore, quite obvious that the epidemiological pattern and geographical distribution differ completely between vampire bats and insectivorous bats, dogs and polar foxes.

The migratory behaviour of several species of insectivorous bats, bringing them into contact with vampire bats in Central America, explains the scattered appearance of foci of bat rabies throughout North America into Canada (Constantine, 1967; Johnston & Beauregard, 1969). Outbreaks of polar madness coincide with periods of high population densities and mass migration (Rausch, 1958; Kantorovich, 1964).

The red fox in the temperate zone covers comparatively short distances. Home ranges are recorded to be between 0.25 and 10 km². There is territorial behaviour during the whelping season. Juvenile foxes disperse in the autumn over quite a large area, but still the median migration distance was only 5.2 km in ear-tagged foxes in Switzerland (Wandeler *et al.*, 1974). A significant number of young females remain in or near the parent territory, while most males emigrate (Jensen, 1973; Storm *et al.*, 1976; Wandeler & von Beust, pers. comm.). The gradual progression of rabies with maximum spread during autumn, winter and early spring and a minimal spread between April and July (Bögel *et al.*, 1976; Steck & Wandeler, 1980) suggests that rabid foxes have an activity range quite similar to that of normal foxes. The incubation period with an average of about 20 days and a range between nine days and several months allows foxes, after exposure to the virus, to move and migrate according to their normal seasonal pattern during the preclinical phase.

Artois (in press) has shown by the radio tracking of three artificially infected wild foxes that they tend to show a period of intensified movement with some, but limited, extension of their excursion early in the clinical phase of the disease, followed by periods of immobility. The progression of the rabies front by 1400 km in 38 years, with an average of 38 km a year, is consistent with these observations. Progression was found by Bögel *et al.* (1976) to be 4.8 km/month, regardless of the relative fox population density above a critical threshold. Maximum monthly spread was 20 km. In Switzerland, there was in one instance a build-up of a new focus 48 km ahead of the nearest rabies front (Steck, Wandeler, Nydegger *et al.*, 1980) but such observations are unusual.

The spread is much influenced by natural barriers such as large rivers, lakes, seashores and areas of low fox population density such as marshland and in particular the high alpine region. Rabies has rapidly moved through many of the main alpine valleys but it took several years for a higher mountain range to be crossed (Krocza, Scharfen & Mendel, 1981; Irsara, 1978; Steck, Wandeler, Nydegger *et al.*, 1980). Our immunization studies in the Valais have also shown that transmission is a chance event. Single rabid foxes which crossed the immunization zone did not necessarily create new foci of infection.

A build-up of fox rabies cases to higher frequencies usually precedes the advance of the rabies front (Bögel *et al.*, 1976). Under marginal conditions this is unlikely to happen. So far, fox rabies has steadily gained new territory. It has recurred in all regions once reached by the epizootic, in cycles of three to six years. Up to the present time there is no indication that rabies might regress. Poland, which was the first country in which this epizootic was recognized, is still infected today, 40 years later. The occurrence of rabies behind the initial front wave is very patchy, closely linked with the regional recovery of a fox population reduced by rabies and control measures.

It is interesting to note that rabies caused a similar epizootic among foxes during the last century and completely disappeared from large areas after several decades (Streiff, 1829). A close surveillance of the present rabies epizootic may give a clue to which factors influence on a continental scale the coming and going of this scourge.

REFERENCES

Artois, M. (In press). Radio-pistage de renards enragés: quelque données applicables aux modèles spatiaux d'étude de la rage. *Comp. immunol., microbiol. infect. diseases.*

Baer, G. M. (Ed.) (1975a). *The natural history of rabies* 1, 2. New York and London: Academic Press.

Baer, G. M. (1975b). Wildlife vaccination. In *The natural history of rabies* 2: 261–266. Baer, G. M. (Ed.). New York and London: Academic Press.

Bigler, W. J., McLean, R. G. & Revino, H. A. (1973). Epizootiologic aspects of racoon rabies in Florida. *Am. J. Epidem.* 98: 326–335.

Bögel, K., Moegle, H., Knorpp, F., Arata, A., Dietz, K. & Diethelm, P. (1976). Characteristics of the spread of a wildlife rabies epidemic in Europe. *Bull. World Hlth Org.* 54: 433–447.

Constantine, D. G. (1967). Activity patterns of the Mexican free-tailed bat. *Univ. New Mex. Publs Biol.* No. 1–79.

Crandell, R. A. (1966). Rabies in Northern Greenland: Some observations on the epizootiology and etiology. *Proc. Natn Rabies Symp.*, CDC, Atlanta, USA: 37–42.

Crick, J., Tignor, G. H. & Moreno, K. (1981). Lagos bat virus in South Africa. *Rabies information exchange, June 1981*: 40–41.

Dean, D. J. (1955). Panel discussion on: Practical problems of rabies control. *Publ. Hlth Reps* 70: 564–569.

Eichwald, C. & Pitzschke, E. (1967). *Die Tollwut bei Mensch und Tier.* Jena: Gustav Fischer Verlag.

Everard, C. O. R., Baer, G. M. & James, A. (1974). Epidemiology of mongoose rabies in Grenada. *J. Wildl. Dis.* 10: 190–196.

Fischer, H., Bögel, K. & Hoppe, K. (1981). *Natural barriers of wildlife rabies in Schleswig-Holstein (FRG).* (WHO consultations on natural barriers of wildlife rabies in Europe, Vienna 28 April–1 May 1981.)

Gessler, M. (1979). *Grenzen der Tollwutausbreitung in Nordrhein-Westfalen.* (WHO consultations on natural barriers of wildlife rabies in Europe, Berne, Oct. 1979: 25–27.)

Gessler, M. & Spittler, H. (1981). *Relationships between fox population density and natural barriers of wildlife rabies in Nordrhein-Westfalen.* (WHO consultations on natural barriers of wildlife rabies in Europe, Vienna, 28 April–1 May 1981).

Irsara, A. (1978). Presenza ed importanza della rabbia in Italia. *Selez. Veterin.* 14: 78–86.

Jensen, B. (1973). Movements of the Red fox (*Vulpes vulpes* L.) in Denmark investigated by marking and recovery. *Dan. Rev. Game Biol.* 8: 1–20.

Johnston, D. H. & Beauregard, M. (1969). Rabies epidemiology in Ontario. *Bull. Wildl. Dis. Ass.* 5: 357–370.

Kantorovich, R. A. (1964). Natural foci of rabies infection. *J. Hyg. Epidem. Microbiol. Immunol.* 7: 100—110.

Kauker, E. (1966). Die Tollwut in Mitteleuropa von 1953 bis 1966. *Sber. heidelb. Akad. Wiss.* 1966 (4): 205—232.

Kauker, E. (1975). Vorkommen unde Verbreitung der Tollwut in Europa von 1966—1974. *Sber. heidelb. Akad. Wiss.* 1975 (2): 49—84.

Krocza, W., Scharfen, E. & Mendel. G. (1981). *Landscape and wildlife rabies in Austria.* (WHO consultations on natural barriers of wildlife rabies in Europe, Vienna 28 April—1 May 1981.)

Lloyd, H. G., Jensen, B., van Haaften, J. L., Niewold, F. J. J., Wandeler, A., Bögel, K. & Arata, A. A. (1976). Annual turnover of fox populations in Europe. *Zbl. Vet. Med.* (B) 23: 580—589.

Macdonald, D. W. (1977). The behavioural ecology of the red fox. In *Rabies — the facts:* 70—90. Kaplan, C. (Ed.). Oxford: Oxford University Press.

Macdonald, D. W. (In press). The effect of habitat on the behaviour and ecology of red foxes, *Vulpes vulpes. Comp. immunol., microbiol. infect. diseases.*

McLean, R. G. (1975). Racoon rabies. In *The natural history of rabies:* 53—57. Baer, G. M. (Ed.). New York and London: Academic Press.

Meredith, C. D., Rossouw, A. P. & Van Praag Koch, H. (1971). An unusual case of human rabies thought to be of chiropteran origin. *S. Afr. Med. J.* 45: 767—769.

Moegle, H. & Knorpp, F. (1978). Zur Epidemiologie der Wildtiertollwut 2. Mitteilung: Beobachtungen über den Dachs. *Zbl. Vet. Med.* (B) 25: 406—415.

Müller, J. (1971). The effect of fox reduction on the occurrence of rabies. Observations from two outbreaks of rabies in Denmark. *Bull. Off. Int. Epiz.* 75: 763—776.

Murphy, F. A. (1977). Rabies pathogenesis. Brief review. *Archs Virol.* 54: 279—297.

Murphy, F. A., Bauer, S. P., Harrison, A. K. & Winn, W. L. (1973a). Comparative pathogenesis of rabies and rabies-like viruses: a) Virus infection and transit from inoculation site to the central nervous system. *Lab. Invest.* 28: 361—376.

Murphy, F. A., Bauer, S. P., Harrison, A. K. & Winn, W. L. (1973b). Comparative pathogenesis of rabies and rabies-like viruses: b) Infection of the central nervous system and centrifugal spread of virus to peripheral tissues. *Lab. Invest.* 29: 1—16.

Parker, R. L. (1975). Rabies in skunks. In *The natural history of rabies*: 41—51. Baer, G. M. (Ed.). New York and London: Academic Press.

Rausch, R. (1958). Some observations on rabies in Alaska, with special reference to wild Canidae. *J. Wildl. Mgmt* 22: 246—260.

Schneider, L. G. (In press). Antigenic variants of rabies virus. *Comp. immunol., microbiol. infect. diseases.*

Schneider, L. G., Dietzschold, B., Dierks, R. E., Matthaeus, W., Enzmann, P. J. & Strohmajer, K. (1973). Rabies group-specific ribonucleoprotein antigen and a test system for grouping and typing of rhabdo-virsuses. *J. Virol..* 11: 748—755.

Selimov, M., Tatarow, A., Ilyasova, R. *et al.* (1980). Problems of natural foci of sylvatic and arctic rabies. Background document, *WHO Consultations on Rabies Prevention and Control. Lyon, March 1980.*

Shope, R. (1975). Rabies virus antigenic relationships. In *The natural history of rabies* 1: 141–152. Baer, G. M. (Ed.). New York and London: Academic Press.

Sikes, R. K. (1962). Pathogenesis of rabies in wildlife. I. Comparative effect of varying doses of rabies virus inoculated into foxes and skunks. *Am. J. vet. Res.* 23: 1041–1047.

Steck, F., Addy, P., Schipper, E. & Wandeler, A. (1968). Der bisherige Verlauf des Tollwutseuchenzuges in der Schweiz. *Schweizer Arch. Tierheilk.* 110: 597–616.

Steck, F. & Wandeler, A. (1980). The epidemiology of fox rabies in Europe. *Epidem. Rev.* 2: 71–96.

Steck, F., Wandeler, A., Nydegger, B., Manigley, C. & Weiss, M., (1980). Die Tollwut in der Schweiz 1967–1978. *Schweizer Arch. Tierheilk.* 122: 605–636.

Steck, F., Wandeler, A., Bichsel, P. & Capt, S. (In press). Oral immunisation of foxes against rabies: Laboratory and field studies. *Comp. immunol., microbiol. infect. diseases.*

Storm, G. L., Andrews, R. D., Phillips, R. L. *et al.* (1976). Morphology, reproduction, dispersal and mortality of midwestern red fox populations. *Wildl. Monogr.* 49: 1–82.

Streiff, M. D. (1829). Ueber das Erscheinen kranker Füchse im Kanton Glarus und die durch den Biss solcher Tiere verursachte Wuthkrankheit. *Arch. Tierheilk.* 4: 140–149.

Toma, B. & Andral, L. (1977). Epidemiology of fox rabies. *Adv. Virus Res.* 21: 1–36.

Wachendörfer, G. (1976). Gegenwärtiger Stand der Vakzination von Füchsen gegen Tollwut. *Der prakt. Tierarzt.* 12: 801–808.

Wachendörfer, G. (1977). Die gegenwärtige Situation und Bedeutung der Tollwut in der Bundesrepublik Deutschland. *Dt-öst. tierärztl. Wschr.* 84: 413–452.

Wandeler, A. (1968). Einige Daten über den bernischen Fuchsbestand. *Rev. suisse Zool.* 75: 1071–1075.

Wandeler, A. & Hörning, B. (1972). Aspekte des Cestodenbefalls bei bernischen Füchsen. *Jb. naturh. Mus. Bern* 4: 231–256.

Wandeler, A., Wachendörfer, G., Förster, U., Krekel, H., Schale, W., Müller, J. & Steck, F. (1974). Rabies in wild carnivores: I. Epidemiological studies. II. Virological and serological studies. III. Ecology and biology of the fox in relation to control operations. *Zentr. Vetmed.* (B) 21: 735–773.

WHO world survey of rabies 1964–1979.

WHO (1979). Rabies in Poland 1940–1977. *Wkly Epidem. Rec.* No. 13. 30 March 1979: 99.

WHO consultations on natural barriers of wildlife rabies, Berne 25–27 Oct. 1979 and Vienna 28 April to 1 May 1981.

Wiktor, T. J. (In press). Antigenic variants of street virus in the Americas. *Comp. immunol., microbiol. infect. diseases.*

Wiktor, T. J., Flamand, A. & Koprowski, H. (1980). Use of monoclonal antibodies in diagnosis of rabies virus infection and differentiation of rabies and rabies-related viruses. *J. virol. Meth.* 1: 33–46.

Wiktor, T. J. & Koprowski, H. (1978). Monoclonal antibodies against rabies virus produced by somatic cell hybridization: Detection of antigenic variants. *Proc. natn. Acad. Sci. USA* 75: 3938–3942.

Winkler, W. G. (1975). Fox rabies. In *The natural history of rabies*: 3—22. Baer, G. M. (Ed.). New York and London: Academic Press.
Wittmann, W. & Kokles, R. (1967). Weitere Untersuchungen zur Frage der latenten Tollwutinfektion bei Füchsen. *Arch. exp. vet. Med.* 21: 165—173.

DISCUSSION

Keymer — I should be grateful if you would say a few words about the possible risks of aerosol infection, because it has been shown that this can occur. For example, a cat was caged in a cave inhabited by bats infected with rabies. Although the cat could not be bitten by the bats it contracted the disease, purely by sharing the same air space.

Steck — I think aerosol exposure is of no epidemiological importance under our conditions here in Europe. You need a very high concentration of rabies virus in the aerosol to be able to transmit the infection. But it is a very dangerous route. For instance, you can infect opossums by the aerosol route when you cannot infect them by the intra-muscular route. But it appears that these conditions are only fulfilled in bat caves in the Americas or in the laboratory with a leaky homogenizer or similar apparatus creating an artificial aerosol. There have been a few human accidents of this sort.

Kaplan — I think Professor Steck will anticipate this comment. It is the concern one has when one introduces a living virus into nature, even though it is an attenuated virus. There is a worry that the strain of virus that is used for the vaccine, the SAD strain which is very similar to the ERA strain, known to be a heterogeneous population of viral particles, may be selected by foxes in serial transmission, and a virulent strain may emerge. This is a concern always when introducing into nature a living agent like that for immunization purposes.

Steck — That is our concern too. There is the question to what extent one should refrain completely from doing these types of experiments because one could say that the risks involved with oral vaccination are too great. We are quite happy that we can now look back on three years' experience during which time we were unable to re-isolate the vaccine virus from the field. Dr Schneider from Tübingen has looked at all critical isolates in the vaccination zone with monoclonal antibodies, and none of them resembled the SAD virus; but we still have more searching to do. We have no laboratory evidence that this virus spreads, for instance among rodents. In the fox it does multiply in the tonsils, but in parallel swabbing or washing of the mouth we were unable to find any vaccine virus. So excretion, at least for the fox, is limited. We have passaged this virus in suckling mice and after 12 passages used it as a vaccine in foxes, and it did not cause disease. Thus there is a certain body of evidence that the fears — which were well justified — that this virus may get loose, are not substantiated.

Zuckerman — Is it correct that the sporadic occurrence of rabies in some of the Channel port towns bears no relation whatever to the advance of rabies across Europe by way of the fox?

Steck — In which species?

Zuckerman — I seem to recall cat and dog.

Steck — The cases of rabies in domestic animals like cats and dogs, which occurred throughout France, had no influence on the spread of the fox rabies epizootic. It could be, of course, that if you have a dog infected with the fox virus, this virus may infect foxes again. If you have a dog from a Mediterranean country with a dog virus, this virus does not seem to be to the same extent adapted to foxes — but, as I said, this adaptation is ill-understood. We have no good markers for adaptation.

Mullaney — Some years ago latent infection was reported in rodents — mice, hamsters and voles in Central Europe. Would you like to comment on the effect of such infection on the perpetuation and spread of rabies? And, if I might ask a second question: there are strict controls, with quarantine regulations for the importation of rabies-susceptible animals into these countries, except for horses, which for competitive and stud purposes are permitted free movement. Since, in France for instance, horses often spend periods at pasture with possible exposure to infection, do you think there is need for some restriction, such as a time "off grass" or in a known clear area prior to export?

Steck — The question concerning infected rodents in Central Europe is a difficult one because my point of view is that these are laboratory artifacts. Nevertheless, there have been several isolates made in Czechoslovakia and in Tübingen. There have been none made with a similar survey by Wachendörfer's group in Frankfurt or by Hannoun in Strasbourg, and only one isolate by ourselves in Switzerland. There is some question about it. Some of the Czech isolates look like CVS virus, which is a laboratory virus as proven by monoclonal antibody. I think so far we do not know in what way these viruses would spread because the isolations from the rodents have been successful only after several successive blind passages and brain inoculations in mice. The question is, was there really originally in the wild mouse that you started from, rabies virus in small quantities which you have been adapting to mice through passage, or have you picked it up on your way? It is very difficult — once you have your isolate, it's there. There is no good information about the epidemiology of such rodent viruses in Europe, but further research is going on, and I hope it will be possible to clarify this point. We don't think that these rodent isolates, and here the people that isolate them agree with our point, have any direct link with the present fox rabies epizootic, which certainly depends on fox-to-fox spread.

The other question, concerning horses: if you look at their population size, they are about as frequently infected with rabies as other grazing animals — domestic animals. And of course the chance of transmitting rabies to a domestic animal by a horse is rather limited.

Nelson — I have a personal interest in this disease because at one time I was bitten by a proved rabid dog and I suffered the painful course of carbolized vaccine which I was told was not very effective! But this was in Uganda, and the pathologist with whom I was working assured me that he had personally examined over 2000 rabid dogs and jackals in Africa (mainly in Nigeria) and that in his opinion the virus in Africa was of extraordinarily low pathogenicity for man. This was very reassuring; and during the next six years as District Medical Officer in the West Nile District of Uganda I saw plenty of rabies in jackals and dogs but I only ever saw one human case. This prompts two questions: (1) Is there any evidence that the rabies virus in Africa, that has been isolated from dogs and jackals, is different from rabies in Europe? (2) Is there any evidence that there is cross-protective immunity between the closely related viruses that have been isolated in Africa and the rabies virus?

Steck — It looks as if the early isolates were of somewhat reduced virulence, but if you take the whole of north Africa and the Near East, there human rabies cases due to dog bites are very frequent today.

Nelson — How about tropical Africa?

Steck — I cannot answer your question concerning isolates of rabies virus south of the Equator or the Sahara.

Nelson — What about the cross-protection?

Steck — There is an outbreak in Tanzania at present where there is a lot of dog rabies and also human cases. Man is relatively resistant to rabies virus of all strains. Compared to a fox it needs something like 10 000 times more to infect man. But I don't know to what extent this difference in virulence has been measured between African strains and, say, north African strains.

Kaplan — Are you referring to what is called oulo fato, as the African strains of low virulence?

Nelson — I don't know. I am asking the question.

Kaplan — That has been shown to be true. Your other question concerns the cross-protection of the African strains: you do have cross-protection between them, although some vaccines will not protect against certain strains in laboratory experiments, as recently shown by Wiktor of the Wistar Institute in Philadelphia.

Symp. zool. Soc. Lond. (1982) No. 50, 77—95

Myxomatosis: The Natural Evolution of the Disease

JOHN ROSS

Agricultural Science Service, Ministry of Agriculture, Fisheries and Food, Worplesdon Laboratory, Guildford, England

SYNOPSIS

Myxomatosis appeared in Great Britain in 1953 and in the following three years killed an estimated 99% of the wild rabbit population. Since then, rabbit numbers have recovered slowly (although they are still well below pre-myxomatosis levels) because of attenuation of myxoma virus strains and more recently the appearance of genetically resistant rabbits. The key factor in the changes in the relationship between the disease and the wild rabbit is almost certainly the method of transmission: mechanical transmission mainly by the rabbit flea. The development of the disease in Britian is compared with developments in Australia and in France. Differences between the three countries may be due to the different main vectors: the rabbit flea in Britain, mosquitoes in Australia and France.

The effects of myxomatosis on the rabbit population in Britain are discussed and it is predicted that although the disease now has a diminished effect, it will continue in the short term at least to be an important factor in determining rabbit numbers.

The main indirect effects of the disease on the environment are outlined. The sudden and virtually complete disappearance of rabbits had far-reaching effects on the vegetation and appearance of the countryside, and on a number of vertebrate species, particularly some predators.

The introduction of myxomatosis into European wild rabbit (*Oryctolagus cuniculus*) populations has given an excellent opportunity to study the natural evolution of an extremely virulent disease, with little interference by man, at least in Great Britain. There have been in recent years a number of excellent reviews of myxomatosis (e.g. Fenner, 1977; Fenner & Myers, 1978; Mead-Briggs, 1977) giving the history of the disease from the discovery of myxoma virus in South America to its introduction into Australia and France and its subsequent spread to Great Britain. This chapter will deal mainly with myxomatosis in Great Britain where developments have been somewhat different from those in Australia. Myxomatosis was first recorded in England in late 1953 and spread to cover almost the whole of the UK by the end of 1956. Though the disease at first killed virtually every rabbit which became infected, the rabbit is still with us — not in the numbers that existed before myxomatosis, but increasing — and the disease is still widespread. Leading to this situation, there have been a number of important changes in the relationship between myxoma virus and its host; these will be outlined, with speculation on the future of that relationship. In view of the title of this Symposium, the environmental effects of myxomatosis will also be briefly considered.

The first major change in the relationship was the appearance of mutant strains of virus with lower virulence. The type of virus which appeared in Kent in 1953 and which was responsible for the first dramatic spread throughout the country was fully virulent, but, within three years of the original outbreaks, weaker strains of virus were found in the field (Hudson & Mansi, 1955). This development was similar to that in Australia where myxomatosis had been introduced in 1950. When weaker strains of myxoma virus appeared there, Fenner & Marshall (1957) devised an arbitrary system of grading virus strains according to virulence with five grades ranging from Grade I causing >99% mortality to Grade V causing <50% mortality. It was shown that survival times were inversely related to mortality rates and a standard method of virulence grading was based on this finding. A small number of rabbits is inoculated with the virus strain to be graded, the mean survival time is calculated and used to assess the virulence grade.

This system of virulence grading was used in large surveys of myxoma virus strains obtained from all over Great Britian in 1962 (Fenner & Chapple, 1965) and again in 1975 (Ross & Sanders, in preparation a). The results of these surveys are shown in Table I, along with results of similar surveys conducted in Australia (Marshall & Fenner, 1960; Fenner & Woodroofe, 1965) and in France (Joubert, Leftheriotis & Mouchet, 1972).

By 1962, the fully virulent strain which caused the original outbreaks in Britain had almost disappeared and the most commonly occurring strains were of virulence Grade III. A similar pattern had developed in Australia and the grade was sub-divided into IIIa and IIIb. The 1975 survey showed that there was a significant change in the virulence of field strains of virus in Britain, with virulence Grades II and IIIa being most common. Tests of much smaller numbers of virus strains from selected areas in other years showed the same tendency towards dominance of Grades II and IIIa since 1974 (Table II).

The trend towards greater attenuation of field strains has proceeded further in Australia and in France than in Britain (Table I)

TABLE I

Virulence of field strains of myxoma virus in Great Britain, Australia and France

Virulence Grade	I	II	IIIa	IIIb	IV	V
Mortality rate (%)	>99	95—99	90—95	70—90	50—70	<50
Mean survival time (days)	<13	14—16	17—22	23—28	29—50	—
Great Britain						
1953	100[a]	—	—	—	—	—
1962	4.1	17.6	38.8	24.8	14.0	0.9
1975	1.6	24.2	51.6	13.3	9.4	0
Australia						
1950—1951	100	—	—	—	—	—
1958—1959	0	25	29	27	14	5
1963—1964	0	0.3	26	33	31	9
France						
1953	100	—	—	—	—	—
1962	11	19.3	34.6	20.8	13.5	0.8
1968	2.0	4.1	14.4	20.7	58.8	4.3

[a] Expressed as percentages.

TABLE II

Virulence of field strains of myxoma virus in Great Britain

	Virulence Grade					
	I	II	IIIa	IIIb	IV	V
1962—1967[a]	0	3[b]	27	6	0	0
1968—1970	0	0	7	2	0	0
1971—1973	0	1	11	17	1	0
1974—1976[a]	0	4	15	1	1	0
1977—1980	0	7	13	2	1	0

[a] Excluding 1962 and 1975 survey figures.
[b] Number of strains in each grade.

and by 1968 in France strains of Grade IV virulence were most common. In Britain, Grade IV strains have always been relatively rare. These differences could be due to the different vectors involved. In Britain, myxomatosis is transmitted mainly by the rabbit flea (*Spilopsyllus cuniculi*), while in France and in Australia the main vectors are mosquitoes. In all cases, transmission is purely mechanical (there is no multiplication of virus within the insect) and is dependent on the amount of virus in the skin lesions of infected rabbits. Attenuated strains of virus quickly displaced the fully virulent strains because they produced high titres of virus for longer periods and therefore were more efficiently transmitted. Fenner, Day & Woodroofe (1956) showed that strains of Grade IV virulence produce lower titres of virus than Grade III strains. Since mosquitoes have longer and larger mouth-parts than rabbit fleas they may be able to transmit strains of Grade IV virulence more effectively than can fleas and this may explain the higher incidence of Grade IV strains in Australia and in France.

The second important change in the evolution of myxomatosis in Britain has been the appearance of genetic resistance to the disease. The results of work on the monitoring of genetic resistance in Britain are summarized in Table III. Young rabbits (between four and eight

TABLE III

Genetic resistance to myxomatosis in wild rabbits in Great Britain

Location	Year	Virulence grade of virus	No. of rabbits tested	Mean survival time of rabbits which died	Mortality rate
Norfolk	1966	IIIa	41	26.7 days	90%
Norfolk	1967	IIIa	34	23.2 days	94%
Norfolk	1968	IIIa	71	25.5 days	86%
Norfolk	1969	IIIa	74	25.9 days	84%
Norfolk	1970	IIIa	27	24.8 days	59%
Norfolk	1974	IIIa	15	—	13%
Norfolk	1976	IIIa	63	35.9 days	21%
Hants	1978	IIIa	16	—	0%
Wilts	1978	IIIa	71	28.6 days	45%
Angus	1979	IIIa	45	30.5 days	44%
Norfolk	1974	I	11	14.1 days	100%
Norfolk	1975	I	11	17.1 days	100%
Wilts	1979	II	53	29.4 days	45%
Wilts	1980	I	44	22.1 days	52%

weeks old) were captured, and held in the laboratory for at least two months. They were then tested for the presence of anti-myxoma antibodies, any immune animals were rejected and the remainder were injected with a standard dose of a virus strain of known virulence. The virulence of the virus strain was checked by injection into a number of laboratory rabbits.

Early results from the Norfolk site were obtained by Vaughan & Vaughan (1968). Up to and including 1969, the mortality rates of rabbits from this site did not differ significantly from that expected with fully susceptible rabbits (90—95%). In 1970 only a small number of rabbits were tested but the mortality rate was lower than expected. In 1974, using a small number of rabbits bred in captivity from immune adults caught at the Norfolk site, the mortality rate was very low. In 1976, a large sample of young rabbits was captured on site and the mortality rate of 21% left no doubt that these rabbits showed a significant degree of resistance to the disease (Ross & Sanders, 1977). It was later shown that rabbits from other widely separated locations also possessed a similar degree of resistance (Ross & Sanders, in preparation b).

The above results show considerable resistance to a moderately attenuated strain of virus (the most commonly occurring virulence grade). Rabbits have also been tested for resistance to more virulent strains. In 1974 and 1975, small numbers of rabbits from the Norfolk study area were tested against a fully virulent strain; all died, though the mean survival time was significantly longer than that of New Zealand White rabbits used as controls (11–12 days). However, rabbits from Wiltshire, tested in 1979 and 1980, showed significant resistance to strains of virulence Grade II (45% mortality) and Grade I (52% mortality).

In Australia, genetic resistance to myxomatosis was detected much earlier: by 1958 rabbits from one area were exhibiting a degree of resistance (Marshall & Douglas, 1961) similar to that shown by rabbits from Norfolk in 1976. More recent work by Douglas, Nolan, Edmonds & Shepherd (1977) has shown that rabbits from one area of Victoria were as resistant to a Grade I strain of virus in 1961–1966 as Wiltshire rabbits were in 1980. This resistance had not increased by 1975 and this is in agreement with the results of breeding experiments with laboratory rabbits which had recovered from infection. Sobey (1969) showed that selection for resistance proceeded rapidly but reached a plateau level within five or six generations (W. R. Sobey, pers. comm.). In Australia, attenuated strains of virus (Grades IIIb, IV and V) quickly became more common than in Britain (Table I) and this may have allowed quicker development of resistance. Also major outbreaks of myxomatosis in Australia, being mosquito-borne, were confined mainly to the summer months and the high summer temperatures experienced in many parts of that country are known to alleviate the effects of the disease (Marshall, 1959).

The change to higher virulence among field strains of virus and the appearance of genetic resistance in the rabbit population in Britain seem to have occurred at around the same time. Research on the transmission of myxomatosis has shown that the appearance of resistance has probably caused the trend towards higher virulence grades. Using fully susceptible rabbits, it has been shown for mosquitoes in Australia (Fenner *et al.*, 1956) and for rabbit fleas in Britain (Mead-Briggs & Vaughan, 1975) that moderately attenuated strains of virus are most efficiently transmitted. Mead-Briggs & Vaughan (1975) have

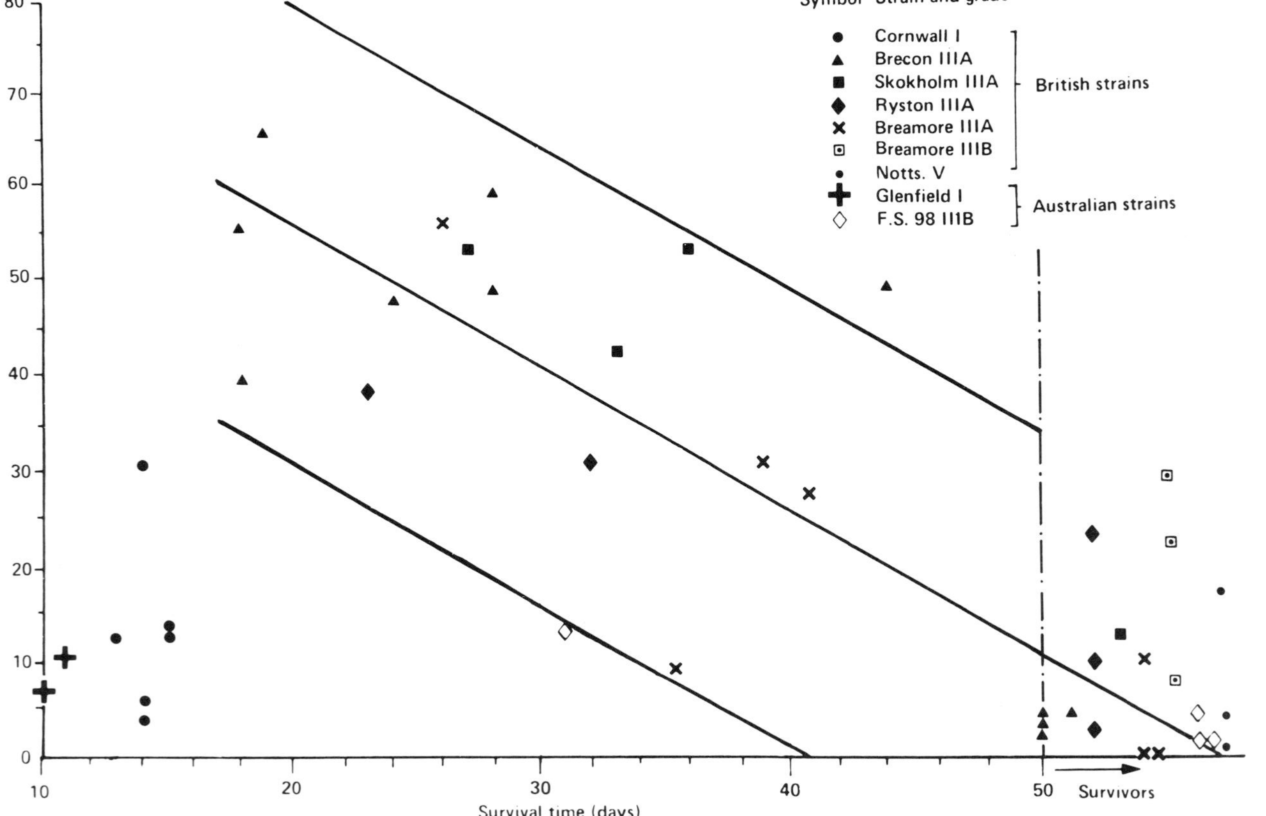

FIG. 1. The relationship between the survival time of a myxoma-infected wild rabbit and the percentage of infective fleas. This regression line (shown with its 95% confidence limits) excludes the results for rabbits infected with the Grade I strains.

shown that the transmissibility of virus by fleas from an infected rabbit depends on the survival time of that rabbit. This is illustrated in Fig. 1 which shows the relationship between the survival time of infected rabbits and the percentage of fleas which have fed on that rabbit and which are then capable of infecting a clean rabbit.

Rabbits which die in 10—15 days (after infection with virulent viruses) yield only a low percentage of infective fleas. Rabbits which survive infection (with any virus) also prove only poorly infective. There is a linear relationship between survival times of 18 days and upwards and transmissibility, with rabbits which die after 18—27 days yielding the highest percentage of infective fleas. With fully susceptible rabbits as used by Mead-Briggs & Vaughan (1975), survival times in this optimal range are produced by moderately attenuated virus strains.

There is evidence (Table III) that, after infection with any given myxoma virus strain, rabbits with a degree of genetic resistance survive longer. Mean survival times (M.S.T.) of resistant rabbits infected with a Grade IIIa strain (29—36 days) are significantly longer than those of fully susceptible laboratory rabbits (17—22 days), and of "wild rabbit base-line" (Vaughan & Vaughan, 1968) (23—25 days). Vaughan & Vaughan used rabbits from the island of Skokholm as a "wild rabbit base-line" in the early work on genetic resistance. The rabbit population on Skokholm was known to be free of myxomatosis since the failure of early experiments with the disease (Lockley, 1940). Myxomatosis reached the island naturally in 1968 and, although it spread only slowly, it was no longer possible to include a "wild rabbit base-line" in tests for genetic resistance. Vaughan & Vaughan (1968), while recording that little genetic resistance had developed as judged by mortality rates, noted that some rabbits from the Norfolk site survived for much longer than Skokholm "base-line" rabbits. They interpreted this as the first sign of resistance entering the population. Similarly, the results of infection of rabbits from Norfolk and Wiltshire with a fully virulent Grade I strain of virus can be interpreted as showing the development of resistance: Norfolk 1974, M.S.T. 14.1 days; Norfolk 1975, 17.1 days; Wiltshire 1980, 22.1 days. Therefore as resistance appears and increases, survival times in the optimal (18—27 days) range will be produced by strains which are more and more virulent to non-resistant rabbits. The Wiltshire resistant rabbits which died after infection with a Grade I virus strain had a mean survival time of 22 days, i.e. within the optimal range. Hence, with the appearance of genetic resistance in wild rabbits, one would expect a trend towards increasing virulence in field strains of virus. Such a change, predicted by Fenner & Ratcliffe

(1965), is taking place now in Great Britain and there is evidence that a similar change may be occurring in Australia (Edmonds, Nolan, Shepherd & Gocs, 1975).

Edmonds *et al.* (1975) tested large numbers of virus strains from the state of Victoria and showed that virus strains collected from the Mallee region in 1970—1974 were of significantly higher virulence than in previous years and also of higher virulence than strains from other parts of Victoria. Rabbits from the Mallee region were found by 1966 to possess a marked degree of resistance to myxomatosis while rabbits from other parts of Victoria showed no resistance against a Grade I strain (Table V) (Douglas *et al.*, 1977). It is surprising that the change towards higher virulence of field strains of virus did not occur earlier in Australia in view of the earlier appearance there of genetic resistance. However, transmission of myxomatosis in Australia has until recently been mainly by mosquitoes which, as stated earlier, may be able to transmit more effectively than fleas virus strains which produce lower titres of virus in skin lesions. Therefore, transmission by mosquitoes would not be as greatly affected as flea transmission by the development of genetic resistance to the disease.

Another factor which may be of importance in explaining the differences in the evolution of the disease in Australia and in Britain is that the virus strains which initiated the outbreaks were from different sources. Both strains were obtained from South America, but the strain introduced into Australia had been isolated from a naturally infected *Oryctolagus* in Brazil in 1910 (Moses, 1911) and had been passaged many times in laboratory rabbits in the 40 years before release in Australia. The Lausanne strain which was released in France and initiated the European (and British) outbreaks, was isolated in 1949 (Bouvier, 1954) and had been passaged only a few times in domestic rabbits. It is possible that selection occurred during the many passages of the "Moses" strain (the "standard laboratory strain" in Australia) resulting in some loss of virulence. This apparent loss is discernible only in rabbits with some resistance. Results obtained by W. R. Sobey (pers. comm.) using fully susceptible and resistant (selected) domestic rabbits (Table IV) show that selected rabbits have some resistance to the "standard laboratory strain" and also to Glenfield, the "hypervirulent" strain described by Sobey (1960), but not to Lausanne strain. Since the Grade I strain used in testing for genetic resistance in Britain was derived from Lausanne strain, it may be that the Wiltshire rabbits tested in 1980 are showing considerably more resistance than the plateau level reached in Australia. Sobey is now conducting trials using Australian

TABLE IV

Mortality rates of unselected and selected adult domestic rabbits infected with highly virulent strains of myxoma virus

Virus strain	Unselected rabbits % mortality	Selected rabbits % mortality
Standard Laboratory	98	80–85
Glenfield	>99	90
Lausanne	>99	>99

and British virus strains and Australian and British resistant wild rabbits. These trials should answer a number of questions about the evolution of the disease and its host in the two countries.

What about the future? Fenner & Ratcliffe (1965) predicted that the relationship between virus virulence and host resistance could take one of two possible courses, leading to two different climax associations.

(1) A disease resembling fibroma may develop in genetically resistant wild rabbits infected with a virus strain which remains very virulent for fully susceptible laboratory rabbits. This is the climax association which has developed in *Sylvilagus* species in South America and California. In these species myxoma virus produces benign tumours which remain infective for weeks before regressing and only rarely produces a fatal disease.

(2) Selection for genetic resistance will continue but not indefinitely, because, after a certain degree of resistance has been acquired, no further rapid changes in virus or host are possible that would result in either a transmissible fibroma (1, above) or a generalized disease that is readily transmitted by biting arthropods. As a result there could be a climax association in which myxomatosis still caused a moderately severe disease with an appreciable mortality. The fecundity of *Oryctolagus* is sufficiently high to counteract such mortality.

There is no evidence at present of the appearance of a fibroma-like infection of resistant rabbits in Britain or Australia. On the other hand, there is evidence from both countries that the processes described in (2) are under way. Results obtained by Douglas *et al.* (1977) (Table V) suggest that in Australia genetic resistance to myxomatosis has reached a steady level. However Fenner & Myers (1978) suggest that this stable situation might be only short-term, representing the limits of change possible by selection acting on

TABLE V

Mortality rates of wild rabbits from two areas of Victoria, Australia after infection with highly virulent strains of myxoma virus

Period	Virus strain	% Mortality	
		Mallee	Gippsland
1961—1966	Standard Laboratory (SLS)	68	94
	Glenfield	98	98
1967—1971	SLS	66	90
	Glenfield	94	99
1971—1975	SLS	67	85
	Glenfield	96	98

existing genetic variability. There may be longer-term evolution based on mutations in virus and host animal.

In Great Britain, the trend towards increased virulence of field strains of virus has been more definite than that detected in Australia and resistance to myxomatosis may have developed further. There is not yet sufficient information to decide whether resistance has "levelled out" and a stable situation has been reached but it is likely to be at a point of balance different from that in Australia.

If the response of rabbits from the Wiltshire site to infection with a fully infective strain of virus (52% mortality rate) is typical of other resistant populations, the effect of myxomatosis will be less than in the past, though the disease will continue to be an important factor in determining rabbit numbers.

However, the mortality rates given in Table III were obtained under laboratory conditions and it is difficult to determine the actual mortality rates in the field. A proportion of rabbits which recover from infection in laboratory cages suffer very severe symptoms and it is doubtful if they would survive in the field because of predators or possibly difficulty in finding food and shelter. In Australia there have been a number of estimates of field mortality rates based on detailed observations of rabbit populations. The proportion of non-immune rabbits which became infected with myxomatosis was known fairly accurately and the proportion of those with the disease which were *not* known to recover gave the survival rate. Estimates of mortality rates vary from 3.7—13.5% (Parer, 1977) during spring and summer epizootics in a population with a high degree of resistance, to virtually 100% (Williams, Fullagar, Davey & Kogon, 1972) during a winter outbreak in a population with no genetic resistance to the disease. High temperatures, such as those that occur in summer in many parts of Australia, ameliorate the effects of moderately attenuated strains of myxoma virus, while low winter temperatures exacerbate the effects (Marshall, 1959). However, it is difficult to disentangle the effects of myxomatosis from those of other mortality factors which may be operating at the same time.

In Britain, a detailed study was made of the original outbreaks of myxomatosis (Armour & Thompson, 1955), and Chapple & Lewis (1964) reported on an outbreak of the disease with an estimated mortality rate of 92%. The best estimate of field mortality rate in this country in recent years was obtained from an intensively studied site between 1971 and 1977 (Ross & Tittensor, in press). In the presence of strains of virulence Grades IIIa and IIIb, 40–60% of infected rabbits died from the disease. It was not possible, without interfering with the long-term intensive study, to determine whether the rabbits on this site had developed genetic resistance, but the results in Table III indicate that genetic resistance was widespread by 1978 and probably earlier: in addition the Hampshire site (Table III) was within 10 miles of the intensive study site. However, despite the difficulties in applying mortality rates determined in the laboratory to the field situation, there can be little doubt that with increasing resistance to the disease, its effect on rabbit numbers will diminish.

Since its appearance in Great Britain in 1953, the disease has been the single most important factor in determining rabbit numbers. It is estimated (Lloyd, 1970) that 99% of rabbits were killed in the first epizootic which swept the country in 1953–1955. Since then rabbit numbers have increased only slowly on a national scale, though in some areas there have been rapid increases causing significant damage to agriculture. Lloyd (1970) reported that, prior to myxomatosis, rabbits were present on 94% of holdings in England and Wales and 47% of holdings were seriously infested. A Ministry of Agriculture, Fisheries and Food survey in 1969 revealed the presence of rabbits on 60% of holdings but only 2% of holdings were seriously infested. Later surveys have shown a gradual increase in the distribution, and therefore, by inference, in the numbers of rabbits in England and Wales, with some acceleration in the late 1970s. This gradual overall increase in the proportion of holdings infested by rabbits disguises very significant changes from year to year in the status of individual holdings from rabbit-infested to rabbit-free and vice versa.

The continued presence of myxomatosis has been a major factor in slowing the recovery of the rabbit population from the disastrous effects of the first epizootic. The results shown in Table III indicate that the disease is probably having a diminishing effect on the rabbit population and therefore the increase in rabbit numbers is likely to continue or even to speed up.

Finally, let us consider briefly the widespread effects of the drastic reduction in rabbit numbers following myxomatosis. The virtual disappearance of such an abundant animal had many direct and indirect effects on the environment. The most dramatic effects were on the appearance of the countryside. Because rabbits had been present in this country for hundreds of years, and because numbers built up relatively gradually, the extent to which they influenced the appearance of the countryside was perhaps not fully appreciated until myxomatosis had almost removed them. Immediately after the first epizootic, there were many reports of the spectacular flowering of many species of wild flowers including species which were thought to be rare or extinct in some areas. On the chalk downland of southern England, rabbits had largely taken over the role of sheep in maintaining the typical close-grazed turf, an early stage of ecological succession. With the coming of myxomatosis the appearance of the downland changed from "smooth-grazed lawn to tussocky hay field" (Nature Conservancy, 1957). There was an increase in palatable grasses and seedlings of trees and shrubs but also of coarse grasses such as *Brachypodium pinnatum*. Details of the changes on Nature Reserves were given by Thomas (1960, 1963) and later changes to one of these areas by Grubb, Green & Merrifield (1969). Chalk heath areas became dominated by *Festuca rubra*, while *Calluna vulgaris* and other woody species grew tall and the areas developed into true heath. Generally, the spectacular show of wild flowers and the initial increase in the number of species recorded were followed by dominance of a few coarse species, formerly held in check by rabbit grazing. The Nature Conservancy made attempts to recover or preserve the former appearance of some areas by mowing twice a year. Merton (1970) reported that some areas of chalk downs, heath lands and sand dunes were becoming woodlands.

In established woodland, myxomatosis had the effect of allowing natural regeneration of trees and other woody species and the reappearance of grasses and herbs (Thomas, 1960). Some of the early benefits to commercial woodland and also to agriculture are outlined by the Agriculture and Forestry Division of FAO (1956). In many coastal areas also there were beneficial effects of myxomatosis, particularly the stabilization of sand dunes previously heavily overgrazed by rabbits (Brown, 1956).

Rabbits have always been an important source of food for a variety of predators and fears were expressed that the loss of rabbits would have serious repercussions for those predators; foxes, buzzards and stoats in particular. All the predators benefited temporarily from the superabundance of dead and dying rabbits during the first outbreaks. Subsequently, foxes (*Vulpes vulpes*) suffered only a temporary decline in numbers because they are very adaptable and were able to compensate for the loss of rabbits by turning increasingly to other prey species (Lever, 1959). Buzzards (*Buteo buteo*) suffered a marked decline in numbers following myxomatosis in the 1950s (Moore, 1957; Parslow, 1967) and the effects of pesticides in the early 1960s (Moore, 1965). Numbers of stoats (*Mustela erminea*) dropped in parallel with the drop in rabbit numbers (Jefferies & Pendlebury, 1968). Buzzard and stoat populations have now stabilized with some fluctuation and have increased from the lowest levels reached, but at lower levels than before myxomatosis. The increased predation on alternative prey had some effect on other predators: in the years following the first outbreaks of myxomatosis, Southern (1970) reported greatly reduced breeding of tawny owls (*Strix aluco*) due to the scarcity of small rodents.

Brown hares (*Lepus europaeus* Pallas) have benefited from the reduction in the rabbit population, but there has not been any dramatic general increase in numbers (Rothschild & Marsh, 1956; Bourlière, 1956), possibly because of increased predation.

The increased height of vegetation after myxomatosis has led to other indirect effects. Voles (*Microtus agrestis*) benefited from the greater protection provided by taller vegetation (Hewson & Kolb, 1973) as did ants from increased shading of ant-hills (Wells, Sheail, Ball & Ward, 1976). On the other hand, changes in vegetation following myxomatosis are said to have contributed to the decline of the sand lizard (*Lacerta agilis*) because the shading provided by taller vegetation caused a loss of bare sand basking areas and delayed hatching of eggs (Jackson, 1979).

It would be a difficult task to judge whether on balance myxomatosis has been a blessing or a curse. Economically, in agriculture and forestry, it has certainly been of great benefit, but it is a most unpleasant disease and has caused great changes to our countryside, though not all of these have been negative. The disease is likely to remain enzootic in the rabbit population and will continue, at least in the short term, to be an important factor in determining rabbit numbers.

ACKNOWLEDGEMENTS

I thank Mr H. V. Thompson for helpful criticism of an earlier draft and Dr W. R. Sobey, C.S.I.R.O. Canberra, for permission to use unpublished data.

REFERENCES

Armour, C. J. & Thompson, H. V. (1955). Spread of myxomatosis in the first outbreak in Great Britain. *Ann. appl. Biol.* **43**: 511—518.

Bourlière, F. (1956). Biological consequences due to the presence of myxomatosis. *Terre Vie* **103**: 123—136 [In French and English].

Bouvier, G. (1954). [Some remarks on myxomatosis]. *Bull. Off. int. Epiz.* **46**: 76—77 [In French].

Brown, P. W. (1956). Story of the rabbit in the North-East of Scotland. *Trans. R. Highld agric. Soc. Scotl.* **67**: 26—35.

Chapple, P. J. & Lewis, N. D. (1964). An outbreak of myxomatosis caused by a moderately attenuated strain of myxoma virus. *J. Hyg., Camb.* **62**: 433—441.

Douglas, G. W., Nolan, I. F., Edmonds, J. W. & Shepherd, R. C. H. (1977). [quoted in Fenner & Myers (1977)].

Edmonds, J. W., Nolan, I. F., Shepherd, R. C. H. & Gocs, A. (1975). Myxomatosis: the virulence of field strains of myxoma virus in a population of wild rabbits with high resistance to myxomatosis. *J. Hyg., Camb.* **74**: 417—418.

F.A.O. (1956). Myxomatosis in rabbits. *Terre Vie* **103**: 248—252.

Fenner, F. (1977). Processes that contribute to the establishment and success of exotic micro-organisms in Australia. (In *Exotic species in Australia — their establishment and success.* Anderson, D. (Ed.)) *Proc. ecol. Soc. Austr.* **10**: 39—61.

Fenner, F. & Chapple, P. J. (1965). Evolutionary changes in myxoma virus in Britain. *J. Hyg., Camb.* **63**: 175—185.

Fenner, F., Day, M. F. & Woodroofe, G. M. (1956). Epidemiological consequences of the mechanical transmission of myxomatosis by mosquitoes. *J. Hyg., Camb.* **54**: 384—303.

Fenner, F. & Marshall, I. D. (1957). A comparison of the virulence for European rabbits (*Oryctolagus cuniculus*) of strains of myxoma virus recovered in the field in Australia, Europe and America. *J. Hyg., Camb.* **55**: 149—191.

Fenner, F. & Myers, K. (1978). Myxoma virus and myxomatosis in retrospect: the first quarter century of a new disease. In *Viruses and environment*: 539—570. Kurstak, E. & Maramorosch, K. (Eds). New York and London: Academic Press.

Fenner, F. & Ratcliffe, F. N. (1965). *Myxomatosis.* Cambridge: University Press.

Fenner, F. & Woodroofe, G. M. (1965). Changes in the virulence and antigenic structure of strains of myxoma virus recovered from Australian wild rabbits between 1950 and 1964. *Aust. J. exp. Biol. med. Sci.* **43**: 359—370.

Grubb, P. J., Green, H. E. & Merrifield, R. C. J. (1969). The ecology of chalk heath: its relevance to the calcicole-calcifuge and soil acidification problems. *J. Ecol.* **57**: 175—212.

Hewson, R. & Kolb, H. H. (1973). Changes in the numbers and distribution of foxes (*Vulpes vulpes*) killed in Scotland from 1948 to 1970. *J. Zool., Lond.* **171**: 134—356

Hudson, J. R. & Mansi, W. (1955). Attenuated strains of myxomatosis virus in England. *Vet. Rec.* **67**: 746–747.

Jackson, H. C. (1979). The decline of the sand lizard (*Lacerta agilis*) population on the sand dunes of the Merseyside coast, England. *Biol. Conserv.* **16**: 177–193.

Jefferies, D. J. & Pendlebury, J. B. (1968). Population fluctuations of stoats, weasels and hedgehogs in recent years. *J. Zool., Lond.* **156**: 513–549.

Joubert, L., Leftheriotis, E. & Mouchet, J. (1972). [*Myxomatosis*] 1. Paris: L'Expansion Scientifique Francaise. [In French].

Lever, R. J. A. W. (1959). The diet of the fox since myxomatosis. *J. anim. Ecol.* **28**: 359–375.

Lloyd, H. G. (1970). Post-myxomatosis rabbit populations in England and Wales. *EPPO Pubs* (A) No. 58: 197–215.

Lockley, R. M. (1940). Some experiments in rabbit control. *Nature, Lond.* **145**: 767–769.

Marshall, I. D. (1959). The influence of ambient temperature on the course of myxomatosis in rabbits. *J. Hyg., Camb.* **57**: 484–497.

Marshall, I. D. & Douglas, G. W. (1961). Studies in the epidemiology of infectious myxomatosis of rabbits. VIII Further observations on changes in the innate resistance of Australian wild rabbits exposed to myxomatosis. *J. Hyg., Camb.* **59**: 117–122.

Marshall, I. D. & Fenner, F. (1960). Studies in the epidemiology of infectious myxomatosis of rabbits. VII The virulence of strains of myxoma virus recovered from Australia between 1951 and 1959. *J. Hyg., Camb.* **58**: 485–488.

Mead-Briggs, A. R. (1977). The European rabbit, the European rabbit flea and myxomatosis. In *Applied biology* 2: 183–261. Coaker, T. H. (Ed.). London and New York: Academic Press.

Mead-Briggs, A. R. & Vaughan, J. A. (1975). The differential transmissibility of myxoma virus strains of differing virulence grades by the rabbit flea *Spilopsyllus cuniculi* (Dale). *J. Hyg., Camb.* **75**: 237–247.

Merton, L. F. H. (1970). The history and status of the woodlands of the Derbyshire limestone. *J. Ecol.* **58**: 723–744.

Moore, N. W. (1957). The past and present status of the buzzard in the British Isles. *Br. Birds* **50**: 173–197.

Moore, N. W. (1965). Pesticides and birds — a review of the situation in Great Britain in 1965. *Bird Study* **12**: 222–252.

Moses, A. (1911). [On the myxoma virus of rabbits.] *Mems Inst. Oswaldo Cruz* 3: 46–53. [In Portuguese].

Nature Conservancy (1957). *Report of The Nature Conservancy, 1957.* London: HMSO.

Parer, I. (1977). The population ecology of the wild rabbit *Oryctolagus cuniculus* in a mediterranean-type climate in New South Wales. *Austr. wildl. Res.* **4**: 171–205.

Parslow, J. L. F. (1967). Changes in status among breeding birds in Britain and Ireland. *Br. Birds* **60**: 2–47.

Ross, J. & Sanders, M. F. (1977). Innate resistance to myxomatosis in wild rabbits in England. *J. Hyg., Camb.* **79**: 411–415.

Ross, J. & Sanders, M. F. (In preparation a). *Further changes in the virulence of myxoma virus in Britain.*

Ross, J. & Sanders, M. F. (In preparation b). *The development of genetic resistance to myxomatosis in Britain.*

Ross, J. & Tittensor, A. M. (In press). Myxomatosis in selected rabbit populations in southern England and Wales 1971—1977. *Proc. 1st World Lagomorph Conference, Guelph, 1979.*

Rothschild, M. & Marsh, H. (1956). Increase in hares (*Lepus europaeus* Pallas) at Ashton Wold, with a note on the reduction in numbers of the Brown rat (*Rattus norvegicus* Berkenhaut). *Proc. zool. Soc. Lond.* **127**: 441—445.

Sobey, W. R. (1960). Myxomatosis: the virulence of the virus and its relation to genetic resistance in the rabbit. *Aust. J. Sci.* **23**: 53—55.

Sobey, W. R. (1969). Selection for resistance to myxomatosis in domestic rabbits (*Oryctolagus cuniculus*). *J. Hyg., Camb.* **67**: 743—754.

Southern, H. N. (1970). The natural control of a population of Tawny owls (*Strix aluco*). *J. Zool., Lond.* **162**: 197—285.

Thomas, A. S. (1960). Changes in vegetation since the advent of myxomatosis. *J. Ecol.* **48**: 287—306.

Thomas, A. S. (1963). Further changes in vegetation since the advent of myxomatosis. *J. Ecol.* **51**: 151—186.

Vaughan, H. E. N. & Vaughan, J. A. (1968). Some aspects of the epizootiology of myxomatosis. *Symp. zool. Soc. Lond.* No. 24: 289—309.

Wells, T. C. E., Sheail, J., Ball, D. F. & Ward, L. K. (1976). Ecological studies on the Porton Ranges: relationships between vegetation, soils and land-use history. *J. Ecol.* **64**: 589—626.

Williams, R. T., Fullagar, P. J., Davey, C. C. & Kogon, C. (1972). Factors affecting the survival time of rabbits in a winter epizootic of myxomatosis at Canberra, Australia. *J. appl. Ecol.* **9**: 399—410.

DISCUSSION

Fielding — It was a very fine survey of the ecological balance existing between the rabbit and the virus of myxomatosis in Great Britain; but in Australia the picture is obviously very different, because the rabbit has been introduced by man into that country and now indeed is causing excessive problems and is upsetting their ecological balance in that it is taking away much of the environment from the smaller marsupial mice and so on. Do you see the possibility of laboratory cultivation of a very virulent, highly pathogenic virus for introduction back into Australasia for the reduction of the rabbit population over there, and if you see this potential possibility in the future, would you say that is a good thing, to try to restore their natural ecological balance?

Ross — There has been talk in Australia of developing a supervirulent strain of virus, but I think there are great dangers in this. If you mess around with the genetics of myxoma virus you may lose its species-specificity, for example, which is something I haven't mentioned, but it is virtually specific to the European rabbit. Experience in Australia has shown that, although (in some states particularly) they have reintroduced fully virulent virus repeatedly to try to maintain the mortality due to myxomatosis, the natural balance has asserted itself regardless, so the current situation in Australia is not so very different from that in this country where we have left myxomatosis strictly alone.

Thompson — One comment and one question. You mentioned the two suggestions about the future by Fenner & Ratcliffe (1965).[1] But they are not of course

---

[1] See list of references, pp. 90—92.

mutually exclusive. We tend to forget, I think, or we overlook, that whereas the *Sylvilagus brasiliensis* in South America has been coping with myxomatosis for many tens of thousands of years, the European wild rabbit has been exposed to it only since 1950 in Australia and 1952 in Europe. So I think that in the mid-1950s we were all imagining that the rabbit would come to terms with myxomatosis, perhaps rather quickly, and that we would see rabbit numbers going up again possibly within a decade. I think it is more obvious to us now that it *will* come to terms, but possibly not within a hundred years. The question I wanted to ask you: as you know, the strain of myxomatosis introduced into Australia deliberately came from the Rockefeller Institute and had been passaged for something of the order of 39 to 40 years, whereas the one introduced more or less accidentally into Europe was recovered in 1949 and had very few passages before release in 1952. Do you think this great difference in the history of the two strains released is likely in the long run to affect the development of the disease in Europe as compared with Australia?

Ross — Yes, there is this big difference between the virus strains used, and I think the answer to your question is, yes, I believe it will make a difference to the eventual outcome in this country as opposed to Australia. There is a difference already showing up. Australian resistant rabbits are showing quite considerable resistance to their Grade I — fully virulent — strains but are showing no resistance at all to the Grade I strains obtained in this country. So already we seem to have overtaken them; our rabbits are resistant to the type of virus that the Australian rabbits are still fully susceptible to.

Keymer — Is there any evidence that mosquitoes transmit myxomatosis in this country, because I have seen the disease in commercial rabbits when there appeared to be no possibility of contact with wild infected rabbits or with their fleas?

Ross — There is no doubt that mosquitoes can transmit myxoma virus in this country as in Australia, but it's of minor importance. The rabbit flea is the important vector in this country. There is a paper by Service (1971)[1] which demonstrates very clearly that mosquitoes in this country are capable of transmitting the disease. But I think it's a local problem rather than a national problem.

Mullaney — How long does maternal immunity to myxomatosis last? The first signs of disease, in my area, in wild rabbits occur about the end of August or early September, when there is conjunctivitis, "dullness" of the eye and loss of vision, and then a little later on lesions appear and they die quickly. The second question I would like to ask you is this: in pox disease like myxomatosis where you get localized lesions, is there an early drop in the secondary viraemia with massive invasion of the skin followed by rapid replication before lesions are noted, to enable the rabbit to survive for 10, 12, up to 20 days, so that fleas can become contaminated?

Ross — There is a lot of work still to be done on maternal antibodies, and there is conflicting evidence in this country and in Australia for the effect that maternal antibodies have. We would say that maternal antibodies do have some protective effect but probably not for more than four to six weeks after birth. The Australians are more pessimistic and think that maternal antibodies have no protective effect whatsoever. So that's an open question.

---

[1] Service, M. W. (1971). A reappraisal of the role of mosquitoes in the transmission of myxomatosis in Britain. *J. Hyg., Camb.* **69**: 105–111.

Mullaney — Why do you think the young rabbits remain normal — apparently normal — until about the end of August?

Ross — I think that the normal pattern of disease in this country is that noticeable epizootics of disease occur in late summer and autumn, and we believe that this is because at that period there are the greatest number of susceptible animals available, and that's why you get the most noticeable outbreaks; and I think that's probably independent of maternal antibody. I think viraemia in myxomatosis, as far as we have discovered, is a very transient phenomenon. It varies between the different strains, but detectable virus only lasts for four to five days in the rabbit, and plays no part in transmission. The transmission is merely from the skin lesions.

Young — In the early days of myxomatosis, there were a number of allegations, if nothing stronger, that it could occur in the hare, and there was even a case, I think, reported of a dog supposedly having myxomatosis, a gamekeeper's dog. I wonder if you would comment on that, and also, would you care to say anything about the possibility of human infection through cuts and sores on the hands of, say farmworkers and gamekeepers?

Ross — There are a number of cases in the literature where hares are reported to have been infected and show symptoms of myxomatosis; some of these are probably well documented, some of them leave some doubt. We have never seen a case (and Dr Keymer will bear me out); there are hares which I have seen that have symptoms like myxomatosis but we haven't been able to isolate any virus. So the answer is perhaps, occasionally, you can get a hare infected. We have also tried injecting large amounts of virus into hares, and although they produce antibodies there are no symptoms. So it must be a very rare event, if it does happen. Dogs: I think the symptoms must have been caused by something else. In humans I think there is no danger whatsoever. Some of the early workers in Australia, before the virus was introduced there, tested virtually every vertebrate in Australia, and found that none of them could be infected with the virus, and also injected themselves with huge doses of virus; there are no symptoms yet, and this was over 30 years ago.

Molyneux — Could an explanation of the absence of myxomatosis in four- to eight-week old rabbits be that in the nest they are infested with fleas from the female and not exposed to fleas from other rabbits which may be more likely to be carrying virus?

Ross — That's true. But I should say that some of the young rabbits that we collected, the four- to eight-week old rabbits that we used for genetic resistance, were shown to be permanently immune, so they must have been in contact with the disease very early on; and at four to eight weeks old they looked perfectly healthy and no symptoms of disease developed subsequently, so they must have had a very mild form of the disease in the early weeks of life. So it does happen.

Henderson — Mr Thompson spoke, perhaps a little euphemistically, of the introduction of myxomatosis in Europe as having been accidental. We heard from Professor Steck about the merits of the barrier of the Channel against the spread of rabies from continental Europe. I would like to ask Dr Ross whether the introduction from France to Great Britain was through natural causes.

Ross — As far as I know, the answer is unknown. It's possible that a mosquito carrying virus crossed the Channel. It's also possible that a farmer went across and brought back a diseased carcass; but we do not know. It was certainly not done officially by the Ministry of Agriculture.

Zuckerman — May I add a point here. It may not have been done officially by the Ministry of Agriculture, but I know that many years ago a report went to the Ministry suggesting that it should be done.

Nelson — I am very worried at the suggestion that people are going to be playing around with the myxomatosis virus by genetic engineering. This could lead to the development of an organism infective to man similar to tanapox which produced an epidemic in east Africa rather like smallpox but with only one or two lesions in the skin. This virus was also thought to be mechanically transmitted by mosquitoes and it turned out to be related to the myxomatosis virus (Downie *et al.*, 1971).[1]

Ross — That's the general feeling in Australia as well, that they just don't want to mess around with it.

---

[1] Downie, A. W., Taylor-Robinson, C. H., Caunt, A. E. *et al.* (1971). Tanapox: a new disease caused by a pox virus. *Br. med. J.* **1971 1**: 363–368.

Symp. zool. Soc. Lond. (1982) No. 50, 97–119

Botulism in Waterfowl*

G. R. SMITH

*Nuffield Laboratories of Comparative Medicine, Institute of Zoology,
The Zoological Society of London, Regent's Park, London NW1, England*

SYNOPSIS

Botulism, a form of food poisoning affecting many vertebrate species, is caused
by ingestion of the lethal neurotoxin of the anaerobic bacterium *Clostridium
botulinum*, of which there are seven toxigenic types (A–G). In animals this paralytic disease is often caused by types C and D — types that rarely affect man.
Types A, B, E and F are mainly important as human pathogens. Botulism is
usually a pure intoxication, but the recent discovery of infant botulism in man
should stimulate further studies on bacterial multiplication and toxigenesis in
the intestine.

Type C kills millions of waterfowl each summer, in the western USA and
many other countries, on lakes and marshes permanently contaminated with
spores. Current studies show that the apparent absence of the disease in the
tropics is probably fallacious. In 1973, 50 000 birds died at the Coto Doñana
Reserve. Outbreaks occur in Britain, especially in the Norfolk Broads; in coastal
areas gulls are often affected, but the source of toxin is obscure.

Research at the Institute of Zoology has shown *C. botulinum* spores to be
surprisingly prevalent in British lakes and waterways. Type B is generally much
more common than C or E, but in the Norfolk Broads all three abound. More
than 70% of London's lakes contain *C. botulinum*. A current survey shows that
in the Mersey estuary, where recent avian mortality is being studied by several
groups, substantial type-C contamination is present. *C. botulinum* is much more
prevalent in lake mud than in soil, but the soil at the site of the former Metropolitan Cattle Market, London, is heavily contaminated. Some mud, such as that
in the Camargue, is inhibitory and might repay further study, with a view to
methods of control.

It is important to distinguish botulism from chemical intoxication. Examination of mud for type C spores can sometimes be useful as a retrospective diagnostic measure, and as an adjunct to wetland management.

BOTULISM AS A DISEASE OF MAN AND ANIMALS

Botulism is a form of food poisoning, caused by ingestion of the
lethal neurotoxin of *Clostridium botulinum*. This bacterium, whose

---

* This paper was originally given under the chairmanship of Dr D. A. J. Tyrrell, after the
session on *General epidemiological principles and policies*.

natural habitat is soil and sediment (freshwater and marine), grows only under conditions of strict anaerobiosis, and its vegetative cells are, moreover, sensitive to oxygen. It readily forms spores that remain alive for long periods under the most adverse natural conditions. The powerful heat-labile toxin, whose structure and biological properties have recently been reviewed by Sugiyama (1980), is released from ageing vegetative cells as they autolyse (Boroff, 1955; Bonventre & Kempe, 1960).

A suitable growth substrate, destined for consumption as food, may become contaminated with spores. Anaerobic conditions and a favourable temperature may then lead to germination, bacterial multiplication and toxigenesis. If the now lethally dangerous foodstuff is eaten, some of the preformed toxin will be absorbed from the small intestine and find its way into the lymph and blood, and thereby to the receptor sites. These are at the endings of the efferent autonomic and somatic nerves that supply muscles and glands. Only the cholinergic nerves are affected, the toxin apparently interfering with the acetylcholine release process (Kao, Drachman & Price, 1976). Botulism is thus a paralytic disease, death in fatal cases eventually resulting from respiratory or cardiac failure. Consumption of contaminated food leads to the ingestion of organisms as well as toxin, but, except under special circumstances (see pp. 109, 110), the disease is generally considered to be a pure intoxication, not an infection.

On the basis of immunological differences between the toxins they produce, there are seven different types (A–G) of *C. botulinum*. Criteria that are satisfactory bacteriologically but less so medically can be used to divide strains into four groups that bear only a slight relation to the toxigenic types (see L. DS. Smith, 1977). The toxins A–G all have similar pharmacological effects, but wide variations occur in their toxicity for different animal species (R. S. Roberts, 1959). They are easily distinguished from each other by neutralization tests in mice with specific antitoxins.

Strains of *C. botulinum* type C consist of two subtypes, Cα and Cβ. They and type D possess, in different proportions, the three toxic components C1, C2 and D (see G. R. Smith, 1976; L. DS. Smith, 1977); the production of these three components is mediated by bacteriophage (Inoue & Iida, 1970, 1971; Eklund, Poysky, Reed & Smith, 1971; Eklund, Poysky & Reed, 1972). *C. botulinum* types C and D can be interconverted by bacteriophage manipulation (Eklund & Poysky, 1974). Some strains of *C. botulinum* produce prototoxin that requires activation by trypsin (Duff, Wright & Yarinsky, 1956; Iida, 1968). These include most type E strains and some non-proteolytic strains of types B and F; some type Cα strains produce C2 toxin that also requires activation (Eklund & Poysky, 1972).

Towards the end of the eighteenth century medical authorities in Europe, especially Germany, became aware of the disease that was later named botulism (from the Latin *botulus*, meaning sausage) because of its frequent association with the consumption of blood sausage. The aetiology was elucidated by van Ermengem (1897) who investigated an illness affecting 23 musicians, three fatally, who ate raw salted ham after performing at a funeral in the Belgian village of Ellezelles. In 1919 two types of the causative organism were designated A and B. Three years later type Cα was isolated in the USA by Bengtson (1922) from larvae of the blowfly *Lucilia caesar*, and Cβ in Australia by Seddon (1922) from "bulbar paralysis" of cattle (see also Pfenninger, 1924; Gunnison & Meyer, 1929). In 1936 type E was isolated from fish in the Ukraine, and in 1960 type F from an outbreak of human botulism in Denmark. Type G was discovered in a sample of Argentinian soil in 1970 but has not yet been incriminated as a cause of natural disease.

Types A, B, E and F are mainly, though not exclusively, of importance as human pathogens. For reasons that are by no means clear, types C and D, which are more often than not responsible for botulism in animals, rarely if ever affect man (see Gilbert, 1974; L. DS. Smith, 1977). Today the human disease is well recognized in Europe, America, Japan and elsewhere, though fortunately it is rare. In Britain the most serious recorded outbreak was that now referred to as the Loch Maree tragedy (Leighton, 1923), in which eight people died after eating duck paste contaminated with type A toxin; the most recent outbreak, which occurred in Birmingham, affected four people, two fatally, and was caused by tinned salmon containing type E toxin (Ball *et al.*, 1979). Botulism is much more common in animals, especially waterfowl, than in man. In South Africa, before vaccination was introduced, about 100 000 cattle died annually from the disease (Sterne & Wentzel, 1952) mainly as a result of the depraved eating of carrion that sometimes occurred on phosphorus-deficient pastures. Cattle also die from "forage poisoning", in which foodstuffs become contaminated with rat and other carcasses containing toxin. Botulism occurs in sheep, horses, mink, ferrets, dogs (mainly foxhounds), turtles, chickens, and many other species, especially of birds. Massive outbreaks, with 10 000 or more deaths, sometimes occur on pheasant farms (Vadlamudi, Lee & Hanson, 1959). Huss & Eskildsen (1974) described botulism on trout farms caused by feeding spoiled marine-fish scraps. Smart, Roberts, McCullagh, Lucke & Pearson (1980) reported 14 fatal cases in an outbreak of type C botulism in captive primates — squirrel monkeys (*Saimiri sciureus*), white-throated capuchins (*Cebus albifrons*), and weeper capuchins (*C. nigrivittatus*) — given contaminated chopped chicken.

BOTULISM IN WATERFOWL

Historical Aspects

From about the year 1910, reports appeared annually of summer mortality, often on a huge scale, in waterfowl on lakes and mudflats in the western USA. The 1910 outbreak on the Great Salt Lake was exceptionally severe. In 1932, fully 250 000 birds died at the northern end of the lake, where "The dead lay at the rate of 8000 to 10 000 to the mile of shore line" (Kalmbach, 1935). Whether outbreaks of such magnitude occurred in the nineteenth century is in doubt. Kalmbach (1968) comments on the lack of evidence in reports written by the observant naturalists who accompanied the various early expeditions into the western regions. The disease, which became known as "western duck sickness", was initially thought to be due to alkaline poisoning, but this explanation lost credibility as outbreaks began to be reported in aquatic environments less highly saline than the Great Salt Lake. Some 20 years after it was first reported, and 33 years after van Ermengem's discovery of *C. botulinum*, western duck sickness was revealed as type C botulism (Giltner & Couch, 1930; Hobmaier, 1930; Kalmbach, 1930; Kalmbach & Gunderson, 1934). In 1952 severe flooding was associated with no less than 4—5 million fatal cases in the western USA (Rosen, 1971).

Present Status

Type C botulism in waterfowl is by no means confined to the USA, and millions of deaths occur annually in lakes, marshes and waterways throughout the world. The countries known to be infected include Canada, Mexico, Argentina, Uruguay, Australia, New Zealand, South Africa, Germany, and Sweden (see G. R. Smith, 1976). Recent serious outbreaks in western Europe have attracted much attention, but the disease has probably existed there for many years (G. R. Smith, 1975). Haagsma, Over, Smit & Hoekstra (1972) reported mass mortality in the Netherlands. Mountfort (1973) described a disastrous outbreak, initially thought to be due to pesticide poisoning, in the Coto Doñana Reserve at the mouth of the Rio Guadalquivir. About 50 000 birds of various species, including 80% of the spoonbill colony, died in the marshes of this famous reserve, jointly created by the World Wildlife Fund and the Spanish Government. There has been further mortality since 1973. The area of the marshes in which most of the deaths took place was shown to be saturated with type C spores (Haagsma & Koeman, 1974; G. R. Smith, 1975, unpublished). Contamination also occurred further afield (G. R. Smith, 1975, unpublished) in the Laguna Dulce de Zorilla (Espera, Cádiz), and in two sites associated with mortality in 1974, namely, L. del Taraje (Las Cabezas, Sevilla) and L. de Medina (Jerez, Cádiz). The problem is therefore probably not strictly localized to Doñana.

In 1969 the disease was confirmed for the first time in Britain, deaths being reported in the Midlands (Blandford, Roberts & Ashton, 1969) and in St. James's Park, London (Keymer, Smith, Roberts, Heaney & Hibberd, 1972). Since then, numerous outbreaks have been recognized. Lloyd, Thomas, Macdonald, Borland, Standring & Smart (1976) stated that, during the unusually hot summer of 1975, at least nine confirmed and 11 unconfirmed outbreaks took place in waterfowl, including gulls, in England, Scotland, Wales, and Northern Ireland. These outbreaks included one that caused the death of several thousand birds on the Norfolk Broads (Borland, Moryson & Smith, 1977) and was followed by a similar episode in 1976.

Natural Distribution of *C. botulinum*

The studies of Meyer & Dubovsky and many others since (see Meyer, 1956; L. DS. Smith, 1977) have given much information on the prevalence and regional distribution of the various types of *C. botulinum* in soil. Type A is, for example, common in the western half of the USA, whereas type B predominates in the east; and type B is much more prevalent than A in Europe. Many studies have also been made of marine sediment, especially in relation to type E in fishery products (Hobbs, 1976). Less work has been done on inland freshwater sediment, but the geographical incidence of botulism in waterfowl demonstrates that type C occurs in many countries.

In Britain studies by Meyer & Dubovsky (1922), Leighton & Buxton (1928), and Haines (1942) showed that 4–8% of soil samples contained type A or B. In relation to botulism in waterfowl, surveys of aquatic environments in Britain have been made at the Nuffield Laboratories of Comparative Medicine, The Institute of Zoology, London. The method, fully described by G. R. Smith & Moryson (1975), consisted of washing a considerable proportion of the bacterial spores from a 50-g mud sample and culturing them, with and without preliminary heating at 60°C for one hour, in cooked meat medium for six to eight days at 30°C. Culture supernate, sterilized by filtration, was then tested biologically, in both the trypsinized and untrypsinized state, for *C. botulinum* toxin. In positive samples the toxin was subsequently identified by neutralization tests. The results are summarized in Table I.

A very high proportion — more than 70% — of London's lakes contained *C. botulinum* (Smith & Moryson, 1975). Of the various

TABLE I

The prevalence of Clostridium botulinum *in different aquatic areas (after G. R. Smith, 1978)*

Area surveyed	Number of mud samples	Percentage of samples containing *C. botulinum* of					
		any type	type B	type C	type D	type E	more than one type
London	69	72[a]	45	17	1	14	6
Norfolk Broads	45	98	62	51	0	60	58
Britain and Ireland (excluding London and Norfolk Broads)	554	35	30	3	1	3	2
Camargue	44	5	0	0	0	5	0

[a] Apparently negative London sites were resampled; this increased the percentage of positive results from 55 to 72.

types found, B was the most common, but type E — thought previously to be absent from this country — and type C were not uncommon. Mud samples taken from the Norfolk Broads during and shortly after a serious outbreak of avian botulism were almost uniformly positive, types B, C and E each being present in more than 50% (Borland *et al.*, 1977); over half the samples contained more than one type. When the survey work was extended (G. R. Smith, Milligan & Moryson, 1978) to other parts of the UK and to Eire it became clear that whereas type B was generally common (30% of all samples), types C and E — although very prevalent in particular localities — were by no means ubiquitous (only 3% of all samples). A small survey made for the purpose of comparison showed that *C. botulinum* was almost entirely absent from the Camargue in France (G. R. Smith, Moryson & Walmsley, 1977), despite the constant seeding of the environment that presumably occurred through the intermediary of migrant birds. This suggested that Camargue mud was inhibitory. Graham (1978) isolated from inhibitory mud samples, including some from the Camargue, numerous strains of *Bacillus* spp. that inhibited the growth of *C. botulinum* type C *in vitro*. In further studies made by their standard technique G. R. Smith & Moryson

(1977) and G. R. Smith & Young (1980) confirmed that in Britain *C. botulinum* is much more prevalent in mud from inland waters than in soil. One exception, however, came to light. The soil in a small grassy park on the site of the former (1855–1939) Metropolitan Cattle Market, London, showed exceptionally heavy contamination (G. R. Smith & Milligan, 1979); 25% of 60 samples contained *C. botulinum* and no less than four types (B, C, D, and E) were demonstrated. However, at three further sites associated for many years with animals — the Zoological Society's premises at Regent's Park and Whipsnade, and the Market Paddocks, Gorgie, Edinburgh — the prevalence of *C. botulinum* in the soil was either very low or nil (G. R. Smith & Young, 1980).

Epidemiology

The disease is almost always caused by *C. botulinum* sub-type Cα. Sub-type Cβ may occasionally be responsible (Pullar, 1934; Reilly & Boroff, 1967) but, as Hariharan & Mitchell (1977) observe, in the routine diagnosis of avian botulism the sub-type is seldom determined. Type E may have played a part in avian botulism on Lake Michigan (see p. 106).

Type C strains grow at 15.6°C but not 10°C (Segner, Schmidt & Boltz, 1971). Large amounts of toxin are produced at temperatures as high as 37°C (Boroff & Reilly, 1959). Strains lacking the bacteriophage that mediates toxigenesis produce no toxin. Some strains have a strong tendency to lose their toxigenicity spontaneously in culture.

The disease occurs not only in freshwater environments but also in marshes surrounding highly saline environments such as the Great Salt Lake. Type C strains have often been isolated from marine sediment, and outbreaks of botulism have been reported on a tidal estuary in New Jersey (Reilly & Boroff, 1967) and in the delta area of the Naka River in Tokyo (Sakai, Itoh, Okazawa, Zen-Yoji & Benoki, 1975).

It is likely that waterfowl spread type C organisms from one lake to another by means of spores carried in the gut or on the body surfaces (Haagsma, 1974). If, having just ingested a lethal dose of toxin, a bird flies to another lake, it will after death seed the new environment with massive numbers of type C organisms produced by invasion of the putrefying carcass from the gut. The new environment may or may not provide conditions that permit the establishment of *C. botulinum* (see G. R. Smith, 1976).

The factors that favour precipitation of an outbreak in an area already contaminated with type C spores are: a prolonged period of warm weather; increasing areas of shallow stagnant water; alkalinity; an abundance of dead aquatic invertebrates; and oxygen depletion associated with rotting vegetation and other organic matter (see G. R. Smith, 1976). According to L. P. Smith (1979), botulism on the Norfolk Broads is closely related to the number of days in summer with a maximum air temperature of 21°C or more, and to the excess of water evaporation over rainfall from April to August. Thermal pollution of water by power stations may have played a part in Dutch outbreaks (Haagsma *et al.*, 1972). A favourable combination of precipitating factors leads to germination, explosive multiplication and toxigenesis.

Bacterial multiplication may occur in sludge or rotting vegetation, or inside particles of organic matter such as decaying aquatic invertebrates (Bell, Sciple & Hubert, 1955). Such particles may protect the organism from an unfavourable macroenvironment. Waterfowl carcasses also provide hospitable surroundings and may become contaminated with massive concentrations of toxin, e.g. 1.4×10^7 intraperitoneal mouse-lethal doses per gram of muscle (Haagsma & Koeman, 1974), a quantity sufficient to kill at least 40 ducks. Dipterous fly larvae that crawl through and ingest the contaminated flesh may become dangerous sources of toxin for other birds (see Rosen, 1971; G. R. Smith, Hime, Keymer, Graham, Olney & Brambell, 1975). In the course of an outbreak environmental contamination increases, thus enhancing the likelihood of future outbreaks. Toxin can persist for considerable periods. In an unusual episode on the Norfolk Broads, several waterfowl died from botulism in March and April 1977 (Graham, Smith, Borland & Macdonald, 1978); the evidence suggested that the toxin had been formed the previous summer.

The Disease in the Tropics

The almost total absence of reports of botulism in waterfowl from the tropics is a matter that requires study. Possibly the disease occurs there but remains undiagnosed. On the other hand outbreaks in temperate climates are precipitated by conditions that are lacking in the tropics, where the temperature of the mud remains comparatively high and fluctuates little.

With the assistance of colleagues at the British Museum (Natural History) and the University of Surrey (Department of Microbiology) we have begun to collect for examination samples of inland freshwater sediment from tropical countries. The study is still in progress, but the early findings are of interest. Of 17 mud samples from

Mauritius, 18 from Nigeria, and eight from tropical Botswana, five, five and one respectively contained *C. botulinum* type C. The ease with which the organism is being found suggests that it multiplies freely in tropical waters; if so, toxin must be released and, although not reported, botulism in waterfowl must surely occur.

Gulls

For more than 120 years occasional outbreaks of disease, resulting in mass mortality ("wrecks") of seabirds, have been known around the shores of the British Isles (NERC, 1971). When such calamities occur, the causes — of which there may be many — often remain obscure.

Since about 1975 there has been a steadily growing awareness that in Britain type C botulism plays a rôle of some importance in gull mortality. In that year Macdonald & Standring (1978) investigated the first recorded outbreak in gulls in this country. It occurred between June and October around the Firth of Forth, causing the deaths of at least 2080 birds belonging to several species of the genus *Larus*. In the same summer further outbreaks in Northern Ireland, Anglesey, and the Wirral (Cheshire) were confirmed by laboratory examination (see Lloyd *et al.*, 1976).

At Walney Island (Cumbria) considerable annual mortality has been observed since 1975 between the months of April and September, and laboratory confirmation of botulism was obtained in 1978, 1980, and 1981; between 26 April and 30 June 1978 approximately 1600 deaths occurred in the breeding colony of 60 000 *L. argentatus* and *L. fuscus* (R. McCleery & A. Hart, pers. comm.). In this laboratory type C toxin was found in deep frozen serum taken in 1981 from three moribund gulls on Walney Island as early as 20 May — at least a month before deaths from summer botulism begin to occur in mallard (*Anas platyrhyncos*) in Britain. This observation is of possible interest in relation to the source of toxin for gulls, at present unknown.

Deaths of *L. argentatus* and *L. ridibundus* from botulism were reported in Motherwell, Scotland, in December 1976, a month in which the local mean maximal air temperature was only 3.5°C (Graham *et al.*, 1978). This temperature was well below the minimum necessary for growth and toxin production. A search for the source of toxin proved fruitless, but the scavenging habits of gulls were suspected to be in some way responsible for the disease.

In a recent study of gull breeding colonies in the inner Bristol Channel, Mudge & Ferns (1980) reported that since 1975 the numbers of *L. argentatus*, *L. fuscus* and *L. marinus* had decreased by 67,

30 and 8% respectively; laboratory findings gave a "strong indication" that botulism was responsible.

As recently as August 1981, botulism in gulls was diagnosed at the Veterinary Investigation Centres in Aberystwyth, Bangor, Leeds and Liverpool (Report, 1981).

G. R. Smith (1976) suggested refuse tips as one possible source of toxin for gulls, and found *C. botulinum* types B and C in a single sample of refuse sludge sent by Macdonald & Standring (1978) during their investigation of the 1975 outbreak on the Firth of Forth. A survey of refuse tips for *C. botulinum* would be of considerable interest.

Evidence regarding the possible occurrence of type E botulism in gulls and other birds on Lake Michigan has been discussed by Fay, Kaufmann & Ryel (1965) and others (see G. R. Smith, 1976).

Diagnosis

When large numbers of birds die, apparently from some form of poisoning, it is important — though in practice not always easy — to distinguish botulism from chemical intoxication. Botulism often occurs unexpectedly, and an experienced diagnostician may not be immediately available. The initial samples are, moreover, often collected by a field worker, who may not understand what kind of material is required. It is conceivable that in an area contaminated with type C spores carcasses produced by chemical poisoning will generate large amounts of botulinal toxin and lead to cases of botulism. There may even be interaction between chemical and botulinal toxins (Jensen & Micuda, 1970).

Clinical signs and post-mortem appearance

Botulism has been recorded in a very wide range of avian species (Kalmbach & Gunderson, 1934; Keymer *et al.*, 1972). The clinical signs vary according to the amount of toxin ingested, the time that has elapsed since ingestion, and perhaps also to individual and species susceptibility. Flaccid paralysis of the limbs and neck (Fig. 1), ocular disturbances and respiratory distress are the main features (G. R. Smith, 1976), but the clinical signs are rarely sufficiently specific to permit a firm diagnosis without supporting evidence. There are no characteristic post-mortem lesions.

Demonstration of toxaemia

Diagnosis is usually based on the demonstration, by mouse inocu-

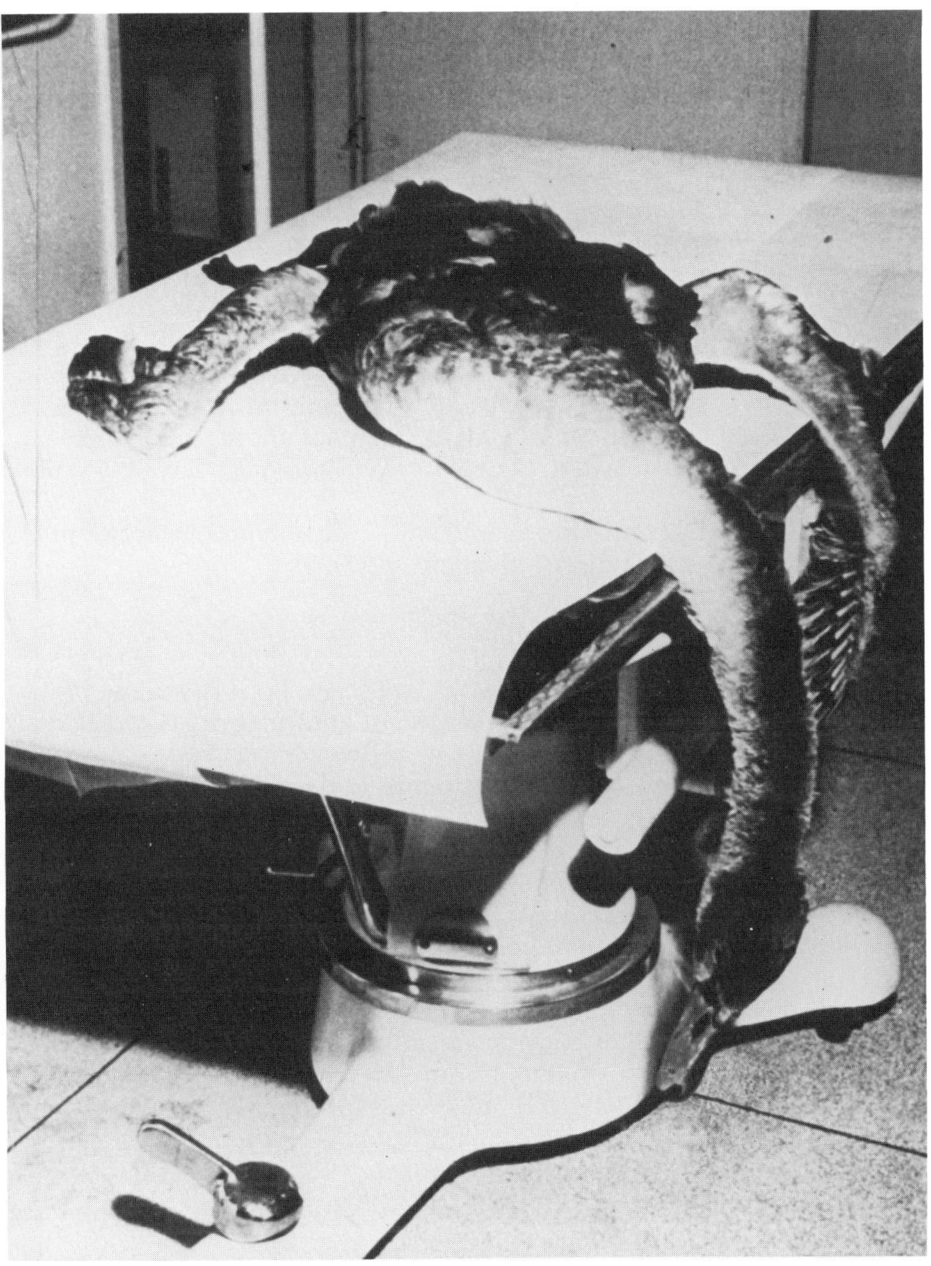

FIG. 1. Flaccid paralysis in a young swan with advanced botulism.

lation and neutralization tests, of type C toxin in the serum of affected birds. Demonstration of toxin in the alimentary tract or in faeces (Jensen, 1981) is usually less successful. The serum of birds with botulism can, on rare occasions, contain several thousand mouse minimal lethal doses (MLD) per millilitre (G. R. Smith, 1980). More often it contains less than 20 MLD, and in many instances is devoid of toxin. It is important, therefore, to examine serum samples from several birds with early and advanced clinical signs, in the hope that toxin can be demonstrated in at least one. The size of the samples submitted should be as generous as possible so that each mouse can be given a large intraperitoneal dose, preferably about 0.5 ml. The successful examination of one positive sample requires at least four mice, and usually more. One-tenth of a unit of C antitoxin will neutralize about 1000 mouse MLD. Samples should not be taken from putrefying carcasses, because toxigenesis may have occurred after death.

The demonstration of toxin in food — a common diagnostic procedure for human botulism — is rarely applicable.

Demonstration of *C. botulinum*

Sometimes a request is made for a retrospective opinion as to whether an outbreak of disease in waterfowl could have been caused by botulism. In such instances the examination of mud for type C spores can be helpful, as is illlustrated by the two following examples.

Towards the end of 1978 a request was received for advice on deaths in flamingoes on the island of Sardinia, strongly suspected by local authorities to be caused by botulism. A few weeks later about a dozen mud samples were received from the aquatic area in which the mortality occurred. Examination by the standard method in use at the Institute of Zoology gave completely negative results. Had an outbreak of botulism occurred, at least some of the samples would surely have been shown to contain type C.

The second example concerns a collaborative study, still in progress, with the North West Water Authority. During the period September to December 1979, at least 2400 birds were found dead or dying along the shoreline in the middle reaches of the Mersey estuary (Head, D'Arcy & Osbaldeston, 1980). Further mortality occurred in August and September 1980. One group has already found a high concentration of lead, a proportion of it in the organic trialkyl form, in the tissues of affected birds (see Head *et al.*, 1980). We are finding that a substantial proportion of mud samples from the Mersey estuary contains *C. botulinum* type C. This is thought-provoking, especially in the light of the earlier survey work (see p. 102;

G. R. Smith, Milligan *et al.*, 1978), which showed that type C — unlike type B — is far from ubiquitous in this country.

The examination of soil samples for type C spores might be used with advantage in wetland management, to distinguish between areas of high and low risk from botulism.

Immunity

Waterfowl that ingest only a small amount of toxin often recover from botulism yet remain susceptible. Some ducks suffer as many as three attacks within a single summer (Rosen, 1971). Haagsma (1974) found that, after recovery from botulism, mallard failed to resist even one MLD of type C toxin, and repeated sub-lethal doses of toxin failed to produce antitoxin. Attempts to demonstrate antitoxin in the serum of normal animals, including man, have generally been unsuccessful (Holdeman, 1970; Lamanna, 1970). Because of the extreme potency of *C. botulinum* toxin, it is likely that sub-lethal doses are too small to stimulate an immune response. Kaufmann & Crecelius (1967) found, however, that repeated sub-lethal doses of type E toxin by mouth produced an increase in the resistance of gulls. Kalmbach (1939) observed that carrion-eating birds such as the turkey vulture (*Cathartes aura*) were resistant to *C. botulinum* toxin. The work of Pates and her colleagues (see Holdeman, 1970) suggested that C antitoxin occurred in vultures. Ohishi, Sakaguchi, Riemann, Behymer & Hurvell (1979) found naturally occurring antibodies against toxins A—F in turkey vultures, coyotes (*Canis latrans*) and crows (*Corvus brachyrhynchos*).

Vaccination of waterfowl is inapplicable except in small valuable collections on contaminated lakes.

Control

The measures available are palliative only. They include various forms of water manipulation, and keeping birds away from dangerous areas (see G. R. Smith, 1976). It is known, however, that some soil organisms (L. DS. Smith, 1975; Graham, 1978) and the sediments from certain aquatic environments (see p. 102; Sugiyama, Bott & Foster, 1970; G. R. Smith, 1978) are inhibitory to *C. botulinum*. Further studies of the factors responsible for the inhibitory nature of certain sediments would seem to offer at least some hope.

BACTERIAL MULTIPLICATION AND TOXIGENESIS *IN VIVO*

Botulism is usually a pure intoxication, but wound botulism in man (see Merson & Dowell, 1973), although rare, shows that under certain circumstances *C. botulinum* can multiply and produce toxin in the tissues. Russian workers such as Shvedov (1960), Minervin (1967) and others (see Dolman, 1961; Petty, 1965) have long maintained that toxigenesis sometimes occurs in the alimentary tract.

The recent discovery of human infant botulism has provided clear evidence of intra-intestinal toxin production (Midura & Arnon, 1976; Arnon, Midura, Clay, Wood & Chin, 1977; Turner, Brett, Gilbert, Ghosh & Liebeschuetz, 1978). Cases caused by *C. botulinum* types A, B and F have been recorded in infants aged from two to 26 weeks. The diets fed to babies of this age group virtually preclude the ingestion of preformed toxin. Organisms and toxin are sometimes found in the faeces many weeks after clinical recovery.

In some outbreaks of botulism in chickens there is no obvious source of toxin, but *C. botulinum* type C is present in considerable numbers in the intestinal contents and litter (T. A. Roberts, Thomas & Gilbert, 1973; Smart & Roberts, 1977). This is suggestive of bacterial multiplication and toxigenesis in the intestine, a possibility now supported by experimental evidence. Miyazaki & Sakaguchi (1978) found that fatal amounts of toxin were produced and absorbed in the caecum of chickens aged 14 days given 100 type C spores directly into the crop.

Some workers have for many years suspected that *in vivo* toxigenesis sometimes occurs in waterfowl, either in the intestine or in internal organs (Boroff & Reilly, 1962). Type C organisms have not infrequently been cultured from the liver of birds dying from botulism (Hobmaier, 1932; Kalmbach & Gunderson, 1934; Bell *et al.*, 1955; Boroff & Reilly, 1962; Dolman, 1964; Haagsma, 1974). Graham & Smith (1977), who used strict aseptic precautions to prevent accidental contamination with spores, found no evidence of invasion of the blood or liver in 27 fresh carcasses of affected waterfowl from four natural outbreaks of botulism. In the light of recent advances, however, studies on possible toxigenesis in the intestines of waterfowl should be made.

CONCLUSIONS

Botulism is an important environmental disease that continues to take a massive toll of water birds. It must be distinguished from chemical poisoning, but in practice diagnosis is not always easy. Indirect methods based on the demonstration of type C spores in mud are sometimes helpful. Increased diagnostic vigilance usually reveals that the disease is more common than was at first suspected. The environmental source of toxin for gulls is a matter requiring further study. Recent work in the medical and agricultural fields shows that toxigenesis occasionally occurs *in vivo*. Control of the disease in waterfowl remains a difficult problem, but further studies of mud known to be inhibitory for *C. botulinum* may offer at least some hope for the future.

REFERENCES

Arnon, S. S., Midura, T. F., Clay, S. A., Wood, R. M. & Chin, J. (1977). Infant botulism: epidemiological, clinical, and laboratory aspects. *J. Am. med. Ass.* **237**: 1946–1951.

Ball, A. P., Hopkinson, R. B., Farrell, I. D., Hutchinson, J. G. P., Paul, R., Watson, R. D. S., Page, A. J. F., Parker, R. G. F., Edwards, C. W., Snow, M., Scott, D. K., Leone-Ganado, A., Hastings, A., Ghosh, A. C. & Gilbert, R. J. (1979). Human botulism caused by *Clostridium botulinum* type E: the Birmingham outbreak. *Q. Jl Med.* (N.S.) **48**: 473–491.

Bell, J. F., Sciple, G. W. & Hubert, A. A. (1955). A microenvironment concept of the epizootiology of avian botulism. *J. Wildl. Mgmt* **19**: 352–357.

Bengtson, I. A. (1922). Preliminary note on a toxin-producing anaerobe isolated from the larvae of *Lucilia caesar*. *Publ. Hlth Rep., Wash.* **37**: 164–170.

Blandford, T. B., Roberts, T. A. & Ashton, W. L. G. (1969). Losses from botulism in Mallard duck and other waterfowl. *Vet. Rec.* **85**: 541–543.

Bonventre, P. F. & Kempe, L. L. (1960). Physiology of toxin production by *Clostridium botulinum* types A and B. I. Growth, autolysis and toxin production. *J. Bact.* **79**: 18–23.

Borland, E. D., Moryson, C. J. & Smith, G. R. (1977). Avian botulism and the high prevalence of *Clostridium botulinum* in the Norfolk Broads. *Vet. Rec.* **100**: 106–109.

Boroff, D. A. (1955). Study of toxins of *Clostridium botulinum*. III. Relation of autolysis to toxin production. *J. Bact.* **70**: 363–367.

Boroff, D. A. & Reilly, J. R. (1959). Studies on the toxin of *Clostridium botulinum*. V. Prophylactic immunization of pheasants and ducks against avian botulism. *J. Bact.* **77**: 142–146.

Boroff, D. A. & Reilly, J. R. (1962). Studies on the toxin of *Clostridium botulinum*. VI. Botulism among pheasants and quail, mode of transmission and degree of resistance offered by immunization. *Int. Archs Allergy* **20**: 306–313.

Dolman, C. E. (1961). Further outbreaks of botulism in Canada. *Can. med. J.* **84**: 191–200.

Dolman, C. E. (1964). Botulism as a world health problem. In *Botulism: proceedings of a symposium*: 5–32. Lewis, K. H. & Cassel, K. (Eds). Cincinnati, Ohio: US Department Health and Educational Welfare.

Duff, J. T., Wright, G. G. & Yarinsky, A. (1956). Activation of *Clostridium botulinum* type E toxin by trypsin. *J. Bact.* **72**: 455–460.

Eklund, M. W. & Poysky, F. T. (1972). Activation of a toxic component of *Clostridium botulinum* types C and D by trypsin. *Appl. Microbiol.* **24**: 108–113.

Eklund, M. W. & Poysky, F. T. (1974). Interconversion of type C and D strains of *Clostridium botulinum* by specific bacteriophages. *Appl. Microbiol.* **27**: 251–258.

Eklund, M. W., Poysky, F. T. & Reed, S. M. (1972). Bacteriophage and the toxigenicity of *Clostridium botulinum* type D. *Nature, New Biol.* **235**: 16–17.

Eklund, M. W., Poysky, F. T., Reed, S. M. & Smith, C. A. (1971). Bacteriophage and the toxigenicity of *Clostridium botulinum* type C. *Science, N.Y.* **172**: 480–482.

Ermengem, E. van (1897). Ueber einen neuen anaeroben Bacillus und seine Beziehungen zum Botulismus. *Z. Hyg. InfectKrankh.* **26**: 1–56.

Fay, L. D., Kaufmann, O. W. & Ryel, L. A. (1965). Mass mortality of waterbirds in Lake Michigan 1963–64. *Publs Gr. Lakes Research Div.* No. **13**: 34–46.

Gilbert, R. J. (1974). Staphylococcal food poisoning and botulism. *Postgrad. med. J.* **50**: 603–611.

Giltner, L. T. & Couch, J. F. (1930). Western duck sickness and botulism. *Science, N.Y.* **72**: 660.

Graham, J. M. (1978). Inhibition of *Clostridium botulinum* type C by bacteria isolated from mud. *J. appl. Bact.* **45**: 205–211.

Graham, J. M. & Smith, G. R. (1977). Observations on the possible invasiveness of *Clostridium botulinum* for waterfowl. *Res. vet. Sci.* **22**: 343–346.

Graham, J. M., Smith, G. R., Borland, E. D. & Macdonald, J. W. (1978). Avian botulism in winter and spring and the stability of *Clostridium botulinum* type C toxin. *Vet. Rec.* **102**: 40–41.

Gunnison, J. B. & Meyer, K. F. (1929). Cultural study of an international collection of *Clostridium botulinum* and *parabotulinum*. XXXVIII. *J. infect. Dis.* **45**: 119–134.

Haagsma, J. (1974). Etiology and epidemiology of botulism in waterfowl in the Netherlands. *Tijdschr. Diergeneesk.* **99**: 434–442.

Haagsma, J. & Koeman, J. H. (1974). *An investigation into the cause of death of birds in the Coto Doñana nature reserve in Spain in the summer and autumn of 1973*. World Wildlife Fund Report.

Haagsma, J., Over, H. J., Smit, T. & Hoekstra, J. (1972). Botulism in waterfowl in the Netherlands in 1970. *Neth. J. vet. Sci.* **5**: 12–33.

Haines, R. B. (1942). The occurrence of toxigenic anaerobes, especially *Clostridium botulinum*, in some English soils. *J. Hyg., Camb.* **42**: 323–327.

Hariharan, H. & Mitchell, W. R. (1977). Type C botulism: the agent, host spectrum and environment. *Vet. Bull.* **47**: 95–103.

Head, P. C., D'Arcy, B. J. & Osbaldeston, P. J. (1980). The Mersey estuary bird mortality autumn-winter 1979: summary report. *NWest Water scient. Rep.* DSS-EST-80-2.

Hobbs, G. (1976). *Clostridium botulinum* and its importance in fishery products. *Adv. food Res.* **22**: 135–185.

Hobmaier, M. (1930). Duck disease caused by the poison of the *Bacillus botulinus*. *Calif. Fish Game* **16**: 285–286.

Hobmaier, M. (1932). Conditions and control of botulism (duck disease) in waterfowl. *Calif. Fish Game* **18**: 5–21.

Holdeman, L. V. (1970). The ecology and natural history of *Clostridium botulinum*. *J. Wildl. Dis.* **6**: 205–210.

Huss, H. H. & Eskildsen, U. (1974). Botulism in farmed trout caused by *Clostridium botulinum* type E. *Nord. VetMed.* **26**: 733–738.

Iida, H. (1968). Activation of *Clostridium botulinum* toxin by trypsin. In *Toxic microorganisms*: 336—340. Herzberg, M. (Ed.). Washington D.C.: UJNR Joint Panels on Toxic Microorganisms, and US Department of the Interior.

Inoue, K. & Iida, H. (1970). Conversion of toxigenicity in *Clostridium botulinum* type C. *Jap. J. Microbiol.* 14: 87—89.

Inoue, K. & Iida, H. (1971). Phage conversion of toxigenicity in *Clostridium botulinum* types C and D. *Jap. J. med. Sci. Biol.* 24: 53—56.

Jensen, W. I. (1981). Evaluation of coproexamination as a diagnostic test for avian botulism. *J. Wildl. Dis.* 17: 171—176.

Jensen, W. I. & Micuda, J. M. (1970). The effect of malathion on the susceptibility of the mallard duck (*Anas platyryhyncos*) to *Clostridium botulinum* type C toxin. In *Toxic microorganisms*: 372—375. Herzberg, M. (Ed.). Washington D.C.: UJNR Joint Panels on Toxic Microorganisms, and US Department of the Interior.

Kalmbach, E. R. (1930). Western duck sickness produced experimentally. *Science, N.Y.* 72: 658—659.

Kalmbach, E. R. (1935). Will botulism become a world-wide hazard to wild fowl? *J. Am. vet. med. Ass.* 87: 183—187.

Kalmbach, E. R. (1939). American vultures and the toxin of *Clostridium botulinum*. *J. Am. vet. med. Ass.* 94: 187—191.

Kalmbach, E. R. (1968). Type C botulism among wild birds — a historical sketch. *Wildlife* No. 110: 1—8. Washington D.C.: Bur. Sport Fisheries Wildlife.

Kalmbach, E. R. & Gunderson, M. F. (1934). Western duck sickness: a form of botulism. *USDA tech. Bull.* No. 411: 1—81.

Kao, I., Drachman, D. B. & Price, D. L. (1976). Botulinum toxin: mechanism of presynaptic blockade. *Science, N.Y.* 193: 1256—1258.

Kaufmann, O. W. & Crecelius, M. S. (1967). Experimentally induced immunity in gulls to type E botulism. *Am. J. vet. Res.* 28: 1857—1862.

Keymer, I. F., Smith, G. R., Roberts, T. A., Heaney, S. I. & Hibberd, D. J. (1972). Botulism as a factor in waterfowl mortality at St. James's Park, London. *Vet. Rec.* 90: 111—114.

Lamanna, C. (1970). Critical comment on research needs in botulism: ecology, nature and action of toxin. In *Toxic microorganisms*: 230—235. Herzberg, M. (Ed.). Washington D.C.: UJNR Joint Panels on Toxic Microorganisms, and US Department of the Interior.

Leighton, G. R. (1923). *Report of the circumstances attending the deaths of eight persons from botulism at Loch Maree (Ross-shire).* (Report Scottish Board Health) Edinburgh: HMSO.

Leighton, G. R. & Buxton, J. B. (1928). The distribution of *Bacillus botulinus* in Scottish soils. *J. Hyg., Lond.* 28: 79—82.

Lloyd, C. S., Thomas, G. J., Macdonald, J. W., Borland, E. D., Standring, K. & Smart, J. L. (1976). Wild bird mortality caused by botulism in Britain, 1975. *Biol. Conserv.* 10: 119—129.

Macdonald, J. W. & Standring, K. T. (1978). An outbreak of botulism in gulls on the Firth of Forth, Scotland. *Biol. Conserv.* 14: 149—155.

Merson, M. H. & Dowell, V. R. (1973). Epidemiological, clinical and laboratory aspects of wound botulism. *New Engl. J. Med.* No. 289: 1005—1010.

Meyer, K. F. (1956). The status of botulism as a world health problem. *Bull. Wld Hlth Org.* 15: 281—298.

Meyer, K. F. & Dubovsky, B. J. (1922). The occurrence of the spores of *Bacillus botulinus* in Belgium, Denmark, England, the Netherlands and Switzerland. *J. infect. Dis.* **31**: 600—609.

Midura, T. F. & Arnon, S. S. (1976). Infant botulism: identification of *Clostridium botulinum* and its toxins in faeces. *Lancet* 1976 (ii): 934—936.

Minervin, S. M. (1967). On the parenteral-enteral method of administering serum in cases of botulism. In *Botulism 1966 (Proc. 5th int. Symp. Food Microbiol.)*: 336—345. Ingram, M. & Roberts, T. A. (Eds). London: Chapman & Hall.

Miyazaki, S. & Sakaguchi, G. (1978). Experimental botulism in chickens: the cecum as the site of production and absorption of botulinum toxin. *Jap. J. med. Sci. Biol.* **31**: 1—15.

Mountfort, G. (1973). Wildlife disaster in Spain. *The Times* October 9th.

Mudge, G. P. & Ferns, P. N. (1980). A census of breeding gulls in the inner Bristol Channel. *Rep. Dept Energy Proj.* No. YD1-40-22.

NERC (1971). The sea bird wreck in the Irish Sea, autumn 1969. *Publs nat. Environm. Res. Counc.*, (C.) No. 4: 1—17.

Ohishi, I., Sakaguchi, G., Riemann, H., Behymer, D. & Hurvell, B. (1979). Antibodies to *Clostridium botulinum* toxins in free-living birds and mammals. *J. Wildl. Dis.* **15**: 3—9.

Petty, C. S. (1965). Botulism: the disease and the toxin. *Am. J. med. Sci.* No. 249: 345—359.

Pfenninger, W. (1924). Toxico-immunologic and serologic relationship of *B. botulinus*, type C, and *B. parabotulinus*, 'Seddon'. XXII. *J. infect. Dis.* **35**: 347—352.

Pullar, E. M. (1934). Enzootic botulism amongst wild birds. *Aust. vet. J.* **10**: 128—135.

Reilly, J. R. & Boroff, D. A. (1967). Botulism in a tidal estuary in New Jersey. *Wildl. Dis.* **3**: 26—29.

Report (1981). Gulls recover from botulism. *Vet. Rec.* **109**: 397.

Roberts, R. S. (1959). Clostridial diseases. In *Infectious diseases of animals: diseases due to bacteria* 1: 160—228. Stableforth, A. W. & Galloway, I. A. (Eds). London: Butterworths.

Roberts, T. A., Thomas, A. I. & Gilbert, R. J. (1973). A third outbreak of type C botulism in broiler chickens. *Vet. Rec.* **92**: 107—109.

Rosen, M. N. (1971). Botulism. In *Infectious and parasitic diseases of wild birds*: 100—117. Davis, J. W., Anderson, R. C., Karstad, L. & Trainer, D. O. Ames: Iowa State University Press.

Sakai, S., Itoh, T., Okazawa, K., Zen-Yoji, H. & Benoki, M. (1975). An outbreak of botulism in waterfowl at the area of the Naka river in Tokyo. *A. Rep. Tokyo metr. res. Lab.* **26**: 14—19.

Seddon, H. R. (1922). Bulbar paralysis in cattle due to the action of a toxicogenic bacillus, with a discussion on the relationship of the condition to forage poisoning (botulism). *J. comp. Path.* **35**: 147—190.

Segner, W. P., Schmidt, C. F. & Boltz, J. K. (1971). Minimal growth temperature, sodium chloride tolerance, pH sensitivity and toxin production of marine and terrestrial strains of *Clostridium botulinum* type C. *Appl. Microbiol.* **22**: 1025—1029.

Shvedov, L. M. (1960). Observations on the production of toxins in the gastro-intestinal tract of animals infected orally with a culture of the pathogen of botulism type A. *J. Microbiol. Epidemiol. Immunobiol.* **31**: 106—111.

Smart, J. L. & Roberts, T. A. (1977). An outbreak of type C botulism in broiler chickens. *Vet. Rec.* **100**: 378—380.

Smart, J. L., Roberts, T. A., McCullagh, K. G., Lucke, V. M. & Pearson, H. (1980). An outbreak of type C botulism in captive monkeys. *Vet. Rec.* **107**: 445–446.

Smith, G. R. (1975). Recent European outbreaks of botulism in waterfowl. *Bull. int. waterfowl Res. Bur.* No. 39/40: 72–74.

Smith, G. R. (1976). Botulism in waterfowl. *Wildfowl* **27**: 129–138.

Smith, G. R. (1978). Botulism, waterfowl and mud. *Br. vet. J.* **134**: 407–411.

Smith, G. R. (1980). Concentrations of toxin in the serum of waterfowl with botulism. *Vet. Rec.* **107**: 513.

Smith, G. R., Hime, J. M., Keymer, I. F., Graham, J. M., Olney, P. J. S. & Brambell, M. R. (1975). Botulism in captive birds fed commercially bred maggots. *Vet. Rec.* **97**: 204–205.

Smith, G. R. & Milligan, R. A. (1979). *Clostridium botulinum* in soil on the site of the former Metropolitan (Caledonian) Cattle Market, London. *J. Hyg., Camb.* **83**: 237–241.

Smith, G. R., Milligan, R. A. & Moryson, C. J. (1978). *Clostridium botulinum* in aquatic environments in Great Britain and Ireland. *J. Hyg., Camb.* **80**: 431–438.

Smith, G. R. & Moryson, C. J. (1975). *Clostridium botulinum* in the lakes and waterways of London. *J. Hyg., Camb.* **75**: 371–379.

Smith, G. R. & Moryson, C. J. (1977). A comparison of the distribution of *Clostridium botulinum* in soil and in lake mud. *J. Hyg., Camb.* **78**: 39–41.

Smith, G. R., Moryson, C. J. & Walmsley, J. G. (1977). The low prevalence of *Clostridium botulinum* in the lakes, marshes and waterways of the Camargue. *J. Hyg., Camb.* **78**: 33–37.

Smith, G. R. & Young, A. M. (1980). *Clostridium botulinum* in British soil. *J. Hyg., Camb.* **85**: 271–274.

Smith, L. DS. (1975). The inhibition of *Clostridium botulinum* by strains of *Clostridium perfringens* from soil. *Appl. Microbiol.* **30**: 319–323.

Smith, L. DS. (1977). *Botulism: the organism, its toxins, the disease.* Springfield, Illinois: Charles C. Thomas.

Smith, L. P. (1979). The effect of weather on the incidence of botulism in waterfowl. *Agric. Meteorol.* **20**: 483–488.

Sterne, M. & Wentzel, L. M. (1952). Botulism in animals in South Africa. *Rep. int. vet. Congr.* **14**: (3): 329–331.

Sugiyama, H. (1980). *Clostridium botulinum* neurotoxin. *Microbiol. Rev.* **44**: 419–448.

Sugiyama, H., Bott, T. L. & Foster, E. M. (1970). *Clostridium botulinum* type E in an inland bay (Green Bay of Lake Michigan). In *Toxic microorganisms*: 287–291. Herzberg, M. (Ed.). Washington D.C.: UJNR Joint Panels on Toxic Microorganisms, and US Department of the Interior.

Turner, H. D., Brett, E. M., Gilbert, R. J., Ghosh, A. C. & Liebeschuetz, H. J. (1978). Infant botulism in England. *Lancet* 1978 (i): 1277–1278.

Vadlamudi, S., Lee, V. H. & Hanson, R. P. (1959). Case report – botulism type C outbreak on a pheasant game farm. *Avian Dis.* **3**: 344–350.

DISCUSSION

Tyrrell (Chairman) – I must admit that I have a very biased view about the clostridia. We are interested in *Clostridium difficile* infection in man, which gives rise to diarrhoeal disease, as you know, and I have been thinking about it again from a clostridial point of view; and I wonder whether it may not be true, as you have already suggested, that some of this wildlife disease actually is due to clostridia replicating in the organism. I even have a fanciful idea that it's difficult to explain the existence of the toxins without there being some sort of survival advantage. And I think that if I were a clever *Clostridium*, I should want to be in that lovely bit of anaerobic protein-rich medium at the time my toxin produced it, so that I could replicate and sporulate and provide a target for all sorts of other organisms to move me around. Am I being too fanciful?

Smith — No, I don't think so. Of course, whenever toxin is ingested, a lot of organisms are usually ingested at the same time. Infant botulism, in recent years, has really hit the news, but in fact it wasn't the beginning of the toxinfection story by a long chalk. People have suggested this sort of thing for many years, particularly, as a matter of fact, in Russia, where certain people like Minervin (1967)[1] and Shvedov (1960)[1] have said for many years that they think the organism sometimes multiplies in the intestine; and the sort of evidence that they bring forward to support that is that people come into hospital with botulism, they are treated successfully, they are discharged; and then ten days later they are brought back into hospital again. It's very suggestive, although it never seemed to impress anybody in this part of the world until infant botulism was discovered. There is also the undoubted case of wound botulism in man; it's a very rare disease, but there must be 20 or so cases, well authenticated cases, on record, in which botulism follows contamination of a wound with *Clostridium botulinum*; clearly the organism multiplies in those tissues. Many years ago, in the 1930s, in relation to botulism in waterfowl, people like Kalmbach & Gunderson (1934)[1] said that you could sometimes isolate the organism from the liver of affected birds, and Boroff and his colleague Reilly (1962)[1] thought that sometimes the organism actually invaded the body before death and produced toxin in sites like the liver. We actually investigated that here a few years ago: Jennifer Graham and I looked at something like 27 waterfowl from four different outbreaks of botulism (Graham & Smith, 1977);[1] we took very, very stringent precautions to avoid any risk of taking the spores in from the external surface because it would be very easy to do with resistant spores — we used about three sets of sterile instruments and so forth — and we failed to find any evidence at all that the organism got into the liver. But the question of whether it multiplies in the intestine is something different again, and it's something that certainly should be looked at.

Nelson — What other animals are killed by botulism? You've got this vast mortality amongst the ducks. When you get that kind of situation, you get foxes wandering all around the mudflats, eating the wild ducks. Does botulism then kill the foxes? Or does it kill dogs?

Smith — Or does it kill human beings? Because there is no doubt — it must have happened on many occasions — that human beings have eaten birds that have died from botulism. Apparently not. Apparently there are no deaths in human beings that have resulted from that cause. As far as human beings are concerned, I have mentioned that types C and D, which occur so often in animals, don't produce the disease in man — there are only about three cases on record of type C botulism in man and not one of those is entirely conclusive. What happens

---

[1] See list of references, pp. 111–115.

with other animals like foxes and so on, which certainly would be susceptible to type C — dogs are, for example — I don't know; I would imagine that if foxes come along and eat carcasses that are decaying after death from botulism, they would be eating large amounts of toxin, and I would think it would kill them. But what happens to them, where they go, I don't know. They presumably creep under a bush and die, and then their carcasses develop large amounts of toxin which may then be eaten by other animals.

Huxley — Is the reason for the different susceptibility of humans to these different types known? Is it a matter of the ease with which the particular toxin is absorbed from the intestine — because it has presumably got to get through without being digested by the various digestive enzymes.

Smith — In fact very little is known about this. Type C toxin is known to be absorbed more quickly and more readily from the gut of gallinaceous birds than toxins of other types. There is that amount of information. I don't know of any information on the ease with which type C toxin is absorbed from the gut of man. We do know, from experimental work, that primates are susceptible to type C toxin, and in 1980 an outbreak of type C botulism was recorded in captive primates in the Bristol area — I think something like 14 died, from having eaten chopped chicken that was contaminated with type C toxin. And three or four species of primates died. But it is a mystery why types C and D don't produce botulism in man more often, or at all; no-one really knows the answer to this.

Payne — We have had some tantalizing glimpses of what possibly might be done to prevent this condition. Could we hear a little more about this bacillus, and is it possible to seed our lakes to prevent the growth of *Clostridium*?

Smith — Well, I think it would be interesting to do this sort of thing. We know areas where the disease readily occurs, and occurs almost every year; we know other areas where it never occurs, and where the organism is apparently almost completely absent. And we know other areas in which the mud contains organisms that inhibit, *in vitro*, the growth of *Clostridium botulinum* type C. Now I would like to see some experiments done with material of this kind. How would one do it? Well, you could envisage beginning by taking a very favourable mud sample — from a favourable environment — and deliberately trying to modify it by adding a small amount of mud from the Camargue, say, which we know is inhibitory. You could then see whether you could make a favourable mud unfavourable by seeding it with isolates of *Bacillus* spp. that produce these antibiotic substances that prevent the growth of the organism. You could then perhaps even go to a small Lake — I have one in particular in mind, in west London, which is heavily contaminated with type C spores (you can demonstrate it in any handful of mud you take, at any time, at any time of the year, and there have been at least two serious outbreaks of botulism in the valuable waterfowl that are on that lake) — and you might perhaps be able to arrange to do some experiments in a situation like that, seeding the environment with organisms and seeing what happened. That is the sort of thing that I have in mind. I don't think it would be very easy — I think it would be an extremely complicated series of experiments to do — but I bet something interesting would come out of it.

Tyrrell — Can I encourage you there by saying that some such experiment may go on in the guts of virtually all of us, all our lives; because my colleagues, Drs Borielli and Larson, found that large numbers of children get *Clostridium difficile* infection in their neonatal period, none of whom seem to suffer very much by it, and yet adults only get *Clostridium difficile* infection later in life, usually when something pretty drastic has been done to their bowel flora, for instance administering antibiotics; normal bowel flora contains organisms which inhibit *Clostridium difficile*. So that's it — we're being protected all through life by just the sort of mechanism which you've suggested for the field.

Plowright — This is really a basic question. You are getting rid of the spores also, aren't you, if there is a process of continual contamination of mud in the Camargue? Does this mean that the spores are germinating and then further growth is inhibited? If this were not so and there was an inhibition of germination then you might still find a very low level of viable spores.

Smith — Yes, I can see the point exactly. I would think that what you say is right. Of course, there is also the point that you have to bear in mind, that even in an environment that is frankly inhibitory, it's always conceivable that in little bits of that environment you may have little patches, pockets of organic matter that may be favourable for the organism to grow in, may even protect it from the surrounding inhibitory environment. That's always possible too, I think. But I would entirely agree with what you say.

Thompson — Can you tell us the extent to which botulism is associated with migratory waterfowl, in for example the USA and Australia? Have there been outbreaks on flyways, for example, in North America?

Smith — Yes, very large outbreaks in North America, and in Australia too, though not on quite such a large scale. There's no doubt about that.

GENERAL DISCUSSION

Bray — I would like to ask a general question, and that is what role there might be in the future for the breeding of cattle in Africa, particularly for better utilization of the coarse grasslands, and of course for resistance, again particularly to trypanosomiasis, looking at such cattle as the N'Dama, and possibly also to piroplasmosis and, who knows, even rinderpest.

Molyneux — As you are well aware, there are three efforts being made in terms of trypanosomiasis control. There is the tsetse vector control aspect; there is chemotherapy; and I think much importance is being attached, particularly by FAO and ILCA, to research on trypanotolerance. And there's certainly a great deal of effort going into research in Togo and, as you are probably aware also, the possibility of a large research activity in Gambia, to investigate the parameters of those qualities you mention of resistance to trypanosomiasis. The basic information on trypanotolerance of livestock, of N'Dama and Muturu and other west African breeds of tolerant livestock, is really not available and this is certainly an area of great potential interest. But one is up against the traditional attitude of many of the nomadic peoples of west Africa, that they prefer the susceptible zebu because of its size and status. I would agree with you, certainly as far as trypanosomiasis is concerned, that this must be investigated, and indeed has been identified by those bodies who are responsible for overseeing activities in this area as being one of major development in the future.

Plowright — I think as far as rinderpest is concerned, there is really little point in developing cattle which are resistant when you can immunize so readily and, as you know, the methods of control in developed countries are extremely effective in any case, without using vaccination. I think that the idea of breeding as a means of obtaining resistance to disease may be applicable in isolated instances, but I do ask myself, if you breed cattle which are resistant to rinderpest, then what would be their reaction to all the other bovine viruses which are present in Africa? It might be quite different, and we might still have many persistent problems.

Henderson — My question concerns population density, which has been mentioned once or twice, and the relationship between cattle and game in east Africa. I can't remember offhand what the cattle population is in Kenya, for example, in Masailand — but did not the use of more efficient vaccines during the 1960s completely eliminate rinderpest from, say half the most susceptible population, and did that not in some measure account for the disappearance of the infection in wild game?

Plowright — I think there is a great deal of evidence that the campaigns of immunization of cattle against rinderpest which had been carried out for many years with goat virus and were then intensified, using culture-attenuated virus, were the factors that led to the elimination of rinderpest from game.

A Speaker — A final point worth noting: there does seem to be some increased susceptibility to rinderpest in cattle which are particularly resistant to trypanosomiasis, and I wonder if this is a general phenomenon or simply a local one?

Plowright — It has been known for a long time that the susceptibility of different breeds of cattle to a particular strain of rinderpest varies very considerably. Therefore, for example, a strain of virus which was very good as a vaccine for most cattle in Africa would kill 40% of Japanese Black cattle. And similarly in Africa the shorthorn Muturus and N'Damas were more susceptible than the zebus. In India hill cattle are regarded as being more susceptible than plains cattle. But if you look for hard evidence about relative susceptibility, experimental evidence, then it is often difficult to find.

Symp. zool. Soc. Lond. (1982) No. 50, 121–135

Influenza in Nature

MARTIN M. KAPLAN

*Director General, Pugwash Conferences on Science and World Affairs,
11a Avenue de la Paix, 1202 Geneva, Switzerland*

SYNOPSIS

For centuries, observers have noted an apparent association between human and animal influenza. It is only with the advent of modern virology some 50 years ago, however, that such an association has been given scientific meaning. Despite the vast amount of work devoted to the "last great plague", the role of animal infections in the origin and spread of human pandemics is still the subject of much speculation.

The influenza virus is highly variable and causes natural infections in a wide variety of domestic and wild animal species, mainly swine, horses, poultry and birds. Work on the molecular biology of the type A influenza virus promises to clarify with precision many of the epidemiological and epizootiological characteristics of the disease. Infections in mammals and birds occupy ecological niches of their own, but inter-species infections can and do occur, although this seems to be the exception rather than the rule. These exceptions appear to be caused by mutations and recombinations of the virus which replicates in segments, thus favouring recombinational events between different strains of the virus affecting man and lower animals.

The disease in man, swine and horses affects chiefly the respiratory system, and is transmitted by the respiratory route. Apart from well-recorded but isolated dramatic episodes of pandemics, chiefly the one in man in 1918, infections in nature usually cause low mortality, but morbidity is high and clinical illness is often severe. Incidents of high mortality in natural infections observed in seals and terns, however, indicate that we may not be dealing here with isolated events in wildlife.

The striking exception with respect to morbidity and mortality is avian infection in wildlife, which is mostly intestinal in character and without clinical signs. Numerous wild avian species circling the globe share a museum of changing viruses, and are probably the evolutionary origin of influenza. This ineradicable reservoir of the chameleon-like virus poses a permanent potential threat to man and to other animal species. Steps are mentioned which might be taken to counter this threat.

INTRODUCTION

The purpose of this chapter is to extract from the vast literature and knowledge of influenza those aspects pertaining to natural infections, particularly in wildlife, and their consequences on the natural history of the disease. There is not much precise information about wildlife influenza, however, so that analogy and conjecture will be in order. This in turn requires consideration of relevant characteristics of the virus and of its effects in human and domestic animal populations.

THE VARIABLE VIRUS

The influenza viruses belong to the group *Orthomyxovirus* and consist of three types categorized as A, B and C. These types are differentiated on the basis of their two major proteins in the core of the virion, the nucleoprotein and the matrix protein. We shall be concerned with influenza type A which is the only type convincingly shown to have caused natural infections in lower animals.

The virion is an irregular sphere approximately 100 nm in diameter, covered with protein spikes attached to its envelope. They are the haemagglutinin (H) which acts to attach the virus to cells (first observed in the clumping of erythrocytes), and the enzyme neuraminidase (N) which is apparently responsible for release of the virus from the cell and its spread to other cells.

The helical RNA core of the virus replicates as eight single-stranded segments, and thus genetic recombination, or reassortment, can readily occur in laboratory-induced or natural infections with the different strains (subtypes) of influenza A (Palese *et al.*, 1980; Webster & Laver, 1975; Barry & Mahy, 1979).

During the past two years great advances have been made in knowledge of the molecular biology of the influenza virus. For example, the nucleic acid composition of complete genes (RNA segments) coding for several H and N proteins has been determined, as well as the structure and amino acid composition of the proteins themselves (Fields, Winter & Brownlee, 1981; Wilson, Skehel & Wiley, 1981; Gething, Bye, Skehel & Waterfield, 1980).

The different H and N antigens are grouped on the basis of several kinds of serological tests, the most widely used ones being the haemagglutinin- and neuraminidase-inhibition tests, and radial immune diffusion reactions (Schild *et al.*, 1980; Advanced laboratory techniques for influenza diagnosis, 1975). Such a grouping serves practical purposes of diagnosis and epidemiological relationships, and is the basis of the official nomenclature recommended by the World Health Organization (WHO).

The two major H and N antigens have been broadly grouped by WHO (World Health Organization, 1980) as H1, H2, H3, . . . , and

N1, N2, N3, ... (Table I). It is the H and N variations mainly that modify the epidemiology and epizootiology of the disease. Minor alterations within the groups, which occur especially in the H antigen, are called antigenic "drift". It is noteworthy that a change in a single amino acid of the H antigen may be responsible for antigenic "drift" (Laver *et al.*, 1980). The second variation, called antigenic

TABLE I

Proposed subtypes of haemagglutinin and neuraminidase antigens of influenza A viruses[a]

H antigens		N antigens	
Proposed subtypes	Previous subtypes (1971 system)	Proposed subtypes	Previous subtypes (1971 system)
H1	H0, H1, Hswl	N1	N1
H2	H2	N2	N2
H3	H3, Heq2, Hav7	N3	Nav2, Nav3
H4	Hav4	N4	Nav4
H5	Hav5	N5	Nav5
H6	Hav6	N6	Nav1
H7	Heq1, Hav1	N7	Neq1
H8	Hav8	N8	Neq2
H9	Hav9	N9	Nav6
H10	Hav2		
H11	Hav3		
H12	Hav10		

[a] Adapted from World Health Organization (1980).

"shift", connotes an abrupt or major change in the composition of previously known H or N antigens within a group, and is accorded a new subscript number in the nomenclature (H1→H2, N1→N2). Until recently subscripts, slashes and abbreviations were used after the letters H and N to denote the animal origin of the strain, the place, strain number and year when isolated (Table II). But this usage is now gradually being replaced by incorporating the many hundreds of animal strains already isolated, and which are constantly increasing, into sequential numbers for the H and N antigens. It must be kept in mind, however, that as our knowledge accumulates of the exact molecular composition of the H and N antigens, the present grouping based on serology may have to be changed. For example, the group H1 contains strains that differ markedly in the amino acids of their haemagglutinin (and therefore in their nucleic acid codons), but this in turn may not be epidemiologically significant if the location(s) of the differing amino acids does not involve mutated "hot" spots

TABLE II

Examples of reference strains for human, swine and horse infections of influenza A viruses according to proposed subtypes of haemagglutinin and neuraminidase antigens[a]

H and N subtypes	Reference strains		
	Human	Swine	Horses
H1N1	A/Puerto Rico/8/34 A/USSR/90/77	A/swine/Iowa/15/30 (formerly Hsw1N1) A/swine/Wisconsin/67 (formerly Hsw1N1)	
H2N2	A/Singapore/1/57 A/Tokyo/3/67		
H3N2	A/Hong Kong/1/68 A/Texas/1/77		
H7N7			A/equine/Prague/1/56 (formerly Heq1 Neq1)
H3N8			A/equine/Miami/63 (formerly Heq2 Neq2)

[a] Adapted from World Health Organization (1980).

of the protein moiety which determine whether or not the virion will escape the immune mechanism of the prospective host (Both & Sleigh, 1981; Wiley, Wilson & Skehel, 1981). When needed for clarity, we shall use here both the old and new nomenclature systems (Table II).

THE DISEASE IN HUMAN POPULATIONS

Influenza epidemics have affected human beings since ancient times. One was recorded by Hippocrates in 412 B.C., and numerous epidemics were described in the Middle Ages. A pandemic in the sixteenth century caused high mortality in some cities, and serious pandemics have occurred since then culminating in the pandemic of 1918–1919 which was estimated to have killed over 20 million people throughout the world (Webster & Laver, 1975; Crosby, 1976; Beveridge, 1977; Kaplan & Webster, 1977). It was not until 1933, in England, that human influenza was found to be caused by a virus (Smith, Andrewes & Laidlaw, 1933) although the influenza virus was first identified as such in pigs three years earlier (Shope, 1931). Fowl plague was shown to be caused by a virus in the early years of this century, but it was not shown to be an influenza strain until half a century later (Schafer, 1955).

The disease in man affects chiefly the respiratory system and is temporarily incapacitating, but mortality is usually low (Kilbourne, 1975; Stuart-Harris & Schild, 1976). The waves of influenza epidemics and pandemics are now known to be due to modifications of virus strains ("drift" or "shift") which by-pass immunological defenses that develop to strains circulating over a period of years. Such changes cannot as yet be predicted. Evidence for rhythmic periodicity of epidemics and pandemics is not convincing, nor does climate or season appear to play a critical role. The person-to-person spread of the virus is through the respiratory route, and is a function of the closeness of human contact (crowds, enclosed places).

When a new strain appears as an epidemic or pandemic it usually replaces the currently circulating strain(s) which is then difficult to isolate. Recent findings, however, show that several influenza A strains can co-exist in the same population, and individuals can be simultaneously affected with two strains and recombinations of the two strains (Bean, Cox & Kendal, 1980). Chronic or latent infections without clinical signs undoubtedly occur, but these are apparently the exception rather than the rule, in contrast to what has been observed in lower animals, particularly in the avian species.

Direct transfers of a swine strain and an avian strain from animals to man, with resultant clinical illness, have been conclusively demonstrated (Beare, Kendal & Craig, 1980; Webster, Geraci, Petursson & Skirnisson, 1981), but such transfers have been observed very rarely. This is surprising in view of the extent of influenza in animals, but we shall return to this point later on.

THE DISEASE IN ANIMAL POPULATIONS

Domestic livestock

Swine

The striking clinical and epizootiological picture described of swine influenza in mid-western states of the USA some 60 years ago is still observed in that region (Easterday, 1975). Herds of pigs in widely separated areas are stricken almost simultaneously with a severe respiratory disease usually following a sudden change of weather in the fall or early winter. The pigs lie prostrate for a few days with fever and coughing, and then recover rapidly. Mortality is between 1% and 5%. Subsequent investigations have shown that pigs throughout the USA and in many other countries are infected with the virus, which can be isolated throughout the year without the pigs showing clinical signs (Pirtle, Hill, Swanson & van Deusen, 1976; Hinshaw,

Bean, Webster & Easterday, 1978; Yamane, Arikawa, Odagiri & Ishida, 1979; Shortridge & Webster, 1979). This indicates that a carrier virus state with shedding episodes can exist in these animals, a fact that was demonstrated experimentally in pigs infected three months previously (Blaskovic *et al.*, 1970).

In a series of experiments carried out in the 1930s, Shope described a complex life cycle of the virus which involved earthworms and swine lung worms as the intermediate hosts of the virus (see Wallace, 1977, for review). While this was a fascinating biological observation, the complex cycle was shown to be unnecessary, and in fact exceptional, in swine influenza as observed today (Easterday, 1975; Blaskovic *et al.*, 1970).

For our purpose it is noteworthy that some strains of human influenza virus can easily infect swine under natural conditions (Kundin, 1970; Pirtle *et al.*, 1976; Hinshaw, Bean *et al.*, 1978; Harkness, Schild, Lamont & Brand, 1972; Shortridge & Webster, 1979), and that the swine virus can infect man (Kendal, Noble & Dowdle, 1977; Beare *et al.*, 1980). The latter occurrence caused great concern in 1976 when soldiers in the Fort Dix, New Jersey, encampment were infected with the swine strain currently infecting pigs in the USA. It was feared that another 1918 pandemic incident might be in the offing, since the prototype virus of the 1918 pandemic was believed to have infected pigs at that time and to have persisted in the animals in a modified form. Also, there was circumstantial evidence that the reverse may have occurred in 1918, i.e. swine to man infection (Kaplan & Webster, 1977). Fortunately the feared pandemic in 1976 and 1977 did not materialize.

Recently, swine have been shown to be infected with an avian strain of the virus (Pensaert, Ottis, Vandeputte, Kaplan & Bachmann, 1981). This was the first clear demonstration that an avian strain can infect mammals in nature, and thus provide a bridge between the vast reservoir of avian infections and man. Subsequently, seals were also found infected with an avian strain (Lang, Gagnon & Geraci, 1981).

Antigenic analysis of the swine virus has revealed significant "drift" in isolations from naturally infected swine during the past 50 years (Kendal *et al.*, 1977).

Equines (for review see Tumova, 1980)

Throughout the centuries influenza-like infections have been reported in horses, but clinical signs cannot be relied upon for assurance that the influenza virus was the cause. Undoubtedly, some of the historical episodes described were caused by the virus, especially the disease around 1890 of which the serological traces have been found in people who were alive at that time.

Equine influenza is characterized by a severe respiratory disorder (conjunctivitis, cough, high fever) with low mortality. The disease is highly disruptive in racing stables. Two strains of the equine virus are involved (Table II), sometimes simultaneously in the same animal. An equine influenza virus was first isolated and identified as such in 1956 in Czechoslovakia. The disease in horses is widely distributed throughout the world. Considerable H "drift" has been detected in recent isolates of equine 2 virus strain (World Health Organization, in press).

Poultry (for review see Easterday, 1975)
Fowl plague (or pest) has long been known as a highly fatal disease of chickens, and its cause by a filtrable virus was demonstrated early in this century. It was not until 1955, however, that the virus was recognized as an influenza virus. Other strains of the virus, including human strains, can also infect chickens, but mortality is much less.

Clinical signs in poultry are respiratory in nature, often accompanied by a severe sinusitis (swelling).

Influenza in duck and turkey flocks is frequently found, and the disease can sometimes cause high mortality. This is also seen in quail raised under battery conditions for commerce. Infection in other domestic poultry is observed occasionally, but often without clinical signs.

Cattle and other domestic ruminants
Using haemagglutination-inhibition tests, influenza antibodies have been reported in sheep, goats and cattle, but such tests alone are undependable unless accompanied by the isolation of the virus. In 1971, a human type virus (H3N2) was isolated in the USSR from calves showing severe respiratory disease. Experimentally, this virus caused similar disease in calves held in isolation and under strict laboratory conditions (Campbell, Easterday & Webster, 1977). There is no question of the validity of the USSR isolate, but it is puzzling why only one such episode has been observed despite extensive observations in virus diseases of cattle in many countries. The proven instance cited, however, perhaps bespeaks only our uncertainty in general about the potential of this versatile virus.

Wildlife

Avians (see Easterday, 1975)
Hundreds of strains of influenza virus have been isolated from

numerous species of wild birds (Table V). Based on serological tests the isolated viruses fall into the antigenic groups shown in Tables III and IV, and we can expect that additional groups will be needed to accommodate new viral isolates which are being obtained in a steady stream. With the exception of one major episode in terns

TABLE III

Examples of reference strains for proposed subtypes of haemagglutinin antigens of influenza A viruses isolated from avian species[a]

Proposed subtype designation	Previous subtype designation	Reference strains
H1	Hsw1	A/duck/Alberta/35/76 (H1N1)
H2	H2	A/duck/Germany/1215/73 (H2N3)
H3	Hav7	A/duck/Ukraine/1/63 (H3N8)
H4	Hav4	A/duck/Czechoslovakia/56 (H4N6)
H5	Hav5	A/tern/South Africa/61 (H5N3)
H6	Hav6	A/turkey/Massachusetts/3740/65 (H6N2)
H7	Hav1	A/fowl plague virus/Dutch/27 (H7N7)
H8	Hav8	A/turkey/Ontario/6118/68 (H8N4)
H9	Hav9	A/turkey/Wisconsin/1/66 (H9N2)
H10	Hav2	A/chick/Germany/N/49 (H10N7)
H11	Hav3	A/duck/England/56 (H11N6)
H12	Hav10	A/duck/Alberta/60/76 (H12N5)

[a] Adapted from World Health Organization (1980).

(see below), it is significant that all isolations to date have been made from birds showing no obvious signs of illness.

We owe this remarkable fund of information to many workers, but particular tribute should be paid to the virologists responsible for the organization of special expeditions designed to clarify the role of migratory birds and fowl in the ecology of influenza. To mention a few: Bernard Easterday and his colleagues for studies in Alaska, the Antarctic and north-central America; Robert Webster and Virginia Hinshaw for studies along the Mississippi flyway and in southern USA; D. Lvov and his co-workers in far-eastern, central and southern parts of the Soviet Union; Graeme Laver and Webster in the Great Barrier Reef off Australia, and in islands near the Peruvian coast; Geoffrey Schild and co-workers in islands off Norway; Claude Hannoun and his associates in France; and Peter Bachmann and Klaus Ottis in the Federal Republic of Germany. It should be noted that some expeditions drew complete blanks, e.g. pelicans in the Antarctic, and wild birds in the Peruvian coastal area. In several instances, expeditions to the same areas were carried out in successive years to

TABLE IV

Examples of reference strains for proposed subtypes of neuraminidase antigens of influenza A viruses isolated from avian species[a]

Proposed new subtype designation	Previous grouping	Reference strains
N1	N1	A/chick/Scotland/59 (H5N1)
N2	N2	A/turkey/Massachusetts/3740/65 (H6N2)
N3	Nav2	A/tern/South Africa/61 (H5N3)
	Nav3	A/turkey/England/63 (H7N3)
N4	Nav4	A/turkey/Ontario/6118/68 (H8N4)
N5	Nav5	A/shearwater/Australia/1/72 (H6N5)
N6	Nav1	A/duck/Czechoslovakia/56 (H4N6)
		A/duck/England/56 (H11N6)
N7	Neq1	A/fowl plague virus/Dutch/27 (H7N7)
N8	Neq2	A/quail/Italy/1117/65 (H10N8)
N9	Nav6	A/duck/Memphis/546/74 (H11N9)

[a] Adapted from World Health Organization (1980).

TABLE V

Partial list of feral avian species infected with influenza virus[a]

Ducks:	mallard	shelduck
	pintail	teal
	sadwall	wigeon
coot	parakeet	shrike
crow	partridge	silverbill
finch	pheasant	snipe
gull	pigeon	starling
macaw	quail	swallow
murre	redstart	tern
mynah	robin	wagtail
nightingale	shearwater	warbler

[a] This list has been compiled for illustrative purposes from numerous sources. A complete registry of virus isolates is maintained by Dr Robert G. Webster, St Jude Children's Hospital, Memphis, Tennessee 38101, USA.

study fluctuations in incidence of infections. Much of this work has been co-ordinated by the World Health Organization (Kaplan, 1980).

Sampling procedures focussed mainly on cloacal swabs, after it was determined that in wild birds and fowl the influenza virus was principally an intestinal infection, rather than a respiratory one (Webster, Yakhno *et al.*, 1978). The virus was easily isolated from duck ponds, and laboratory studies showed that the virus persisted in the water for four days at 22°C and for over 30 days at 0°C (Hinshaw, Webster & Turner, 1979).

The striking exception to the lack of clinical signs of the disease in wild birds was the death of large numbers of terns infected with the virus in South Africa; a closely related strain of virus was isolated in chickens in Scotland, and was believed to have been introduced by terns (Becker, 1966).

Thus we have a picture of a commensal infection in birds travelling widely throughout the world, producing a steady flow of "drifts" and "shifts" of influenza strains. The tern episode in South Africa should alert us to the possibility that this was not an isolated event, and that in nature large numbers can die when a "new" strain appears of a highly pathogenic character.

Other animals

Ferrets, mink and mice are highly suceptible to laboratory infections with the virus, but infections of these animals in nature have not been detected. If infections of ferrets with certain strains do occur in nature, high mortality can be expected.

Gibbons, baboons and monkeys are known to be susceptible to infection with the Hong Kong (H3N2) human strain of virus. Under laboratory conditions baboons showed no clinical signs, but four out of 36 gibbons died after infection contracted by contact with gibbons infected two to three weeks earlier (Johnsen, Wooding, Tanticharoenyos & Karnjanaprakorn, 1971).

Little is known about infections of *ruminants* or *Equidae* in the wild. One interesting observation was the detection of antibodies to Heq2Neq2 (H3N8) in sera of Mongolian horses in the early 1960s (Tumova, 1980). These horses constitute an isolated animal population, which presumably has had no contact with horses from other countries. The Heq2Neq2 strain of virus was first isolated from racehorses in Miami, Florida, USA, in 1963. It is therefore curious how an apparently long-standing infection of horses in Mongolia reached Miami, if this in fact happened. A possible explanation is that it was carried by migratory birds.

Influenza viruses have been isolated by Russian workers from the lungs and livers of apparently healthy *striped whales* in the Pacific (Lvov *et al.*, 1978). Genetic and antigenic studies showed that the virus is antigenically similar to Hsw1N1 strains isolated from avian sources, and falls intermediate between avian and mammalian strains (R. G. Webster, pers. comm.). An influenza A-like strain was isolated from migratory *Pacific salmon* in 1976 (D. K. Lvov, pers. comm.).

From December 1979 and during the next year some 500 harbor seals (*Phoca vitulina*) were found dead on Cape Cod, Massachusetts, USA. Post-mortem examination revealed acute hemorrhagic pneumonia. A virus antigenically similar to that of classical fowl plague (A/fowl plague/Dutch/27) was isolated from the lungs and brain tissues (Lang *et al.*, 1981). Workers associated with the seals contracted a conjunctivitis; virus was isolated from one worker suffering from conjunctivitis who handled an experimentally infected seal which sneezed directly into his eye (Webster, Geraci *et al.*, 1981). The seal episode thus provides another proved instance of an avian strain of virus infecting and causing disease in mammals, including man. The bridge between avian virus infections and man is now definitely known to be in place.

One can only conjecture the extent of morbidity and mortality in wildlife. The high morbidity and mortality sometimes observed in domestic poultry, the great susceptibility of ferrets, the severe clinical disease in calves on one farm in the USSR, and the gibbon and seal incidents described above should warn us that the chameleon-like virus we are dealing with may be causing — and will cause — great havoc in wildlife.

WHENCE? WHITHER?

I shall now indulge briefly in crystal-ball gazing with time flows backward and forward.

Whence?

The origin of the avian species in evolution antedated that of placental mammals by over 100 million years. Influenza viruses appear to have adapted themselves harmoniously to avian hosts in the wild as intestinal infections. It is reasonable to assume therefore that influenza viruses originated in birds and spread to other species in the course of time. The great variety of strains inhabiting birds throughout the world, many of them closely related to human and mammalian strains, have provided a vast reservoir of the antigens required for new infective strains in man and lower animals. Cross-species infections were caused by direct transfer of "new" strains which were produced by minor antigenic mutations ("drift"), or by recombination of the genetic RNA segments of the virus ("shift"). The "shift" mechanism incorporating virus genes derived from animals was operative in causing at least some of the major pandemics which have afflicted man in the past, and can be expected to do so again in the future.

The importance of animal reservoirs for human influenza is not accepted by some research workers who believe that we need not look beyond the human species to explain the origin of epidemics and pandemics. Most investigators, however, have now come to accept that strains of the virus found in lower animals contribute importantly to the epidemiology of human influenza. We should have a clearer idea about who is right within the next few years when molecular analyses of different strains have accumulated.

Whither?

It is safe to predict that in the near and distant future there will continue to be influenza epidemics and pandemics and, in a more limited sense, epizootics and panzootics (more limited because of animal husbandry practices and wildlife patterns). But we cannot predict when and where these outbreaks will occur because of the variables associated with a mutant-prone virus, and the ecological differences within human and animal populations (e.g. mobility, concentration, immune status).

We are, however, far from powerless or helpless in this situation. Precise molecular comparisons can now be made between virus strains that will give important insights into the origin and epidemiology of influenza outbreaks, at least of the recent past where virus strains as originally isolated have been preserved. The attributes of infectivity, transmissibility and pathogenicity, which are largely verbal formulations at present, will become clarified with the improved techniques now at our disposal. Molecular analysis will continue to expose the anatomy of the virus genome and its protein products. Mutation "hot spots" and loci important in neutralization reactions are being revealed. The latter should enable the eventual synthesis of highly specific and pure antigens and antibody preparations capable of preventing infections.

We face, however, formidable difficulties when we take into account the large number of different strains already identified in the animal kingdom which are in the process of change through mutation and recombination. We are indeed fortunate that very few of these strains appear to be infective for man or to cross over to other animal species. We must keep in mind that this observation, however, is based upon only 50 years' scientific observation, the finer details of which have been elucidated only during the past 20 years. Thus we may well have missed many episodes in past history when species cross-overs occurred with resultant devastating outbreaks. The 1918 pandemic, for example, may have been one of them. What, then, could and should be done?

Every opportunity should be taken to isolate and identify new animal strains. All novel animal influenza strains should be freeze-dried and "banked". Recombinants should be prepared of their H antigens with other strains which have been adapted to eggs so that high yields can be readily obtained for vaccine production. Analysis of the molecular structure with respect to strains and segments of genomes of particular interest should also be done whenever possible.

One clear conclusion emerges from the above account. Influenza, which has been called the last great plague, provides perhaps the most fascinating territory for exploration by those interested in human—animal disease relationships of infective agents.

REFERENCES

Advanced laboratory techniques for influenza diagnosis (1975). Immunology Series No. 6. U.S. Department of Health, Education and Welfare, Public Health Services, Center for Disease Control, Bureau of Laboratories, Atlanta, Georgia 30333.

Barry, R. D. & Mahy, B. W. J. (1979). The influenza virus genome and its replication. *Br. med. Bull.* 35: 39—46.

Bean, W. J., Cox, Nancy J. & Kendal, A. P. (1980). Recombination of human influenza A viruses in nature. *Nature, Lond.* 284: 638—640.

Beare, A. S., Kendal, A. P. & Craig, J. W. (1980). Further studies in man of Hsw1N1 influenza viruses. *J. med. Virol.* 5: 33—38.

Becker, W. C. (1966). The isolation and classification of tern virus: influenza virus A/tern/South Africa/1961. *J. Hyg., Camb.* 64: 309—320.

Beveridge, W. I. B. (1977). *Influenza: the last great plague.* London: Heinemann.

Blaskovic, D., Jamrichova, O., Rathova, V., Kociskova, D. & Kaplan, M. M. (1970). Experimental infection of weanling pigs with A/swine influenza virus. 2. The shedding of virus by infected animals. *Bull. Wld Hlth Org.* 42: 767—770.

Both, W. B. & Sleigh, M. J. (1981). Conservation and variation in the hemagglutinins of Hong Kong subtype influenza viruses during antigenic drift. *J. Virol.* 39: 663—672.

Campbell, C. H., Easterday, B. C. & Webster, R. G. (1977). Strains of Hong Kong influenza virus in calves. *J. infect. Dis.* 135: 678—680.

Crosby, A. W. (1976). *Epidemic and peace 1918.* London: Greenwood Press.

Easterday, B. C. (1975). Animal influenza. In *The influenza viruses and influenza:* 449—481. Kilbourne, E. D. (Ed.) London: Academic Press.

Fields, S., Winter, G. & Brownlee, G. G. (1981). Structure of the neuraminidase gene in human influenza virus A/PR/8/34. *Nature, Lond.* 290: 213—217.

Gething, M. J., Bye, J., Skehel, J. J. & Waterfield, M. D. (1980). Cloning and DNA sequence of double-stranded copies of haemagglutinin genes from H2 and H3 strains elucidates antigenic shift and drift in human influenza virus. *Nature, Lond.* 287: 301—306.

Harkness, J. W., Schild, G. C., Lamont, P. H. & Brand, C. M. (1972). Studies on relationships between human and porcine influenza. *Bull. Wld Hlth Org.* **46**: 709–719.

Hinshaw, Virginia, S., Bean, W. J., Webster, R. G. & Easterday, B. C. (1978). The prevalence of influenza viruses in swine and the antigenic and genetic relatedness of influenza viruses from man and swine. *Virology* **84**: 51–62.

Hinshaw, Virginia S., Webster, R. G. & Turner, B. (1979). Water-borne transmission of influenza A viruses. *Intervirology* **11**: 66–68.

Johnsen, D. O., Wooding, W. L., Tanticharoenyos, P. & Karnjanaprakorn, C. (1971). An epizootic of A2/Hong Kong/68 influenza in gibbons. *J. infect. Dis.* **123**: 365–370.

Kaplan, M. M. (1980). The role of the World Health Organization in the study of influenza. *Phil. Trans. R. Soc. Lond.* (B) **288**: 417–421.

Kaplan, M. M. & Webster, R. G. (1977). The epidemiology of influenza. *Scient. Am.* **237**: 88–106.

Kendal, A. P., Noble, G. R. & Dowdle, W. R. (1977). Swine influenza viruses isolated in 1976 from man and pig contain two coexisting subpopulations with antigenically distinguishable hemagglutinins. *Virology* **82**: 111–121.

Kilbourne, E. D. (1975). Epidemiology of influenza. In *The influenza viruses and influenza*: 483–538. Kilbourne, E. D. (Ed.). London: Academic Press.

Kundin, W. D. (1970). Hong Kong A2 influenza virus infection among swine during a human epidemic in Taiwan. *Nature, Lond.* **288**: 957–958.

Lang, G., Gagnon, A. & Geraci, J. R. (1981). Isolation of an influenza A virus from seals. *Archs Virol.* **68**: 189–195.

Laver, W. G., Air, G. M., Webster, R. G., Gerhard, W., Ward, C. W. & Dopheide, T. A. A. (1980). The mechanism of antigenic drift in influenza virus: sequence changes in the haemagglutinin of variants selected with monoclonal hybridoma antibodies. *Phil. Trans. R. Soc. Lond.* B **288**: 313–326.

Lvov, D. K., Zdanov, V. M., Sazonov, A. A., Braude, N. A., Vladimirtceva, E. A., Agafonova, L. V., Skljanskaja, E. I., Kaverin, N. V., Reznik, V. I., Pysina, T. V., Oserovic, A. M., Berzin, A. A., Mjasnikova, I. A., Podcernjaeva, R. Y., Klimenko, S. M., Andrejev, V. P. & Yakhno, M. A. (1978). Comparison of influenza viruses isolated from man and from whales. *Bull. Wld Hlth Org.* **56**: 923–930.

Palese, P., Racaniello, V. R., Desselberger, U., Young, J. & Baez, M. (1980). Genetic structure and genetic variation of influenza viruses. *Phil. Trans. R. Soc. Lond.* (B) **288**: 299–305.

Pensaert, M., Ottis, K., Vandeputte, J., Kaplan, M. M. & Bachmann, P. (1981). Evidence for the natural transmission of influenza A virus from wild ducks to swine and its potential importance for man. *Bull. Wld Hlth Org.* **59**: 75–78.

Pirtle, E. C., Hill, H. R., Swanson, M. R. & van Deusen, R. A. (1976). Haemagglutination-inhibiting antibodies against swine influenza and Hong Kong influenza viruses in swine sera in the USA. *Bull. Wld Hlth Org.* **53**: 7–11.

Schafer, W. (1955). *Z. Naturforsch.* (B) **10**: 81 (cf. Easterday, 1975).

Schild, G. C., Newman, R. W., Webster, R. G., Major, Diane & Hinshaw, Virginia S. (1980). Antigenic analysis of influenza A surface antigens: considerations for the nomenclature of influenza virus. *Archs Virol.* **63**: 171–184.

Shope, R. E. (1931). Swine influenza. III. Filtration experiments and etiology. *J. exp. Med.* **54**: 373–385.

Shortridge, K. F. & Webster, R. G. (1979). Geographical distribution of swine (Hsw1N1) and Hong Kong (H3N2) influenza virus variants in pigs in southeast Asia. *Intervirology* **11**: 9–15.

Smith, W., Andrewes, C. H. & Laidlaw, P. P. (1933). A virus obtained from influenza patients. *Lancet* 1933 (ii): 66–68.

Stuart-Harris, C. H. & Schild, G. C. (1976). *Influenza, the viruses and the disease*. London: Edward Arnold Ltd.

Tumova, Bela (1980). Equine influenza — a segment in influenza virus ecology. *Comp. Immun. Microbiol. infect. Dis.* **3**: 45–59.

Wallace, G. W. (1977). Swine influenza and lungworms. *J. infect. Dis.* **135**: 490–492.

Webster, R. G., Geraci, J., Petursson, G. & Skirnisson, K. (1981). Conjunctivitis in human beings caused by influenza A virus of seals. *New Engl. J. Med.* **304**: 911.

Webster, R. G. & Laver, W. G. (1975). Antigenic variation of influenza viruses. In *The influenza viruses and influenza*: 269–311. Kilbourne, E. D. (Ed.). London: Academic Press.

Webster, R. G., Yakhno, Maya, Hinshaw, Virginia S., Bean, W. J. & Murti, K. G. (1978). Intestinal influenza: replication and characterization of influenza viruses in ducks. *Virology* **84**: 268–278.

Wiley, D. C., Wilson, I. A. & Skehel, J. J. (1981). Structural identification of the antibody-binding sites of Hong Kong influenza haemagglutinin and their involvement in antigenic variation. *Nature, Lond.* **289**: 373–378.

Wilson, I. A., Skehel, J. J. & Wiley, D. C. (1981). Structure of the haemagglutinin membrane glycoprotein of influenza virus at 3 Å resolution. *Nature, Lond.* **289**: 366–373.

World Health Organization (1980). A revision of the system of nomenclature for influenza viruses: a WHO memorandum. *Bull. Wld Hlth Org.* **58**: 585–591.

World Health Organization (In press). The ecology of influenza viruses: a WHO memorandum. *Bull. Wld Hlth Org.*

Yamane, N., Arikawa, J., Odagiri, T. & Ishida, N. (1979). Annual examination of influenza virus infection among pigs in Miyagi Prefecture, Japan: the appearance of the Hsw1N1 virus. *Acta virol.* **23**: 240–248.

DISCUSSION

Steck — Is there any relationship between these virulent strains from birds and classical fowl plague?

Kaplan — Once in a while you will isolate a classical fowl plague strain that does not seem to be virulent. The one in the United States from seals was close to fowl plague, but it doesn't seem to kill chickens.

Symp. zool. Soc. Lond. (1982) No. 50, 137–179

Leishmanial Parasites of Mammals in Relation to Human Disease

RALPH LAINSON

Director, The Wellcome Parasitology Unit, Department of Parasitology, Instituto Evandro Chagas, Fundação Serviços de Saúde Pública, Caixa Postal 3, 66.000 Belém, Pará, Brazil

SYNOPSIS

Species and subspecies of the protozoal parasite *Leishmania* have a wide distribution among wild and domestic animals throughout most tropical and subtropical countries of the world, where they are transmitted by phlebotomine sandflies (Diptera: Psychodidae: Phlebotominae). Some leishmanias, such as those of neotropical porcupines and armadillos, seem to be severely host-restricted, owing either to a high degree of specificity on the part of the sandfly vector, or to an incapacity to survive in other vertebrate hosts. In other cases there is a notable lack of host specificity on the part of the sandfly vectors, and man may frequently become infected when he intrudes into the enzootic foci. Whereas infection in the natural mammalian host is mostly benign and inapparent, it usually produces severe disease in man in the form of skin lesions or generalized visceral involvement. The traditional grouping of the leishmanias into "cutaneous" or "visceral" parasites is unsound: a number of parasites responsible for cutaneous leishmaniasis of man in the Americas are predominantly viscerotropic in the wild mammalian hosts, and *Leishmania tropica* (a causative agent of Old World cutaneous leishmaniasis) is now known to occasionally produce visceral leishmaniasis in man. Among the protozoal diseases of man, leishmaniasis probably ranks second only to malaria in medical and economic importance, and the incrimination of wild or domestic animal reservoirs becomes an important necessity for the establishment of base-lines for the control of human infection. Current methods for the detection of *Leishmania* in the vertebrate host are discussed. Table I lists the known species and sub-species of *Leishmania*, together with their known geographical distribution and recorded mammalian hosts.

INTRODUCTION

The Systematic Position of *Leishmania* (after Levine *et al.*, 1980)

Kingdom PROTISTA Haeckel, 1866
Sub-Kingdom PROTOZOA Goldfuss, 1817
Phylum SARCOMASTIGOPHORA Honigberg & Balamuth, 1963

Sub-Phylum MASTIGOPHORA Diesing, 1866
Class ZOOMASTIGOPHOREA Calkins, 1909
Order KINETOPLASTIDA Honigberg, 1963. emend. Vickerman, 1976
Sub-Order TRYPANOSOMATINA Kent, 1880
Family TRYPANOSOMATIDAE Doflein, 1901 emend. Grobben, 1905
Genus *Leishmania* Ross, 1903

The genus *Leishmania* may be defined as a protozoal parasite, closely related to the trypanosomes, and loosely grouped with these organisms in the "haemoflagellates". It is generally felt that the leishmanias form an ancient group of parasites, originating from flagellates which were at one time peculiar to insects: with the development of a haematophagous habit by the primitive insect host, and migration of the flagellates to the biting mouthparts, the leishmanias gained entrance into certain vertebrates. Parasites located in the cells of the skin or the blood of these animals now act as a reservoir of infection for certain sandflies (Diptera: Psychodidae: Phlebotominae) (Fig. 3). Like other members of the Trypanosomatidae, *Leishmania* appears to be without sex, and multiplication in both the sandfly and the vertebrate is limited to asexual binary fission.

The Morphology and Life-cycle of *Leishmania*

In the vertebrate host
The vertebrate hosts of *Leishmania* include many different species of mammals (Table I) and some reptiles, particularly Old World lizards. Infection is very common in rodents, canids, edentates and marsupials. No infections have been recorded in birds or amphibians, possibly because these animals have rarely been examined for this parasite.

In the mammalian host the parasite is in the amastigote form (Fig. 1). This is a round, oval or cigar-shaped body, about 1.5×2.5 to 3.0×6.5 μm, depending on the species or sub-species of *Leishmania*. Although a rudimentary flagellum is present, within the flagellar pocket, it does not protrude beyond the surface of the parasite. The cytoplasm stains a pale blue with Romanowsky stains and usually contains a number of small vacuoles: the single nucleus and rod-shaped kinetoplast both stain an intense reddish-purple. Within the macrophage the amastigotes divide by binary fission, often packing the cell. Spread of infection within the host may take place when an infected macrophage divides and shares its parasites between the daughter cells, or after the rupture of heavily parasitized cells and the ingestion of liberated amastigotes by further macrophages. It is a strange fact, then, that the organism inhabits the cells which should be protecting the host against their invasion. In the natural host, how-

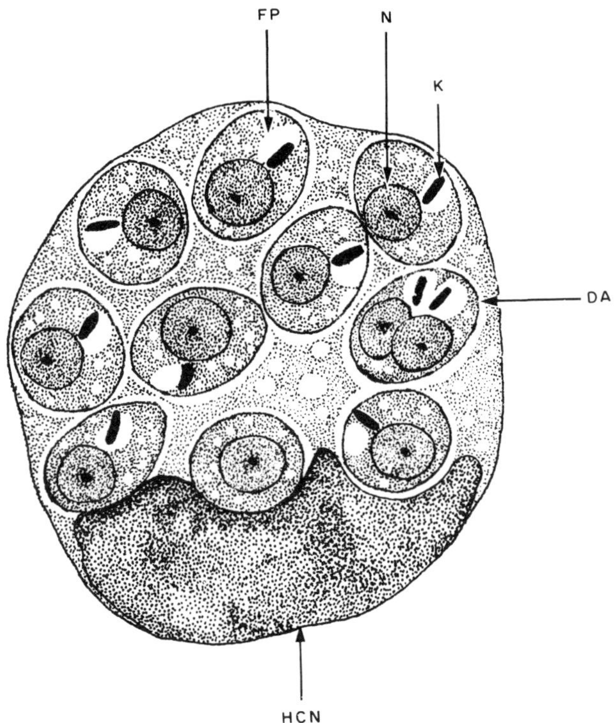

FIG. 1. Amastigotes of *Leishmania* in a macrophage cell of the vertebrate host. DA, dividing amastigote; FP, flagellar pocket; HCN, host-cell nucleus; K, kinetoplast; N, nucleus.

ever, a balance is usually struck, with parasites reaching only moderate numbers, under control of the animal's immune responses.

In the insect host

Phlebotomine sandflies are the only known vectors of *Leishmania*. They are "pool-feeders", piercing the skin of the vertebrate host with their cutting mouthparts and creating a tiny pool of blood which is sucked up into the stomach of the fly. If the host is infected with *Leishmania*, free or intra-cellular amastigotes may be taken up with the blood-meal. During the first 72 h in the insect gut, the parasite elongates and the rudimentary flagellum grows out into a long whip-like structure extending from the flagellar pocket. The resulting promastigote (Fig. 2) is highly motile, and undergoes profuse multi-plication by longitudinal binary fission. The size of the flagellate is very variable and, again, depends on the species or sub-species of *Leishmania*: it varies from about 16.0–40 μm long (including the flagellum) by about 1.5–3.0 μm wide. The flagellum and kinetoplast

TABLE I

Mammalian hosts of Leishmania

Parasite	Known geographical distribution	Recorded mammalian hosts
Section PERIPYLARIA: The *L. braziliensis* Complex: *L. b. braziliensis* Vianna, 1911	Type locality Minas Gerais State, east Brazil: distribution ill-defined	Man:[3] wild animal hosts uncertain: possibly some of those listed for the "unidentified parasites" of the *L. braziliensis* Complex, below
L. b. guyanensis Floch, 1954	The Amazon basin, north of the River Amazonas: i.e. the Guyanas; Amapa, north Pará and Amazonas States in Brazil	The sloth *Choloepus didactylus*[1] and the anteater *Tamandua tetradactyla*;[1] the opossum *Didelphis marsupialis*[2] and the rodent *Proechimys guyannensis*;[2] man[3]
L. b. panamensis Lainson & Shaw, 1972	Panama and Costa Rica. Similar parasites have been reported in man from Honduras and Belize (Cen. Am.) but nothing is known about their reservoir hosts	The sloths *Choloepus hoffmanni*[1] and *Bradypus infuscatus*;[2] the procyonids *Nasua nasua*,[2] *Potos flavus*[2] and *Bassaricyon gabbii*;[2] the monkeys *Aotus trivergatus*[2] and *Saguinus geoffroyi*;[2] the dog[3] and man[3]
Unidentified parasites, probably subspecies of *L. braziliensis*	Pará State, north Brazil	The rodents *P. guyannensis, Rattus rattus,* and *Rhipidomys leucodactylus*; the opossum *D. marsupialis*; the sloth *C. didactylus*; the armadillo *Dasypus novemcinctus*: man[3]
	Mato Grosso State, Central Brazil	The rodent *Oryzomys concolor*: man[3]
	São Paulo State, South Brazil	The rodents *O. capito, O. nigripes* and *Akodon arviculoides*: man[3]
	Panama	The rodent *Diplomys labilis* (Herrer, Telford *et al.,* 1971): man[3] (Walton *et al.,* 1977)
	Costa Rica	Man[3]
L. b. peruviana Velez, 1913	Peruvian and Argentinian Andes	The dog[3] and man[3]
Section SUPRAPYLARIA: The *L. mexicana* Complex: *L. mexicana mexicana* Biagi, 1953	Yucatan peninsula, Belize, Guatemala	The rodents *Ototylomys phyllotis,*[1] *Heteromys desmarestianus,*[2] *Nyctomys sumichrasti*[2] and

L. mexicana amazonensis Lainson & Shaw, 1972	Amazon basin of Brazil and probably in neighbouring countries	The rodents *P. guyannensis*,[1] *O. capito*,[2] *O. concolor*,[2] *O. macconnelli*,[2] *Neacomys spinosus*,[2] *Nectomys squamipes*,[2] and *Dasyprocta* sp.[2] The marsupials *Marmosa murina*,[2] *M. cinerea*,[2] *Metachirus nudicaudatus*,[2] *D. marsupialis*[2] and *Philander opossum*;[2] the fox *Cerdocyon thous*;[2] man[3]
L. mexicana pifanoi Medina & Romero, 1959	Venezuela	The rodent *Heteromys anomalus*; man[3]
L. mexicana aristedesi Lainson & Shaw, 1979	Eastern Panama	*O. capito*,[1] *Proechimys semispinosus*[2] and *Dasyprocta punctata*;[2] *Marmosa robinsoni*[2]
L. mexicana garnhami Scorza *et al.*, 1979	Venezuelan Andes	*D. marsupialis*; man[3]
L. mexicana enriettii Muniz & Medina, 1948	Paraná State, Brazil	The domestic guinea pig, *Cavia porcellus*: wild animal host unknown
L. mexicana subsp.	Mato Grosso and Minas Gerais States, Brazil	The dog;[3] man[3]
L. mexicana subsp.	Trinidad, West Indies	*O. capito*,[1] *P. guyannensis*,[2] *H. anomalus*;[2] the marsupials *Caluromys philander*,[2] *Marmosa fuscata*[2] and *M. mitis*[2]
The *L. hertigi* Complex *L. hertigi hertigi* Herrer, 1971	Panama and Costa Rica	The tree-porcupine, *Coendou rothschildi*[1]
L. hertigi deanei Lainson & Shaw, 1977	Piauí and Pará States, Brazil	The tree-porcupines, *C. prehensilis*[1] and an undescribed *Coendou* sp.[1]
The *L. tropica* Complex: *L. tropica tropica* (Wright, 1903)	In Asia: Israel, Jordan, Iraq, Iran, Afghanistan, Turkmen and Usbek S.S.R., Syria and India. In Africa:[a] Morocco, Algeria, Tunisia and Libya. In Europe:[a] Bulgaria, Crete, S. France, Greece, Italy, Portugal, Sicily, Spain and Yugoslavia	The disease caused by *L. t. tropica* is regarded by some as anthroponotic, in which case man is the primary host. Infections recorded in the dog and *Rattus rattus* are of doubtful epidemiological significance

TABLE I (*continued*)

Parasite	Known geographical distribution	Recorded mammalian hosts
L. tropica aethiopica Bray, Ashford & Bray, 1973	Ethiopian Highlands and Mt. Elgon in Kenya	The hyraxes *Procavia habessinica* and *Heterohyrax brucei* in Ethiopia: the hyraxes *Procavia johnstoni* and *Dendrohyrax arboreus*, and the giant-rat *Cricetomys* sp., in Kenya: man[3]
L. tropica subsp.?	Namibia, S. Africa	Man[3]
L. tropica subsp.?	Namibia, S. Africa	The hyrax, *P. capensis*[1]
L. tropica subsp.?	Ethiopia	The rodent *Arvicanthis* sp. See Chance (1979)
The *L. major* Complex: *L. major* Yakimoff & Schockor, 1914	N.W. China; Thar desert of N.W. India and Pakistan; deserts of Turkmenia, Usbekistan, S. Tadjikistan and Kazakhstan S.S.R.; N. Afghanistan; E. Saudi Arabia; S.W. Iran; Iraq; Kuwait; Jordan; Libya; Israel. In Africa: Algeria; Tunisia; United Arab Republic; the Sahara; Senegal; Sudan; Kenya and Mali. Probably Chad; Niger; Upper Volta	Mainly rodents: in N.W. China, Turkmenian and Usbek S.S.R., N. Afghanistan and N.E. Central Iran, the great gerbil *Rhombomys opimus*,[1] with the gerbil *Meriones libycus*[2] in some areas. In the USSR the following rodents[3] are occasionally infected: *Meriones tamariscinus, Meriones meridianus, Spermophilopsis leptodactylus, Citellus fulvus, Hystrix leucura, Allactaga severtzovi, A. elator, Alactagulus acontion, Dipus sagitta, Nesokia indica, Mus musculus, Cricetulus migratorius* and *Microtus afghanus*: non-rodents[3] include the hedgehog, *Hemiechinus auritus*, the hare *Lepus tolai*, the mustelids *Mustela nivalis* and *Vormela peregusna*. Rodent hosts elsewhere include *Meriones hurrianae* in N.W. India (Rajasthan) and Pakistan (Baluchistan); *Meriones* spp. in S.W. Iran, E. Saudi Arabia and Iraq; *Psammomys obesus*[1] and *Meriones* spp. (?) in Israel and Libya; *Mastomys erythroleucus*,[1] *Tatera gambiana*[1] and *Arvicanthis niloticus* in Senegal; *Tatera* and *Xerus* spp. in Kenya. Many infections in dogs[3] throughout this geographic range are probably due to *L. major* rather than *L. t. tropica*.

L. major subsp.?	N.W. India	Rodents (*Meriones* sp.?): man[3] (See Chance, 1979)
L. gerbilli Wang, Qu and Guan, 1964	S. Mongolia	*Rhombomys opimus*.[1] Apparently not infective to man
The *L. donovani* Complex: *L. donovani donovani* (Laveran & Mesnil, 1903)	India: Assam, Bihar, Bengal, Orissa, Tamil Nadu, Gujarat: E. Pakistan: Nepal: N. China Plain[b] and E. China[b]: in Hopeh, Shansi, Shantung, Kiangsu, Anhwei, Honan, Hupeh, Szechwan, Tsinghai and Sinkiang	No hosts other than man yet discovered, and the disease appears to be an anthroponosis. Exact distribution in China is uncertain due to the coexistence of *L. d. infantum* (see below)
L. donovani infantum Nicolle, 1909	Central Asia; Turkmenia, Usbekistan, Tadjikistan, Kirgizia, S. Kazakhstan, Azarbaidjan, Georgia and Armenia, S.S.R., N. and N.W. China, in Liaoning, Hopeh, Shansi, Shensi, Ningsia, N. Szechwan, N. Tsinghai and Kansu. In S.W. Asia: Iraq, Saudi Arabia, Yemen, S. Yemen, Iran and Afghanistan. In Africa: N. Algeria, N. Tunisia, Libya, Egypt, Central African Republic, Chad, Gabon, Nigeria, Malawi, Niger, Upper Volta, Zaire, and Zambia. Canine infections only, in Congo, Somalia and Senegal. All countries of the Mediterranean littoral, extending into Hungary and Rumania	Primitive hosts in Central Asia are the jackal *Canis aureus*,[1] the wolf *Canis lupus*[1] and foxes, *Vulpes vulpes*[1] and *V. corsak*:[1] the dog[3] has become the major reservoir for human infection in urban areas, extending the disease into S.W. Asia, the Mediterranean and China. The epidemiological significance of infections found in the domestic rats, *Rattus rattus* and *R. norvegicus*, in Yugoslavia and Italy has still to be assessed. Man[3]
L. donovani archibaldi Castellani & Chalmers, 1919. (synonym for *L. d. infantum*?)	S. and central Sudan	The rodents *Acomys albigens, Arvicanthis niloticus* and *Rattus rattus*: the carnivores *Genetta genetta* and *Felis serval*: man[3]. Dogs have *not* been incriminated
L. donovani subsp.? (= *L. d. infantum*?)	Kenya: possibly S. Ethiopia and Somalia	The dog[3] and man[3]

TABLE I (*continued*)

Parasite	Known geographical distribution	Recorded mammalian hosts
L. donovani chagasi Cunha & Chagas, 1937 (synonym for *L. d. infantum, pro parte*?)	Argentina, Bolivia, Brazil, Colombia, Ecuador, Paraguay, Surinam, Venezuela: Guatemala, Honduras, Mexico, El Salvador: Guadeloupe	The fox, *Lycalopex vetulus*: the dog:[3] man.[3] Parasites isolated from the fox *Cerdocyon thous* in the Amazonian Region of Brazil may be *L. d. chagasi*, or another viscerotropic parasite of man (see below)
L. donovani subsp. (= *L. d. chagasi* s.l.?)	Pará State, Amazonian Region of Brazil	The fox, *Cerdocyon thous*:[1] man:[3] the dog[3]

Where sufficient information is available, mammalian hosts are indicated as: primary, natural host,[1] secondary natural host,[2] or "accidental" victim.[3] [a]Considerable doubt now exists as to the presence of true oriental-sore due to *L. tropica tropica* in both Europe and Africa: skin lesions of man in these geographic regions are thought to be due either to variants of *L. d. infantum* (Europe and Africa) or *L. major* (Africa). [b]Evidence for the presence of *L. d. donovani* in China is still inconclusive, but strongly suggested by the absence of canine infection in some endemic areas, clinical features, and the differing response of *L. d. donovani* and *L. d. infantum* to anti-leishmanial drugs.

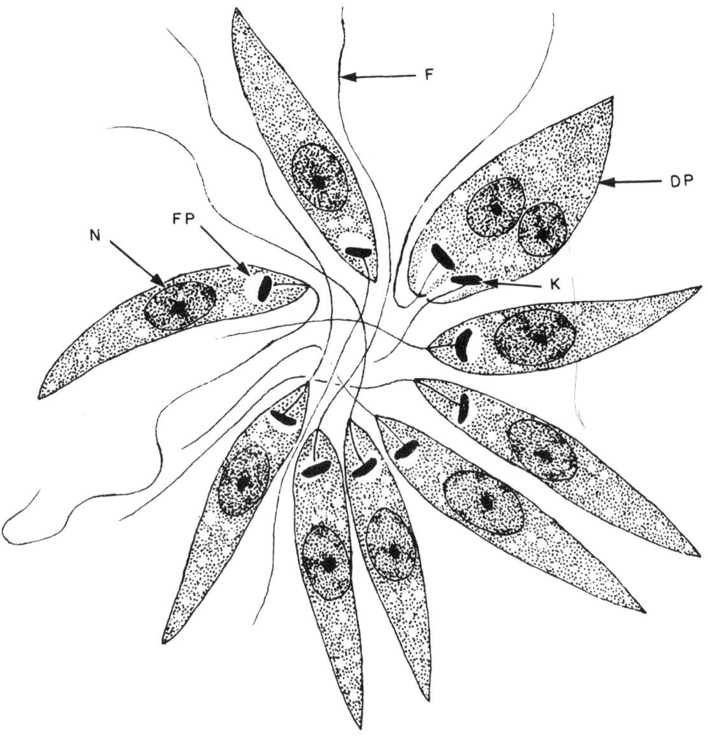

FIG. 2. Promastigotes of *Leishmania* in the midgut of the insect vector. DP, dividing promastigote; F, flagellum; FP, flagellar pocket; K, kinetoplast; N, nucleus.

are situated at the anterior end of the organism, and the nucleus is more or less centrally placed.

Subsequent development of the infection in the sandfly varies among the different leishmanias, principally in the time needed for the invasion of the biting mouthparts, and in the location of attached forms of the flagellates in different parts of the alimentary tract (Killick-Kendrick, 1979; Lainson & Shaw, 1979). Transmission takes place by the bite of the infected insect, when flagellates are injected, *via* the proboscis, into the skin of the new host. Finally the mode of development in the sandfly vector is used as an important character for the basic taxonomic grouping of the leishmanias (Lainson & Shaw, 1979).

Geographical Distribution and Medical Importance

Species and sub-species of *Leishmania* have a remarkably wide distribution in most tropical and sub-tropical countries, extending

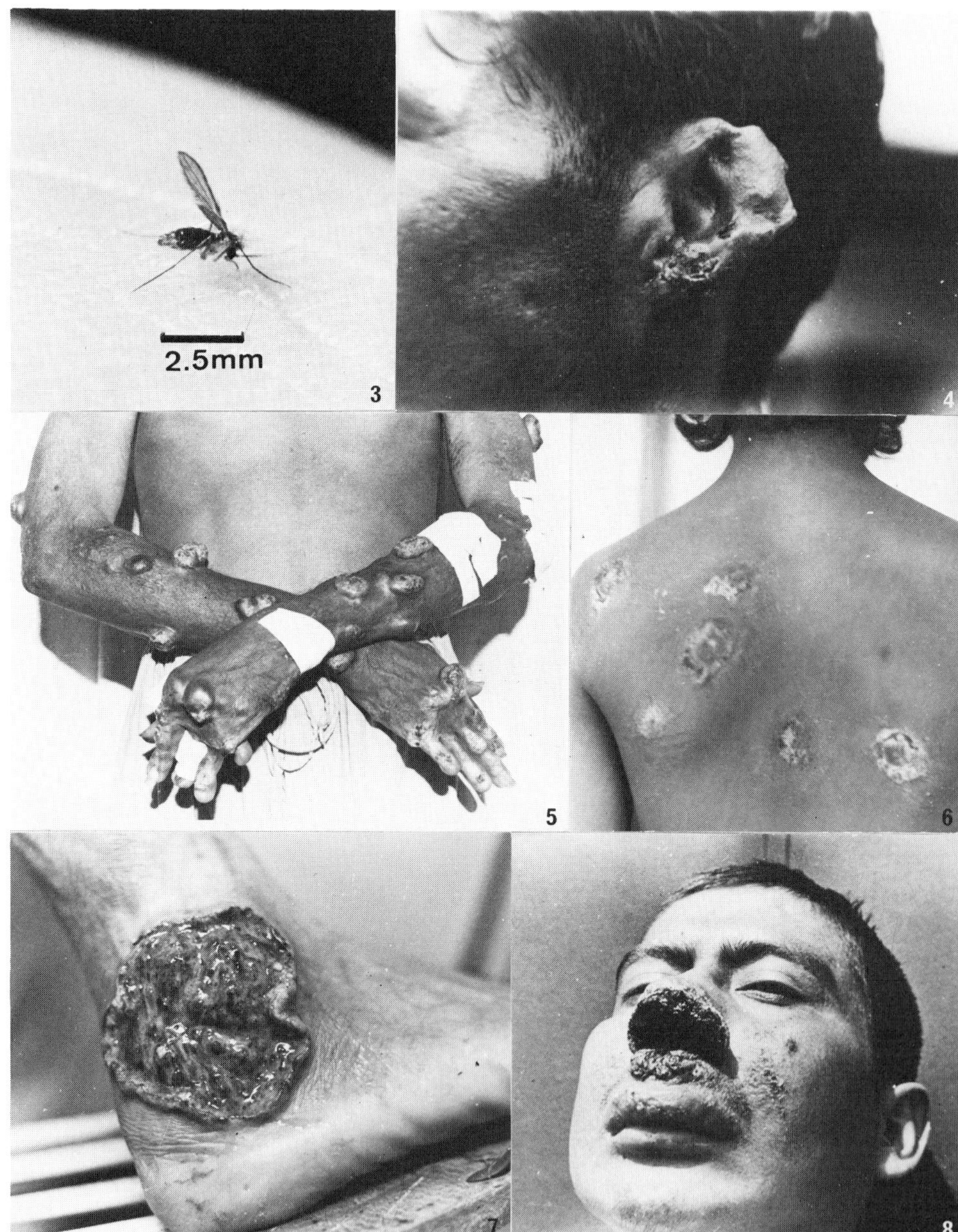
2.5mm
3
4
5
6
7
8

throughout most of Central and South America, the Mediterranean countries, Africa, Central Asia, India and China. Interestingly, *Leishmania* was until recently thought to be absent from North America, north of Mexico: three cases of human cutaneous leishmaniasis have now been reported from Texas, however (Simpson, Mullins & Stone, 1968; Shaw *et al.*, 1976), and canine visceral leishmaniasis has been recorded among foxhounds from a kennel in Oklahoma (Anderson *et al.*, 1980). The canine disease may originally have been imported in dogs coming from Europe, but spread to other dogs in the same kennel remains a mystery in the apparent absence of mammal-biting phlebotomine sandflies. There are no records of *Leishmania* in Australia, although phlebotomine sandflies do occur in certain parts of that country: a single case of mucocutaneous leishmaniasis in Japan remains of doubtful origin.

With the possible exceptions of "kala azar" (Indian visceral leishmaniasis) and cutaneous leishmaniasis due to *L. tropica tropica*, man is usually regarded as an "accidental victim" who plays a small rôle, if any, in the maintenance of the different leishmanial parasites in nature. He becomes infected when he intrudes into the enzootic cycle and is bitten by those sandfly vectors which have a broad host-spectrum. As an unaccustomed host, man usually reacts violently to the intruding organism, with the production of skin lesions (Figs 4–8) or pathological changes in the viscera. As a human disease, leishmaniasis is probably second in importance only to malaria among protozoal infections, and a global incidence of about 400 000 new cases per year has been estimated (Anon., 1981): everything suggests that this figure is still on the increase. Visceral leishmaniasis has exacted an enormous toll of human life since ancient times, and recent, massive outbreaks of the disease in India have been the subject of great concern: American cutaneous and mucocutaneous leishmaniasis have debilitated and mutilated man for centuries.

As most forms of leishmaniasis are zoonoses, it is clearly of considerable importance to indicate the mammalian hosts of the different leishmanias, to provide base-lines for the control of the human

---

FIG. 3. A phlebotomine sandfly, *Lutzomyia* sp., biting man (from Lainson & Shaw, 1971).

FIG. 4. Typical erosion of the ear, by *L. m. mexicana* (from Lainson & Strangways-Dixon, 1963).

FIG. 5. Long-standing diffuse cutaneous leishmaniasis due to *L. m. amazonensis*.

FIG. 6. "Pian-bois" due to *L. b. guyanensis*.

FIG. 7. Large ulcer on the foot, due to *L. b. braziliensis* s.l., from Pará State, Brazil.

FIG. 8. Mucocutaneous leishmaniasis due to *L. b. braziliensis* s.l. (from Lainson & Shaw, 1971).

disease. This chapter deals with the recorded hosts and geographical distribution of the recognized species and sub-species of *Leishmania* (Table I), and current methods for detection and isolation of the parasite. It is beyond the scope of this article to discuss the complicated process of identification and taxonomy of the leishmanias: accent has been placed on epidemiological situations in the Americas, simply because it is here that the author has had most practical experience.

HISTORICAL EVENTS IN THE INCRIMINATION OF WILD AND DOMESTIC ANIMALS AS RESERVOIRS

In the Old World

The dog

Leishmania was first seen by Cunningham (1885) in histological sections of a skin lesion ("Delhi boil") of man in India, and was shown to be the cause of human visceral leishmaniasis ("kala azar") in the same country by Leishman (1903). For some time, therefore, leishmaniasis was studied predominantly by clinicians as an essentially human disease, and simply classified as "cutaneous" due to *Leishmania tropica* (Wright, 1903), or "visceral" due to *Leishmania donovani* (Laveran & Mesnil, 1903).

Nicolle & Comte (1908), however, made the important discovery of a visceral infection in a dog, in Tunisia, and by 1912 canine leishmaniasis had been found in all those Mediterranean countries where the human disease had been recorded. It is generally agreed that the causative organism, now referred to as *L. donovani infantum* Nicolle, 1909, was primarily a parasite of wild Canidae (jackals, wolves and foxes) in rural Central Asia: in time the domestic dog became secondarily involved, and the disease spread from urban areas in Transcaucasia to south-west Asia, Africa and the Mediterranean Basin in one direction, and to China in the other.

The dog has never been shown to be a reservoir of kala azar due to *L. donovani donovani* in India, where the disease appears to have become an anthroponosis.

Attempts to incriminate the dog as a reservoir of cutaneous forms of human leishmaniasis have been less successful. There are many reports of amastigotes found in canine skin lesions, but their true identity has seldom been ascertained. In many instances they are likely to have been *L. d. infantum*, which is known to invade the skin during the evolution of the visceral disease in dogs. Or they may have been another parasite, *Leishmania major*, which is more commonly found in man and wild rodents, as discussed below.

Wild animals

While canids are recognized as the principal hosts of *L. d. infantum*, some evidence is accumulating to suggest that other animals may be involved as reservoirs of this or other leishmanias responsible for human visceral leishmaniasis. In the Sudan, for example, infections with *L. donovani sensu lato* have been found in wild rodents and carnivores (Hoogstraal & Heyneman, 1969) (Table I). The parasite appears to be identical with that causing both cutaneous and visceral leishmaniasis of man in that region. In Yugoslavia, Petrović, Borjośki & Savin (1975) found domestic rats, *Rattus rattus* and *R. norvegicus*, infected with *L. d. infantum*; and Bettini, Gradoni & Pozio (1978) have confirmed the existence of natural infections with this parasite in *R. rattus* in Italy. Whether or not these non-canine hosts have any real epidemiological significance is debatable: they may merely represent "victims", like man.

From very early on it was clear that there were two forms of cutaneous leishmaniasis of man, variously distributed in the principal endemic areas of Central Asia and the Indian continent, and Russian clinicians referred to them as "wet rural" and "dry urban" forms. The "wet rural" disease was found to be associated with the edges of arid or semi-desert regions, whereas the "dry urban" form occurred in the vicinity of towns and ancient settlements. Yakimoff & Schockor (1914) used the sub-specific names of *Leishmania tropica major* and *Leishmania tropica minor* for the causative agents of the two diseases, respectively, on the basis of clear-cut differences in the size of the amastigotes. Present-day specialists, however, refer to them simply as *Leishmania major* and *Leishmania tropica*.

Working in the desert regions of Turkmenia, S.S.R., Latyšev & Krjukova (1941) found up to 25.0% of the burrowing rodents *Rhombomys opimus* (the great gerbil) to harbour *L. major* in their skin, and this animal was subsequently shown to be the principal reservoir of the parasite in north-west China, Iran and Afghanistan. A variety of other rodents play a secondary rôle in these and other countries, including lesser gerbils, jirds, ground-squirrels, jerboas, Nile-rats, sand-rats, voles, and multi-mammate rats (Table I). Non-rodent hosts include hedgehogs, porcupines, hares and mustelid carnivores, but these are probably of minor importance.

Chinese workers (Wang, Qu & Guan, 1964) isolated another parasite from *R. opimus*, in southern Mongolia, and called it *Leishmania gerbilli*. It has large amastigotes, like *L. major*, to which it is probably related. It differs from that parasite, however, in its failure to infect inoculated volunteers.

In Ethiopia, Ashford *et al.* (1973) found hyraxes, *Procavia* and *Heterohyrax*, to be natural hosts of *L. tropica aethiopica* Bray, Ashford & Bray, 1973, the causative agent of cutaneous leishmaniasis in that country: other hyraxes, and the "great rat" *Cricetomys*, have been found infected with what appears to be the same parasite on neighbouring Mt. Elgon, Kenya (Mutinga, 1975).

Leishmaniasis was unrecorded in South Africa until recently, when Grové (1970) reported four cases of the cutaneous disease in Namibia. Grové & Ledger (1975) isolated a *Leishmania* from the skin of hyraxes, *Procavia*, from the same region: surprisingly, however, biochemical studies (Chance, 1979) suggested that this organism was not the same as that infecting man. Both probably represent subspecies of *L. tropica*.

In the New World

Human visceral leishmaniasis
The existence of the disease in the New World was first established in Paraguay by Migone (1913), and since that time has been recorded in almost every Latin American country (Table I). Major endemic areas occur in the semi-arid regions, where transmission is peri-domestic (Deane, 1956) and involves foxes and dogs (Fig. 9) and the sandfly *Lutzomyia longipalpis*. The incrimination of dogs and foxes and clinical similarities between the Mediterranean and American visceral leishmaniases, have led to the suggestion by some that both diseases are due to *L. d. infantum* which was introduced into Latin America by settlers (or their dogs) in relatively recent times.

Lainson & Shaw (1979) discussed the ecological and epidemiological situation in the Amazon Region of Brazil (Pará), where sporadic cases of visceral leishmaniasis have been recorded, and suggested that another parasite might be involved, with a silvatic reservoir. Such a parasite had been isolated from the fox, *Cerdocyon thous* (Lainson, Shaw & Lins, 1969) (Fig. 10), but loss of the stock had not permitted comparison with *L. d. infantum*.

Biochemical comparison of stocks of *L. d. infantum* with a limited number of isolates of *Leishmania* from cases of American visceral leishmaniasis (none of which was from the Amazon Region) have, it is true, so far failed to separate the two parasites (Schnur *et al.*, 1981; Lainson, Miles & Shaw, 1981). The latter authors pointed out, however, that striking differences in the growth of Latin American stocks in both hamsters and NNN culture (see p. 166), have led to an artificial selection of the easily handled stocks for biochemical studies, and a complete failure to characterize the more fastidious parasites. Mention was made of a recent, solitary case of Amazonian visceral leishmaniasis on the island of Marajo, Pará, on the Amazon delta. The man had never left the island; he was a forest-worker, in the same general area where the last, single case had occurred 14 years ago; and members of the victim's family, and neighbours, claimed that sandflies (which they well knew, as "tatuquiras") did not occur in their houses.* These features suggested the rare contact of man with some *Leishmania* of silvatic (feral) origin (Fig. 30): it is particularly unfortunate that no material could be obtained for isolation of the parasite and its biochemical comparison with stocks from the classical endemic areas of visceral leishmaniasis in north-east Brazil.

Human cutaneous leishmaniasis

Whatever the reasons for regarding American visceral leishmaniasis as due to *L. d. infantum*, there is no doubt concerning the indigenous nature of the various forms of the American cutaneous disease, due to species or sub-species of the *L. mexicana* and *L. braziliensis* complexes. Ancient Peruvian and Ecuador pottery from the era 400—900 A.D. depicts human faces with mutilations similar to those caused by cutaneous and mucocutaneous leishmaniasis, and Spanish historians, at the time of the Conquest, described ugly sores on the faces of Peruvian Indians. The aetiology of the disease was finally recognized when the Brazilian parasitologists Lindenberg (1909) and Carini & Paranhos (1909) independently demonstrated *Leishmania* in human skin lesions in south Brazil. Their observations were closely followed by that of a Dr A. L. Barton, in Lima, Peru, who also saw amastigotes in the skin lesion of a patient suffering from "uta", in 1910 (Townsend, 1915).

Both the Brazilian and Peruvian parasites were at first considered as *L. tropica*, but were later named *Leishmania braziliensis* Vianna, 1911 and *Leishmania peruviana* Velez, 1913, respectively. *L. braziliensis* was first associated with the highly mutilating nasopharyngeal lesions of man (Fig. 8), in Brazil, by Splendore (1912). The geographic range of New World dermal leishmaniasis was slowly extended to include almost every country of the continent, from south Mexico to Argentina, and new specific or sub-specific names appeared — such as *Leishmania mexicana* Biagi, 1953 for the causative agent of "chiclero's ulcer" in the Yucatan, Belize and Guatemala (Fig. 4), and *L. braziliensis guyanensis* Floch, 1954 for that of "pian bois" in the Guyanas and north Brazil (Fig. 6).

---

* Since the preparation of this chapter the author and colleagues have, in fact, found abundant *Lutzomyia longipalpas* in chicken-houses by the side of the victim's house, and have isolated *Leishmania donovani chagasi* from a fox, *Cerdocyon thous*, caught in the same locality.

American cutaneous and mucocutaneous leishmaniasis was shown to be most prevalent in forested areas, suggesting the existence of a wild animal reservoir. The first concerted effort to unravel the epidemiology of any one form of the disease was not in the forest, however, but in the endemic foci of "uta", on the relatively barren slopes of the Peruvian Andes. Early studies by the Harvard School of Tropical Medicine expedition (Strong *et al.*, 1913, 1915) had established the peridomestic nature of the disease, at altitudes between 900 and 3000 m, in non-forested regions. Later epidemiological studies by Herrer (1951) concentrated, therefore, on dogs and other domestic animals. Amastigotes were found in skin lesions, and sometimes in apparently normal skin, of 46 out of 513 dogs examined (8.9%), but insufficient material from wild animals was examined to exclude the likely possibility of a feral reservoir.

Muniz & Medina (1948) discovered the enigmatic *Leishmania enriettii* in domestic guinea pigs in Paraná State, Brazil: one suspects its origin in some wild animal, although this has yet to be discovered. Close similarities between this parasite and *L. mexicana*, in terms of DNA buoyant densities and enzyme electrophoretic profiles, led Chance (1979) to feel that it would be better regarded as a subspecies of *L. mexicana*: I think this view is correct (Table I), especially in view of other similarities shared by the two parasites in the large size of their amastigotes and promastigotes, and their ease of *in vitro* culture.

Hertig and colleagues (Anon., 1957, 1959) obtained the first evidence that silvatic animals acted as the reservoir of American cutaneous leishmaniasis, when they isolated *Leishmania* in cultures of heart-blood from forest rodents in Panama: Forattini (1960) made similar isolations, in south Brazil, and also demonstrated amastigotes in smears from skin lesions of some animals. Unfortunately, the precise identity of both the Panamanian and Brazilian parasites was not determined (Table I).

FIG. 9. Emaciated dogs suffering from visceral leishmaniasis due to *L. d. chagasi*, in Ceará State, north-east Brazil (from Lainson & Shaw, 1971).

FIG. 10. The fox, *Cerdocyon thous*, a host of *L. m. amazonensis*. It has also been shown to be the host of a viscerotropic *Leishmania* which is possibly *L. d. chagasi* or a hitherto unidentified parasite responsible for human visceral leishmaniasis in the Amazon Region (from Lainson, Shaw & Lins, 1969).

FIG. 11. The "big-eared tree rat", *Ototylomys phyllotis*, an important reservoir of *L. m. mexicana* in Belize (from Lainson & Strangways-Dixon, 1964).

FIG. 12. The principal host of *L. m. amazonensis* in the Brazilian Amazon Region, the echimyid rodent *Proechimys guyannensis*.

FIG. 13 and 14. Some secondary hosts of *L. m. amazonensis*: the cricetid rodent *Oryzomys capito*, and the opossum *Philander opossum*.

Lainson & Strangways-Dixon (1962, 1964) discovered the reservoir hosts of *L. mexicana mexicana* among silvatic rodents in Belize (Fig. 11): this prompted similar studies on the epidemiology of cutaneous leishmaniasis in north Brazil, where Lainson & Shaw (1973, 1979, and unpublished observations) found a similar parasite of man, *L. mexicana amazonensis* (Fig. 28), in no less than 13 different mammalian species (Table I; Figs 10 and 12—15).

There is increasing evidence indicating the existence of a number of other sub-species of *L. mexicana* in the Americas and, like *L. m. mexicana* and *L. m. amazonensis*, they appear principally to be parasites of rodents, less frequently of marsupials (Table I). Thus the causative agent of diffuse cutaneous leishmaniasis of man in Venezuela, originally named *L. braziliensis pifanoi*, is clearly closely related to *L. mexicana* and should be referred to as *L. mexicana pifanoi* Medina & Romero, 1959 (Lainson & Shaw, 1972, 1979): it is biochemically distinct from both *L. m. mexicana* and *L. m. amazonensis* (Miles *et al.*, 1980).

Parasites very similar to *L. m. pifanoi* have been isolated from the rodent *Heteromys* in Venezuela (Torrealba, Gómez-Núñez & Ulloa, 1972), and seen in smears of skin lesions of *Proechimys* (Convit, 1968) and *Zygodontomys* (Kerdel-Vegas & Essenfeld-Yahr, 1966). Pons (1968) described leishmaniotic lesions of dogs and donkeys in the same country: the causative *Leishmania* was not identified, however, and it is most likely that both the dog and the donkey are "accidental" hosts, like man, and become infected when they accompany him to the forest.

Tikasingh (1974) isolated *L. mexicana* from a number of rodents and marsupials in Trinidad: the parasite has not been identified biochemically, but from the geographical location it may well be *L. m. pifanoi*.

In Panama, Herrer, Telford & Christensen (1971) isolated a sub-species of *L. mexicana* from rodents and marsupials: this parasite was shown to be biochemically distinct from all known sub-species (Chance, Peters & Shchory, 1974; Gardener, Chance & Peters, 1974) and was named *L. mexicana aristedesi* Lainson & Shaw, 1979. No human infection has yet been recorded.

The causative agent of human cutaneous leishmaniasis in the Caratinga and Rio Doce valleys, Minas Gerais, Brazil (Dias, 1975) is considered by Chance (1979) as yet another sub-species of *L. mexicana*. Wild animals have not yet been incriminated as reservoirs, and canine infections recorded in the same areas may be just as accidental as those of man. Finally, Scorza *et al.* (1979) gave the specific name of *L. garnhami* to a *Leishmania* causing cutaneous leishmaniasis of man in the Venezuelan Andes. From the parasite's behaviour in hamsters and NNN culture, and its biochemistry, however, it is very close to *L. m. amazonensis* (Lainson, Shaw & Miles, unpublished observations), and it would be better referred to as *L. mexicana garnhami*: a single opossum was found infected.

Continuing studies by workers in Panama revealed occasional infections with *Leishmania* in a variety of silvatic mammals, including monkeys and procyonids. It was finally concluded, however (Herrer, Christensen & Beumer, 1973), that the high rate of infection in the two-toed sloth, *Choloepus hoffmanni*, indicated this animal as the principal reservoir host of *L. braziliensis panamensis* Lainson & Shaw, 1972.

Lainson, Shaw & Póvoa (1981) and Lainson, Shaw, Ready *et al.* (1981) studied the epidemiology of *L. b. guyanensis* in north Pará State, Brazil (Fig. 31). The major reservoir was found to be the sloth, *Choloepus didactylus* (Fig. 16), with an infection rate as high as 46.0%. The anteater, *Tamandua tetradactyla* (Figs 17, 18) is also commonly infected (22.2%). It was felt that this animal may be of importance in the spread of foci of infection: it is much more mobile than the sloth, and is frequently encountered crossing major highways from one area of forest to another. Secondary hosts were occasionally found among rodents and marsupials and, in primary forest, they probably play a minor epidemiological rôle. However, Arias *et al.* (1981) isolated *L. b. guyanensis* from three out of 15 opossums, *Didelphis* (Fig. 15), from forest much frequented by man, and "poor in game": it is possible that with the disappearance of the primary edentate hosts the sandfly vectors must now turn their attention to animals such as *Didelphis*, which are not unduly perturbed by man's proximity. A high infection rate has also been found in sloths (*C. didactylus*) in French Guyana (Gentile *et al.*, 1981).

Attempts to locate reservoirs of *L. braziliensis braziliensis* s.l. have met with little success (Fig. 32), for a number of reasons. First, Vianna's type material is no longer available for comparison and his description is inadequate by today's standards; secondly, there is clearly a complex of closely related sub-species of *L. braziliensis*, each of which is likely to have different sandfly vectors and vertebrate hosts; thirdly, many of these parasites are very difficult to maintain in culture or laboratory animals. Parasites which may represent *L. b. braziliensis* of man have occasionally been isolated from wild animals (Table I: Unidentified parasites). Forattini *et al.* (1972, 1973) isolated *Leishmania* from the rodents *Akodon* and *Oryzomys* in São Paulo State, south Brazil; and isolations have also

been made from the rodents *Proechimys, Oryzomys, Rattus* and *Rhipidomys*, the opossum *Didelphis* and the sloth *Choloepus*, in forested areas of Pará State, north Brazil, and Mato Grosso State, central Brazil (Lainson & Shaw, 1979; Lainson, Shaw, Ready *et al.*, 1981 and unpublished observations). These parasites all shared certain characteristics with those from man in the same geographic regions — namely, their poor growth in hamster skin and reluctance to adapt to *in vitro* culture media that are highly suitable for other leishmanias: in many cases their maintenance has proved so difficult that the parasites have been lost, and in no case has it been possible to make a biochemical comparison with the few stocks of *L. b. braziliensis* s.l. of man that have been successfully maintained. Working in the Serra dos Carajás, Pará, an area highly endemic for cutaneous and mucocutaneous leishmaniasis, Lainson, Shaw, Ward & Fraiha (1973) were able to pinpoint the major sandfly vector transmitting infection to man, but failed to locate the reservoir among a variety of predominantly terrestrial mammals (unpublished observations). It is possible, then, that arboreal endentates may also prove to be involved in the epidemiology of *L. b. braziliensis* (Fig. 32).

The discovery of a parasite of the *braziliensis* complex commonly infecting armadillos (Lainson, Shaw, Ward, Ready & Naiff, 1979) (Fig. 19) raised some hopes that a major reservoir of *L. b. braziliensis* in man had at last been located: enzyme electrophoresis of 13 isolates has separated the two parasites, however, although they are clearly related organisms. The parasite of *Dasypus novemcinctus* differs most strikingly from *L. b. braziliensis* in its tiny amastigotes (Fig. 26) and the extreme ease with which it is cultured.

The significance of the above-mentioned *L. b. braziliensis*-like parasites of wild mammals in relation to human leishmaniasis has still to be assessed, and will depend on their careful comparison with large numbers of isolates made from man in the same localities. Some leishmanias may prove to be entirely restricted to certain wild animals — by a degree of specificity on the part of their sandfly

---

FIG. 15. Another secondary host of *L. m. amazonensis*: the opossum *Didelphis marsupialis*.

FIG. 16. The two-toed sloth, *Choloepus didactylus*, primary reservoir-host of *L. b. guyanensis*.

FIGS 17 and 18. The lesser anteater, *Tamandua tetradactyla*, another important host of *L. b. guyanensis*. Figure 17 shows the black variety, commonly found infected, north of the Amazon River. Figure 18 shows the more usual black and cream variety, south of the river, and as yet unassociated with any known *Leishmania*.

FIG. 19. The nine-banded armadillo, *Dasypus novemcinctus*, common host of a *Leishmania*, as yet not known to infect man.

FIG. 20. The tree-porcupine, *Coendou prehensilis prehensilis*, host of *L. hertigi deanei*.

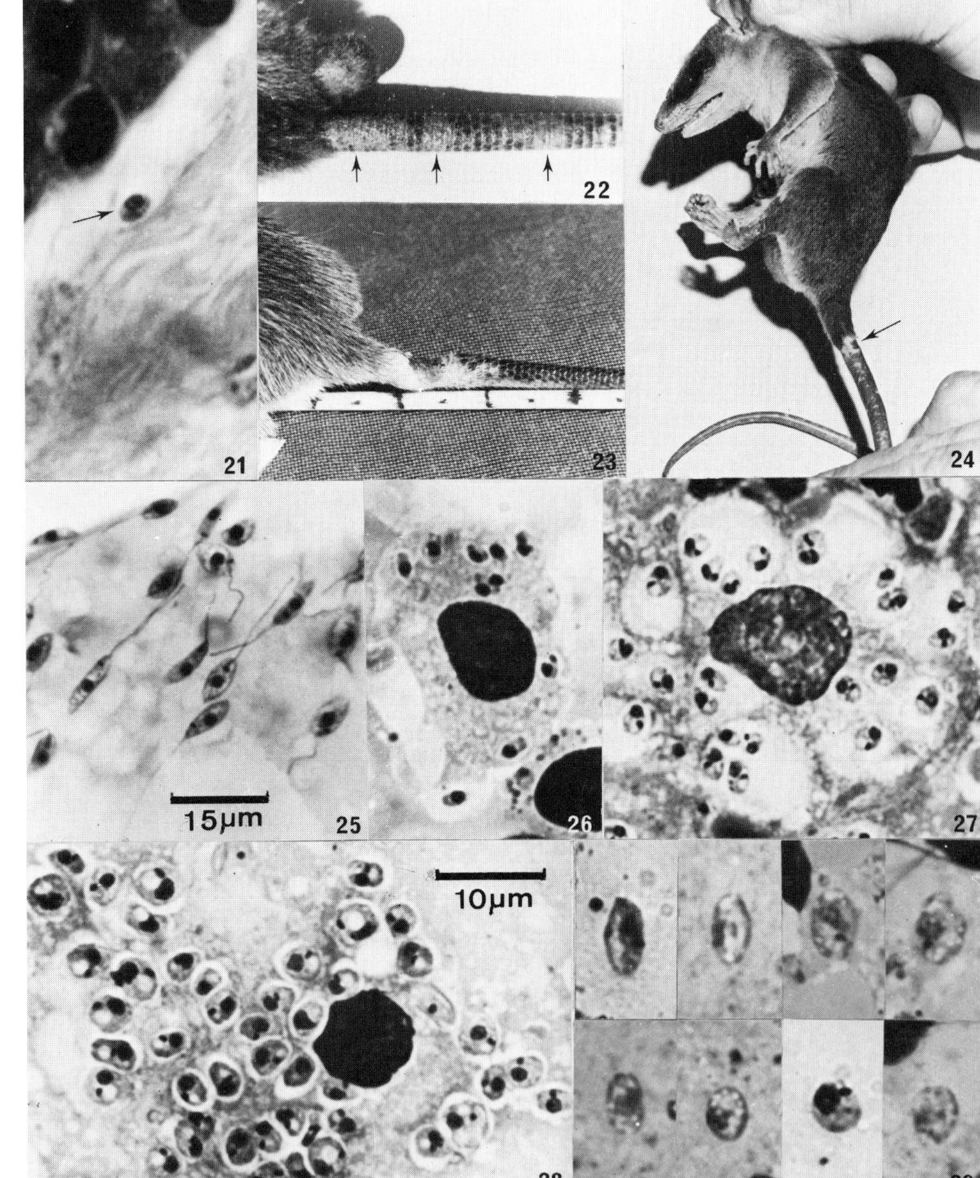

vectors, or an incapacity to establish themselves in other vertebrate hosts, including man. This is probably true for *L. hertigi hertigi* Herrer, 1971 and *L. hertigi deanei* Lainson & Shaw, 1977 (Fig. 29) which are parasites of neotropical porcupines, *Coendou* (Fig. 20). These organisms have biological and biochemical characteristics which are unassociated with any *Leishmania* yet isolated from man.

TECHNIQUES FOR THE INCRIMINATION OF RESERVOIR HOSTS

Base-line Epidemiological Information

Association of proven sandfly vectors with their mammalian hosts

In a given endemic area, the first major step is clearly the isolation and characterization of a large number of stocks of *Leishmania* from man. More than one parasite may be involved, and selected stocks must serve as markers for any future search for sandfly and wild mammalian hosts.

This achieved, information on the biting-habits, resting-sites and other biological features of the local sandflies may provide clues as to the mammalian reservoirs. In this respect, it is particularly important to concentrate on those sandfly species shown to feed frequently on man. In some epidemiological situations the presence of a single *Leishmania* parasite, associated with the intimate contact of strictly limited sandfly and mammalian faunas, gives an immediate line of approach. The best example of this is rural cutaneous leishmaniasis due to *L. major*, discussed above, where the sandfly vectors live in

---

FIG. 21. Histological section of skin of *C. p. prehensilis* showing a single amastigote of *L. h. deanei* in the dermis: note complete absence of host-cell reaction. The parasite appears much smaller than those seen in smears (Fig. 29), owing to shrinkage in preparation of the tissue. Section at 5.0 μm: Giemsa stain.

FIG. 22. Lesions on the tail of the rodent *O. phyllotis*, due to *L. m. mexicana* (from Lainson & Strangways-Dixon, 1964).

FIG. 23. More deeply ulcerated lesion on the tail of the spiny mouse, *Neacomys spinosus*, due to *L. m. amazonensis*.

FIG. 24. Similar lesion, due to *L. m. amazonensis*, on the tail of the murine opossum *Marmosa murina* (from Lainson & Shaw, 1969).

FIG. 25. Promastigotes of *L. m. mexicana* in NNN culture. Giemsa-stained smear.

FIGS 26—29. Amastigotes of some neotropical leishmanias in impression smears made from skin of infected hamsters: all Giemsa-stained, and scale in Fig. 28 applies to all. Note striking size range, which dispenses with the often repeated myth that all leishmanias are morphologically indistinguishable. Figure 26, *Leishmania* sp., from the armadillo, *D. novemcinctus* (from Lainson, Shaw, Ward, Ready & Naiff, 1979); Fig. 27, *L. b. guyanensis*; Fig. 28, *L. m. amazonensis*; Fig. 29, *L. hertigi deanei*.

the burrows of the reservoir hosts. At the opposite extreme, there
may be numerous leishmanias associated with a very wide range of
sandfly species and a very diverse mammalian fauna. Thus, at least
four different leishmanias have been shown to be circulating in study
areas of Amazonian forest which are no bigger than 1 km^2 (Lainson,
Shaw, Ready *et al.*, 1981), and where there are approximately 94
known species of sandflies and about 70 recognized species of wild
mammals (not including bats)!

In Panama, and in the Amazonian forests of the Guyanas and
north Brazil, it was the arboreal nature of the major vectors,
Lutzomyia trapidoi and *Lu. umbratilis*, respectively, that led to the
incrimination of arboreal endentates as the primary reservoirs of *L.
b. panamensis* and *L. b. guyanensis* (Floch, 1957; Johnson,
McConnell & Hertig, 1963; Wijers & Linger, 1966; Lainson, Ward &
Shaw, 1976; Ready *et al.*, in preparation) (Fig. 31). On the other
hand, the sandflies *Lutzomyia olmeca olmeca* and *Lu. flaviscutellata*,

FIG. 30. Ecology of American Leishmaniasis. *Amazonian visceral leishmaniasis.* Sporadic
canine, vulpine and human cases occur in forested areas of north Brazil. Some attribute
them to *L. donovani chagasi*, transmitted by *Lutzomyia longipalpis*, as in the semi-arid
endemic areas of Latin America. Marked ecological differences, however, suggest a different
epidemiology. Hypothetical silvatic cycles are shown on the left: sandflies (V1) feeding on
arboreal reservoirs (1a?) descend to oviposit (O), and ascend again for next blood-meal.
Alternately, there may be a cycle among terrestrial reservoirs (1t?). Man is a "victim" host
(3) when penetrating forest. Early studies, on right, suggested a peridomestic cycle
originating from marauding silvatic animals (1t?) and transmitted among dogs and man (3)
by a secondary, domesticated sandfly claimed to be *Lu. longipalpis* (V2?). Recent investigations of a case tend to support acquisition of infection in the forest.

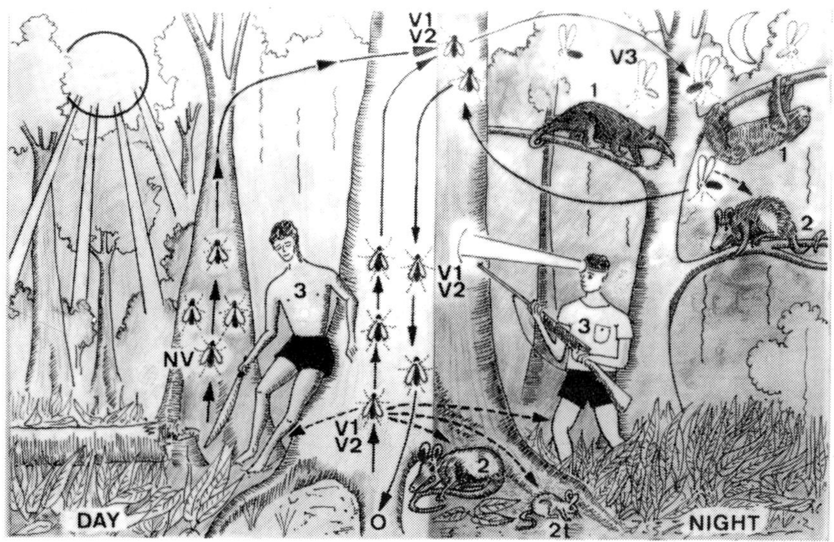

FIG. 31. Ecology of American Leishmaniasis. *Cutaneous leishmaniasis due to L. braziliensis guyanensis in primary forest of north Pará, Brazil*: nulliparous sandflies (NV) ascend tree-trunks to canopy, where *Lutzomyia umbratilis* (V1) and *Lu. whitmani* (V2) feed, at night, on the primary reservoirs (1), the sloth *Choloepus didactylus* and the anteater *Tamandua tetradactyla*. Opossums (2) and a third sandfly, *Lu. anduzei* (V3) may sometimes be involved. Engorged flies migrate down trees to oviposit (O). During the day they move back to the canopy for the next blood-meal, when infected flies transmit. If disturbed at ground-level during daytime or early night-time, some flies will feed on the nearest mammal, and terrestrial species (2t) are occasionally found infected. Man is a common "victim" host (3) due to his excessive disturbance of the environment. Broken lines = transmission to subsidiary, or "victim" hosts.

vectors of *L. m. mexicana* and *L. m. amazonensis*, respectively, are rarely found more than a few feet from the forest floor: reservoir hosts of these parasites are consequently limited to terrestrial or semi-terrestrial mammals, particularly rodents (Fig. 33).

Analysis of blood-meals from wild-caught sandflies would at first sight appear to be the most direct indication of the mammalian reservoir hosts. Unfortunately, many sandfly species are opportunistic feeders: thus, workers in Panama (Tesh, Chaniotis, Aronson & Johnson, 1971; Tesh, Chaniotis & Johnson, 1972) showed the feeding habits of different species to change — not only from site to site, but from one time of the year to another. Lainson & Shaw (1979) collected blood-engorged *Lu. umbratilis* from trees in areas highly endemic for cutaneous leishmaniasis, and it was shown that 51.3% had fed on primates (monkeys), 28.0% on rodents and 18.0% on anteaters: on another occasion, in the same area, 35 out of 36 flies had

FIG. 32. Ecology of American Leishmaniasis. *Cutaneous and mucocutaneous leishmaniasis due to* **L. braziliensis braziliensis** *s.l., in Serra dos Carajás, Pará, Brazil.* The sandfly *Psychodopygus wellcomei* is the major vector to man, feeding on him by day and night. Reservoir hosts are unknown and possibilities are as follows. Nulliparous sandflies (NV) may ascend tree-trunks to feed on arboreal reservoirs (1a?), descend to lay eggs (O) and re-enter canopy (V1) for next blood-meal. Man may be infected in daytime (3d) when disturbing flies on tree-trunks, or at night (3n) by free-ranging flies. Alternatively, there may be terrestrial reservoirs (1t?), and infected sandflies may attack man from ground-level foliage by day (3d) or night (3n).

fed on primates and one on a porcupine (information by courtesy of Dr P. F. L. Boreham). The major reservoir hosts of *L. b. guyanensis* were later shown, however, to be sloths and anteaters. If precipitin tests are to be used, then the engorged flies must clearly be collected from numerous, widely dispersed areas, over an extensive period, and related to the detection of *Leishmania* in dissected flies from the same areas, at the same time.

Screening of Suspected Mammalian Reservoir Hosts

To date there has been developed no really reliable serological screening method for the examination of wild animals, and if it is developed it must in any case be accompanied by isolation of the parasite(s), as discussed below, so that the organisms may be compared with those infecting man.

There have developed two erroneous impressions regarding infection with *Leishmania*: first, that it always results in disease, and

FIG. 33. Ecology of American Leishmaniasis. *Cutaneous leishmaniasis due to L. mexicana amazonensis in Amazonian Brazil.* Principal reservoir host is the echimyid rodent *Proechimys* spp. (1): secondary hosts in other terrestrial or semi-terrestrial mammals (2), mainly rodents, marsupials and foxes. Sandfly vector *Lutzomyia flaviscutellata* (V1) is low-flying, nocturnal and not much attracted to man. Human infection rare and largely restricted to night-time hunters, fishermen or others penetrating low-lying "varzea" forest at night.

secondly that there is a clear-cut division between "cutaneous" and "visceral" parasites. Neither is true, and both originated largely from past observations on the human infection, or those of other "accidental" hosts such as the dog.

Apparent and inapparent infections

There is much evidence to suggest that the leishmanias belong to a very ancient group of parasites. As is usually the case in long-established host—parasite relationships, disease does not normally result in the natural hosts: the infection remains inapparent, with scanty parasites scattered in the dermis or the viscera, where they provoke little or no host-cell response (Fig. 21). In earlier studies, in areas endemic for cutaneous leishmaniasis, workers tended to look only for active skin lesions among the wild animals they were screening for such parasites as *L. tropica, L. major, L. mexicana* and *L. braziliensis.* While the occasional production of discrete skin lesions of rodent or marsupial hosts did lead to the incrimination of certain species as reservoirs (Figs 22—24), a large proportion of inapparent infections clearly were missed, and the true prevalence of infection in the different host species not correctly assessed. Thus, *L. m. amazonensis* was originally described from lesions seen on the tails of the rodent *Oryzomys capito*, which was thought to be the principal reservoir. Subsequent examination of apparently *normal* skin from a wide range of other animals, however, revealed another rodent, *Proechimys guyannensis*, as the major host, with an infection rate of up to 16.8% in some study areas (Lainson & Shaw, 1973; Lainson, Shaw, Ready *et al.*, 1981).

Visceral and cutaneous infections

Although it is still customary to refer to "visceral" and "cutaneous" leishmanias, the viscerotropic and dermatropic characters are by no means always retained in either man or (more particularly) wild animals. Thus, primary cutaneous lesions may appear several months before the onset of visceral leishmaniasis in east African and Sudanese patients, and sometimes infection appears not to proceed past the cutaneous phase: *L. d. donovani* (Indian kala azar) may produce a massive invasion of the skin in cases of "post kala azar dermal leishmanoid"; dogs infected with *L. d. infantum* or *L. d. chagasi* usually show very abundant parasites in the skin, rendering them excellent (if temporary) reservoirs of infection for man; recently, isolates from cases of visceral disease in India and Israel have been identified as *L. t. tropica* by Chance (1979), and *L. major* is commonly found in the viscera of both naturally and experimentally infected animals; finally, in the Americas, isolates of *L. b. guyanensis* from sloths and ant-eaters have almost always been from the viscera and very rarely from the skin (Lainson, Shaw & Póvoa, 1981), although the disease caused by this parasite in man is only known in the form of dermal lesions. Interestingly, parasites of the *L. mexicana* complex do seem to be largely restricted to the skin of their wild animal hosts.

Obtaining material for culture and the inoculation of laboratory animals

Whenever possible, animals should be returned alive to the laboratory, where the examination can be made under sterile conditions with tissues from freshly killed animals. Sometimes this is not possible, and the next best thing is to send the removed, entire organs and skin-snips in sealed, sterile containers kept on ice in a firmly closed polystyrene container or a vacuum flask. Amastigotes of *Leishmania* will survive for at least 48 h under such conditions: the tissues will serve perfectly well for the inoculation of laboratory animals, and even sterile cultures may be obtained if tissue is taken from deep within such organs as the spleen and liver.

The low-grade nature of most natural infections renders the search of tissue smears or sections largely a waste of time, for the parasite can rarely be found. Detection of infection usually depends, therefore, on two methods: incubation of tissue in a suitable *in vitro* medium enabling the parasite to multiply to a detectable level as promastigotes (Fig. 25); or inoculation of triturated tissue into highly susceptible laboratory animals, such as the hamster or mouse, in which the organism will proliferate as amastigotes (Figs 26—29).

Tissues for culture must be taken under conditions of maximum sterility. Carcases of small- to medium-sized animals should be washed thoroughly with soap and water and then well rinsed under a running tap. As much as possible of the subsequent routine should be carried out in some sort of sterile cabinet (the "lamina-flow" type is ideal), the whole work area having been swabbed down with 70.0% spirit and subjected to ultra-violet light for a number of hours previously (overnight). Areas of naked skin such as the nose, ears and feet (the tail may need shaving) are rigorously cleaned with 70.0% alcohol containing iodine, again with pure 70.0% spirit to remove *every trace* of iodine, and finally with sterile physiological saline.

Small snips of skin from the selected areas are pooled in a small quantity of sterile saline in a suitable tissue-grinder: the saline should contain antibiotics (usually approximately 200 IU penicillin and 2 mg streptomycin per ml). The triturated skin may be cultured at once, or kept in the refrigerator overnight for maximum effect of the antibiotics. The rubbery texture of some skins may make trituration in a glass tissue-grinder a difficult task: a sterile pestle and mortar can be used, with the addition of a pinch of fine, sterile sand to facilitate trituration.

Blood for culture is best taken from the heart, with a sterile Pasteur pipette, after the carcase is opened: the ventral surface of the animal is previously cleaned, as described above. Pieces of liver and spleen are treated in the same way as the skin-snips.

Triturated tissues will serve for both culture and the inoculation of laboratory animals. Some workers prefer to culture small, entire pieces of the tissues rather than triturates.

In Vitro *Culture*

There are an extraordinary number of media concocted for the cultivation of *Leishmania* and trypanosomes: they may be diphasic, semi-solid or liquid. Some are very complex mixtures, but the most widely used are relatively simple variations of the time-honoured "NNN" medium of Novy, MacNeal and Nicolle (see Wenyon, 1926).

Basically, this is simply a mixture of a commercially prepared agar-base, with or without certain nutrient substances, and blood (usually from a rabbit). Instructions for use are usually provided and all that is required is to add the agar powder to the appropriate quantity of glass-distilled water. The mixture is heated, in a water-bath, until the agar has dissolved, and the hot fluid dispensed, in about 5.0-ml amounts, into tubes or small bottles of choice: screw-cap bottles of the "McCartney" or "Universal" type are particularly useful. Leaving the caps loose, the bottles are autoclaved at 15 p.s.i. for about 20—30 min: on removal, the caps are tightened and the bottles allowed to cool to about 50—56°C in a water-bath. The desired amount of sterile, defibrinated rabbit blood (usually 0.5 ml) is added to each bottle, which is firmly re-capped. The mixture of blood and agar is made uniform by gently rotating the bottle, which is then sloped until the mixture sets: the slopes should not reach more than half-way up the bottle.

When all the slopes have firmly set, the bottles are left, upright, at room temperature for 24 h. This will enable a check for contamination, and allows the formation of a small quantity of water of condensation at the bottom of each bottle. It is into this fluid that the material to be cultured is introduced, again under strictly aseptic conditions. Care should be taken not to put in too much tissue suspension (one or two drops are sufficient), and if pieces of tissue are used they should be small enough to be completely covered by the condensation fluid.

Seeded culture bottles should be maintained at about 23—26°C and examined for the growth of promastigotes after seven to 10 days. A tiny drop of fluid is removed on a flamed bacteriological loop and examined microscopically under a small cover-slip, using the x 40 objective. It is best to use phase-contrast optics: failing this, the microscope condenser should be racked well down.

If cultures appear negative, they must be re-examined after a further seven days and then weekly for four to six weeks. On successful isolation, the parasite should be transferred to the same medium *without* antibiotics, as there is some evidence that continued exposure to these substances has an adverse effect on the growth of promastigotes.

In spite of much experimentation, a number of leishmanias remain far from easy to culture, and it is very likely that no single medium will ever be devised to cater for the nutritional idiosyncrasies of all these parasites. Individual laboratories have to decide which medium (or media) is best for the parasite(s) of their endemic area, and stick to it. Although most textbooks do not say so, many stocks of *L. donovani* s.l. may require very careful attention if they are to be maintained in culture. The situation is most difficult and complicated, however, among parasites of the *L. braziliensis* complex which exhibit a whole range of cultural characteristics. Thus, *L. b. panamensis* and *L. b. guyanensis* grow well in certain blood—agar mixtures, such as Senekji's medium (Herrer, Thatcher & Johnson, 1966) and NNN with Difco Agar Base (Walton, Shaw & Lainson, 1977), respectively. On the other hand, many isolates of the parasite referred to as *L. b. braziliensis* grow poorly in these media, and some (other sub-species?) have ultimately been lost owing to their failure to adapt to any culture medium so far tried (Walton *et al.*, 1977; Lainson, Shaw, Ready *et al.*, 1981).

Known sub-species of *L. mexicana*, and *L. major*, share the common feature of extreme ease of isolation and maintenance in a wide variety of very simple NNN media. So much so that care must be taken to see that stocks are not lost by *over*-growth of the flagellates, which may occur in as little as 10 days. Asian stocks of *L. tropica* do not offer any serious problems in culture: those from cases of cutaneous leishmaniasis in Europe, however, may prove difficult, probably because they are not, in fact, *L. tropica* (see Table I).

Inoculation of Laboratory Animals
The hamster (*Mesocricetus auratus*) still remains unsurpassed in the isolation and maintenance of most of the leishmanias, while the mouse is less susceptible, notably to infection with parasites of the *L. braziliensis* complex. The guinea pig is useful for work with *L. m. enriettii* but, contrary to most of the literature, this parasite can be held in hamsters in which it produces much more discrete skin lesions.

If it is impossible to return wild animals to the base laboratory, hamsters may be taken to the field (they will survive very rough, tropical conditions) and inoculated on the spot.

Tissue suspensions are inoculated intradermally into the skin of the nose and/or feet: if studies are on reservoirs of parasites of the *L. donovani* complex inoculation is best made by the intraperitoneal route, although the visceral infection will eventually be established if parasites are inoculated into the skin. After inoculation the hamsters should, if possible, be kept in an animal house where the temperature can be maintained at about 23—26°C. High temperatures (unfortunately only too common in most endemic areas) are to be avoided, as the development of cutaneous lesions may be curtailed or even prevented by excessive heat.

The resulting infection *usually* becomes apparent by the development of a nodule or ulcer at the site of inoculation, and the presence of amastigotes can be verified by the examination of Giemsa-stained smears. Once infection is confirmed, *in vitro* cultures may be prepared by the methods given above.

Some leishmanias undergo a rapid proliferation in hamster or mouse skin, resulting in huge, non-vascular, tumour-like growths containing enormous numbers of amastigotes: there is little if any host-cell response. This is the case with the better known sub-species of *L. mexicana* — *L. m. mexicana*, *L. m. amazonensis*, *L. m. pifanoi* and *L. m. aristedesi* — and with *L. major*. The lesions appear quickly, within a few months, and *continue to grow, unchecked*, until the animal's death. Metastases usually occur, and may involve all the extremities of the body: they have the same characteristics as the primary lesion: sub-species of *L. tropica* grow well, but much more discretely, in hamster and mouse skin.

The growth of sub-species of *L. braziliensis* is very different from that of the suprapylarian leishmanias just discussed. Although appearance of a primary nodule due to *L. b. guyanensis* may be within a few weeks (depending a great deal on the number of parasites inoculated), *the growth of the lesion becomes checked*, and it remains relatively small for the hamster's life-span. Metastases may occur to extremities other than the site of inoculation, but they, too, share the same characteristics as the primary lesion — small, vascular and (compared with *L. mexicana* infections) with moderate numbers of amastigotes (Fig. 27) which are associated with a marked host-cell reaction. Parasites become quite numerous in the viscera of hamsters infected with *L. b. guyanensis*, and this is reminiscent of the visceral nature of infections in the natural hosts, the sloth and the anteater. Infections with *L. b. panamensis* produce similar skin lesions in hamsters.

In the case of some sub-species of *L. braziliensis*, loosely referred to as *L. b. braziliensis*, growth in hamster skin may be extremely slow, and a visible lesion may not be detectable for as long as nine to 12 months after the inoculation. The lesion remains small, usually non-ulcerative, and amastigotes are very scanty: there is marked host-cell reaction, and the infection may "die out".

Certain leishmanias which, as far as is known, are limited to specific wild animal hosts, produce asymptomatic infections in hamster skin. Often, the parasites are so scanty that they can only be detected following culture of the skin from the site of inoculation. This is the case with the strange leishmanias of neotropical porcupines, *L. hertigi hertigi* and *L. h. deanei*, and the *Leishmania* of armadillos. Finally, instances have been noted (Lainson & Shaw, 1979) in which *no* infection could be detected in hamsters inoculated with material from human lesions (considered due to *L. b.*

braziliensis s.l.) in which amastigotes had been seen microscopically. The same authors also failed to infect hamsters which had been inoculated with liver and spleen suspensions from an opossum (*Didelphis marsupialis*), although the same suspensions had produced a fleeting growth of promastigotes in NNN culture.

It follows that inoculated hamsters should be kept under observation for at least one year before being sacrificed for a more detailed study. Because of the possible development of occult infections, this examination should include the *in vitro* culture of skin, liver and spleen, as well as a search of stained smears for possible amastigotes: the isolation of viscerotropic parasites (such as members of the *L. donovani* complex) must be remembered, and animals with a generally sickly appearance should be promptly killed and examined with this in mind.

As discussed earlier, the nature of infection with *L. donovani* s.l. in the hamster may be very variable. Some stocks from South America produce a rapid, fulminating infection leading to death of the animal in a few months: the liver and spleen are much enlarged, and the parasite load enormous. In other cases the infection is much more protracted and the hamster slowly becomes more and more wasted in appearance. The liver and spleen remain of normal size, however, and contain only moderate numbers of amastigotes. Apart from their poor growth in blood—agar media, some stocks may also "die out" after a number of sub-passages (liver and spleen suspensions, intraperitoneally) in hamsters.

Differential diagnosis must be made between *Leishmania, Histoplasma* and *Toxoplasma* infections, and the latter two parasites are not uncommonly met with in hamsters inoculated with the tissues of wild animals: they should not, however, offer much difficulty to the experienced parasitologist.

CONCLUDING REMARKS

With the methods presently available, then, isolates of *Leishmania* do tend to be artificially selected out into those organisms which grow well, or reasonably well, in the hamster or *in vitro* culture. From our experience, however (Lainson & Shaw, 1979), briefly touched on here, it is certain that there are many more leishmanias in existence than we previously imagined. Possibly many are host-restricted and of no medical importance: the on-going characterization of isolates from man, however, may produce some surprises (and complications!) in our over-all picture of *Leishmania* and the leishmaniases.

ACKNOWLEDGEMENTS

My thanks are due to the Zoological Society of London for the opportunity to participate in this most interesting Symposium, and for financial assistance to do so: to the Wellcome Trust, London, and the Instituto Evandro Chagas, Fundação SESP of Brazil, under whose joint auspices we have studied leishmaniasis during the past 15 years: to my colleagues, Dr J. J. Shaw, R. D. Ward, P. D. Ready and H. Fraiha, who have participated in many of the epidemiological studies discussed in this paper: and to all the staff of the Section of Parasitology, Instituto Evandro Chagas, for their unstinted technical assistance. Permission to use certain illustrations was kindly given by the editors of the *Transactions of the Royal Society of Tropical Medicine and Hygiene* (Figs 4, 10, 11, 22, 24 and 26) and the *University of Toronto Press* (Figs 3, 8 and 9).

REFERENCES

Anderson, D. C., Buckner, R. G., Glenn, B. L. & Macvean, D. W. (1980). Endemic canine leishmaniasis. *Vet. Path.* 17, 94–96.

Anon. (1957). *The twenty-ninth annual report of the work and operation of the Gorgas Memorial Laboratory, covering the fiscal year ended June 30, 1956.* Washington: United States Government Printing Office.

Anon. (1959). *Thirty-first annual report of the work and operation of the Gorgas Memorial Laboratory, covering the fiscal year ended June 30, 1958.* Washington: United States Government Printing Office.

Anon. (1981). *Report of the Steering Committee of the Scientific Working Group on the Leishmaniases to the STRC.* Geneva: W.H.O.

Arias, J. R., Naiff, R. D., Miles, M. A. & Souza, A. A. (1981). The opossum, *Didelphis marsupialis* (Marsupialis: Didelphidae), as a reservoir-host of *Leishmania braziliensis guyanensis* in the Amazon Basin of Brazil. *Trans. R. Soc. trop. Med. Hyg.* 75: 537–541.

Ashford, R. W., Bray, M. A., Hutchison, M. P. & Bray, R. S. (1973). The epidemiology of cutaneous leishmaniasis in Ethiopia. *Trans. R. Soc. trop. Med. Hyg.* 67: 588–601.

Bettini, S., Gradoni, L. & Pozio, E. (1978). Isolation of *Leishmania* strains from *Rattus rattus* in Italy. *Trans. R. Soc. trop. Med. Hyg.* 72: 441–442.

Carini, A. & Paranhos, U. (1909). Identificação de úlceras de Baurú ao botão do Oriente. *Rev. Med., São Paulo* 12: 111–116.

Chance, M. L. (1979). The identification of *Leishmania*. (In *Problems in the identification of parasites and their vectors.*) *Symp. Br. Soc. Parasit.* 17: 55–74.

Chance, M. L., Peters, W. & Shchory, L. (1974). Biochemical taxonomy of *Leishmania*. 1. Observations on DNA. *Ann. trop. Med. Parasit.* 68: 307–316.

Convit, J. (1968). [cited in Pifano, *et al.* (1968): q.v.].

Cunningham, D. D. (1885). On the presence of peculiar parasitic organisms in the tissue culture of a specimen of Delhi boil. *Scient. Mem. med. Offrs Army India* 1: 21–31.

Deane, L. M. (1956). *Leishmaniose visceral no Brasil.* Rio de Janeiro, Brasil. Serviço Nacional de Educação Sanitário.

Dias, M. (1975). *Contribuição ao estudo da leishmaniose tegumentar no Municipio de Caratinga, M. Gerais, Brasil.* M.Sc. Thesis: Universidade Federal de Minas Gerais.

El-Adhami, B. (1976). Isolation of *Leishmania* from a black rat in the Baghdad area, Iraq. *Am. J. trop. Med. Hyg.* **25**: 759—761.

Floch, H. (1957). Epidémiologie de la Leishmaniose forestière américaine en Guyane française. *Riv. Malariol.* **36**: 233—242.

Forattini, O. P. (1960). Sôbre os reservatórios naturais de leishmaniose tegumentar americana. *Revta Inst. Med. trop. São Paulo* **2**: 195—203.

Forattini, O. P., Pattoli, D. B. G., Rabello, E. X. & Ferreira, O. A. (1972). Infecções naturais de mamíferos silvestres em área endêmica de *Leishmaniose tegumentar* do estado de São Paulo, Brasil. *Revta Saúde Públ.* **6**: 255—261.

Forattini, O. P., Pattoli, D. B. G., Rabello, E. X. & Ferreira, O. A. (1973). Nota sôbre infecção natural de *Oryzomys capito laticeps* em foco enzoótico de leishmaniose tegumentar no estado de São Paulo, Brasil. *Revta Saúde Públ.* **7**: 181—184.

Gardener, P. J., Chance, M. L. & Peters, W. (1974). Biochemical taxonomy of *Leishmania*. II. Electrophoretic variation of malate dehydrogenase. *Ann. trop. Med. Parasit.* **68**: 317—325.

Gentile, B., Le Pont, F., Pajot, F. X. & Besnard, R. (1981). Dermal leishmaniasis in French Guiana: the sloth (*Choloepus didactylus*) as a reservoir host. *Trans. R. Soc. trop. Med. Hyg.* **75**: 612—613.

Grové, S. S. (1970). Cutaneous leishmaniasis in South West Africa. *S. Afr. Med. J.* **44**: 206—207.

Grové, S. S. & Ledger, J. A. (1975). *Leishmania* from a hyrax in South West Africa. *Trans. R. Soc. trop. Med. Hyg.* **69**: 523—524.

Herrer, A. (1951). Estudos sobre leishmaniasis tegumentaria en el Peru. V. Leishmaniasis natural en perros procedentes de localidades utógenas. *Revta Med. exp. Lima* **8**: 119—137.

Herrer, A., Christensen, H. A. & Beumer, R. J. (1973). Reservoir hosts of cutaneous leishmaniasis among Panamanian forest mammals. *Am. J. trop. Med. Hyg.* **22**: 585—591.

Herrer, A., Telford, S. R. & Christensen, H. A. (1971). Enzootic cutaneous leishmaniasis in eastern Panama. I. Investigation of the infection among forest mammals. *Ann. trop. Med. Parasit.* **65**: 349—358.

Herrer, A., Thatcher, V. E. & Johnson, C. M. (1966). Natural infections of *Leishmania* and trypanosomes demonstrated by skin culture. *J. Parasit.* **52**: 954—957.

Hoogstraal, H. & Heyneman, D. (1969). Leishmaniasis in the Sudan Republic. *Am. J. trop. Med. Hyg.* **18**: 1091—1210.

Johnson, P. T., McConnell, E. & Hertig, M. (1963). Natural infections of leptomonad flagellates in Panamanian *Phlebotomus* sandflies. *Expl Parasit.* **14**: 107—122.

Kerdel-Vegas, F. & Essenfeld-Yahr, E. (1966). American leishmaniasis in a field rodent. *Trans. R. Soc. trop. Med. Hyg.* **60**: 563.

Killick-Kendrick, R. (1979). Biology of *Leishmania* in phlebotomine sandflies. In *Biology of the Kinetoplastida* **2**: 395—460. Lumsden, W. H. R. & Evans, D. A. (Eds). London and New York: Academic Press.

Lainson, R., Miles, M. A. & Shaw, J. J. (1981). On identification of viscerotropic leishmanias. *Ann. trop. Med. Parasit.* **75**: 251—253.

Lainson, R. & Shaw, J. J. (1969). Leishmaniasis in Brazil. III. Cutaneous leishmaniasis in an opossum, *Marmosa murina* (Marsupialia, Didelphidae) from the lower Amazon region. *Trans. R. Soc. trop. Med. Hyg.* **63**: 738—740.

Lainson, R. & Shaw, J. J. (1971). Epidemiological considerations of the leishmanias, with particular reference to the New World. In *Ecology and physiology of parasites*: 21—57. Fallis, A. M., (Ed.). Toronto: University of Toronto Press.

Lainson, R. & Shaw, J. J. (1972). Leishmaniasis of the New World: taxonomic problems. *Br. med. Bull.* **28**: 44—48.

Lainson, R. & Shaw, J. J. (1973). Leishmanias and leishmaniasis of the New World, with particular reference to Brazil. *Bull. PanAm. Hlth Org.* **7**(4): 1—19.

Lainson, R. & Shaw, J. J. (1979). The rôle of animals in the epidemiology of South America leishmaniasis. In *Biology of the Kinetoplastida* 2: 1—116. Lumsden, W. H. R. & Evans, D. A. (Eds). London and New York: Academic Press.

Lainson, R., Shaw, J. J. & Lins, Z. C. (1969). Leishmaniasis in Brazil. IV. The fox, *Cerdocyon thous* (L) as a reservoir of *Leishmania donovani* in Pará State, Brazil. *Trans. R. Soc. trop. Med. Hyg.* **63**: 741—745.

Lainson, R., Shaw, J. J. & Póvoa, M. (1981). The importance of edentates (sloths and anteaters) as primary reservoirs of *Leishmania braziliensis guyanensis*, causative agent of "pian-bois" in north Brazil. *Trans. R. Soc. trop. Med. Hyg.* **75**: 611—612.

Lainson, R., Shaw, J. J., Ready, P. D., Miles, M. A. & Póvoa, M. (1981). Leishmaniasis in Brazil. XVI. Isolation and identification of *Leishmania* species from sandflies, wild mammals and man in north Pará State, with particular reference to *Leishmania braziliensis guyanensis*, causative agent of "pian-bois". *Trans. R. Soc. trop. Med. Hyg.* **75**: 530—536.

Lainson, R., Shaw, J. J., Ward, R. D. & Fraiha, H. (1973). Leishmaniasis in Brazil. IX. Considerations on the *Leishmania braziliensis* complex: importance of sandflies of the genus *Psychodopygus* (Mangabeira) in the transmission of *L. braziliensis* in north Brazil. *Trans. R. Soc. trop. Med. Hyg.* **67**: 184—196.

Lainson, R., Shaw, J. J., Ward, R. D., Ready, P. D. & Naiff, R. D. (1979). Leishmaniasis in Brazil. XIII. Isolation of *Leishmania* from armadillos (*Dasypus novemcinctus*), and observations on the epidemiology of cutaneous leishmaniasis in north Pará State. *Trans. R. Soc. trop. Med. Hyg.* **73**: 239—242.

Lainson, R. & Strangways-Dixon, J. (1962). Dermal leishmaniasis in British Honduras: some host reservoirs of *L. braziliensis mexicana*. A preliminary note. *Br. med. J.* **1**: 1596—1598.

Lainson, R. & Strangways-Dixon, J. (1963). *Leishmania mexicana*: the epidemiology of dermal leishmaniasis in British Honduras. *Trans. R. Soc. trop. Med. Hyg.* **54**: 242—265.

Lainson, R. & Strangways-Dixon, J. (1964). The epidemiology of dermal leishmaniasis in British Honduras. Part II. Reservoir-hosts of *Leishmania mexicana* among the forest rodents. *Trans. R. Soc. trop. Med. Hyg.* **58**: 136—153.

Lainson, R., Ward, R. D., & Shaw, J. J. (1976). Cutaneous leishmaniasis in north Brazil: *Lutzomyia anduzei* as a major vector. *Trans. R. Soc. trop. Med. Hyg.* **70**: 171—172.

Latyšev, N. I. & Krjukova, A. P. (1941). On the epidemiology of the cutaneous leishmaniosis. The cutaneous leishmaniosis as a zoonotic disease of wild rodents in Turkmenia. *Trav. Acad. Milit. Med. Armée Rouge U.R.S.S. Moscow* 25: 229–242.

Leishman, W. B. (1903). On the possibility of the occurrence of trypanosomiasis in India. *Br. med. J.* 1: 1252–1254.

Levine, N. D., Corliss, J. O., Cox, F. E. *et al.* (1980). A newly revised classification of the Protozoa. *J. Protozool.* 27: 37–58.

Lindenberg, A. (1909). A úlcera de Baurú e seu micróbio. *Revta Med., Univ. São Paulo* 12: 116–120.

Migone, L. E. (1913). Un caso de kala-azar a Assuncion (Paraguay). *Bull. Soc. Path. exot.* 6: 118–120.

Miles, M. A., Póvoa, M., de Souza, A. A., Lainson, R. & Shaw, J. J. (1980). Some methods for the enzyme characterization of Latin-American *Leishmania*, with particular reference to *Leishmania mexicana amazonensis* and subspecies of *Leishmania hertigi*. *Trans. R. Soc. trop. Med. Hyg.* 74: 243–252.

Muniz, J. & Medina, H. (1948). Leishmaniose tegumentar do cobaio (*Leishmania enriettii* n.sp.). *Hospital (Rio de Janeiro)* 33: 7–25.

Mutinga, M. J. (1975). The animal reservoir of cutaneous leishmaniasis on Mount Elgon, Kenya. *E. Afr. Med. J.* 52: 142–151.

Nicolle, C. & Comte, C. (1908). Origine canine du kala-azar. *C.r. hebd. Séanc. Acad. Sci.* 146: 789–791.

Petrović, Z., Borjoški, A. & Savin, Z. (1975). Les résultats de recherches sur le réservoir de *Leishmania donovani* dans une région endémique du Kala-azar. *Proc. Europ. Multicoll. Parasit. Trogir.* 2: 97–98.

Pifano, C. F., Kerdel-Vegas, F. & Romero, M. J. (1968). Nidos de zoonosis en la leishmaniasis tegumentaria americana. *Derm., Venez.* 11: 520–529.

Pons, A. R. (1968). Leishmaniasis tegumentaria americana en el Asentamiento Campesino de Zipayare. Aspectos epidemiológicos, clinicos e inmunológicos. Su importancia en la reforma agraria. *Kasmera* 3: 5–59.

Ready, P. D., Lainson, R., Shaw, J. J. & Ward, R. D. (In prep.). *The ecology of Lutzomyia umbratilis Ward & Fraiha, 1977 (Diptera: Psychodidae) in relation to its rôle as principal vector of Leishmania braziliensis guyanensis, causative agent of "pian-bois" in north Pará State, Brazil.*

Schnur, L. F., Chance, M. L., Ebert, F., Thomas, S. C. & Peters, W. (1981). The biochemical and serological taxonomy of visceralizing *Leishmania. Ann. trop. Med. Parasit.* 75: 131–144.

Scorza, J. V., Valera, M., Scorza de, C., Carnevali, M., Moreno, E. & Lugo-Hernandez, A. (1979). A new species of *Leishmania* parasite from the Venezuelan Andes region. *Trans. R. Soc. trop. Med. Hyg.* 73: 293–298.

Shaw, P. K., Quigg, L. T., Allain, D. D., Juranek, D. D. & Healy, G. R. (1976). Autochthonous dermal leishmaniasis in Texas. *Am. J. trop. Med. Hyg.* 25: 788–796.

Simpson, M. H., Mullins, J. F. & Stone, O. J. (1968). Disseminated anergic cutaneous leishmaniasis. *Arch. Dermatol.* 97: 301–303.

Splendore, A. (1912). Leishmaniosi con localizzazione nella cavita mucose (nuova forma clinica). *Bull. Soc. Path. exot.* 5: 411–438;

Strong, R. P., Tyzzer, E. E., Brues, C. T., Sellards, A. W. & Gastiaburu, J. C. (1913). Verruga Peruviana, Oroya Fever and Uta. *J. Am. Med. Assoc.* 8: 1713–1716.

Strong, R. P., Tyzzer, E. E., Brues, C. T., Sellards, A. W. & Gastiaburu, J. C. (1915). *Harvard School of Tropical Medicine. Report of the first expedition to South America,* 1913, Cambridge, Mass.

Tesh, R. B., Chaniotis, B. N., Aronson, M. D. & Johnson, K. M. (1971). Natural host preferences of Panamanian phlebotomine sandflies as determined by precipitin test. *Am. J. trop. Med. Hyg.* **20**: 150–156.

Tesh, R. B., Chaniotis, B. N. & Johnson, K. M. (1972). Further studies on the natural host preferences of Panamanian phlebotomine sandflies. *Am. J. Epidemiol.* **95**: 88–93.

Tikasingh, E. S. (1974). Enzootic rodent leishmaniasis in Trinidad, West Indies. *Bull. PanAm. Hlth Org.* **8**: 232–242.

Torrealba, J. W., Gómez-Núñez, J. C. & Ulloa, G. (1972). Isolation of *Leishmania braziliensis* by intra-peritoneal inoculation of blood from a reservoir host into hamsters. *Trans. R. Soc. trop. Med. Hyg.* **66**: 361.

Townsend, C. H. T. (1915). The insect vector of uta, a Peruvian disease. *J. Parasit.* **2**: 67–73.

Walton, B. C., Shaw, J. J. & Lainson, R. (1977). Observations on the *in vitro* cultivation of *Leishmania braziliensis*. *J. Parasit.* **63**: 1118–1119.

Wang, J., Qu, J. & Guan, L. (1964). A study on the *Leishmania* parasite of the big gerbil in northwest China. *Acta parasit. sin.* **1**: 105–117 (with English summary).

Wenyon, C. M. (1926). *Protozoology* 2. London: Baillière, Tindall & Cox. (Reprinted 1966 by Baillière, Tindall & Cassell.)

Wijers, D. J. B. & Linger, R. (1966). Man-biting sandflies in Surinam (Dutch Guiana): *Phlebotomus anduzei* as a possible vector of *Leishmania braziliensis*. *Ann. trop. Med. Parasit.* **60**: 501–508.

Yakimoff, W. L. & Schockor, N. I. (1914). Recherches sur les maladies tropicales humaines et animales au Turkestan. II. La leishmaniose cutanée spontanée du chien au Turkestan. *Bull. Soc. Path. exot.* **7**: 186–187.

DISCUSSION

Clarke (Chairman) — In animals that are infected by the parasite, is there any evidence that the sex ratio of the offspring is upset? I ask this because in man the virus which causes hepatitis B results in women who are carriers having more boys than girls. There is also an example in butterflies and in *Drosophila* where there is an upset in the sex ratio due to a parasite. I therefore wondered if any work had been done on upsets in the sex ratio in the animals that harbour leishmaniasis.

Lainson — I cannot answer this question very adequately, as I am unaware of any research done on this subject in reference to wild animals infected with *Leishmania*, and we seldom have the opportunity to study their reproduction either in their natural habitat or in the laboratory. All I can say is that we have not observed any effect on the sex ratio of offspring in experimentally infected laboratory animals, such as the hamster.

Zuckerman — You showed a slide of two mangy sick dogs suffering from leishmaniasis, so I imagine that those creatures suffer physically to the same extent as does man. Is there any indication that the various hosts that harbour the parasite also suffer?

Lainson — With regard to the parasites associated with cutaneous leishmaniasis in the New World, we can say the following. Among the very large number of wild animal hosts we have examined, the most pronounced evidence of pathogenicity found has been the production of ulcers of the skin, particularly on the tail, as I have shown you. Visible lesions are relatively rare, however, and, in general, infection is inapparent, benign and chronic. Rare records exist of fulminating, generalized infections (Disney, 1964),[1] but these are exceptional and may have been exacerbated by concomitant bacterial infections. Examples of the good host—parasite relationship between *Leishmania* species and their *natural* wild mammalian hosts are to be found in the inapparent infections of *L. mexicana amazonensis* in the skin of rodents and marsupials (Lainson & Shaw, 1973),[2] *L. braziliensis panamensis* in the skin and viscera of sloths (Herrer & Christensen, 1975),[3] *L. hertigi hertigi* and *L. hertigi deanei* in the skin and viscera of porcupines (Herrer, 1971;[4] Herrer & Christensen, 1975;[3] Lainson & Shaw, 1977)[5] and *L. braziliensis guyanensis* predominantly in the viscera of sloths and anteaters (Lainson, Shaw & Póvoa, 1981).[2]

The frequency with which we find infections with sub-species of *L. braziliensis* in the viscera of the natural host is remarkable in view of the fact that they are only known to produce skin infections in man. Thus, among 27 isolates of *L. braziliensis guyanensis* made from sloths, 25 (92.5%) were from the viscera *only*, one from the viscera and skin, and one from the skin only. This is one indication of how erroneous is our old concept of dermatropic and viscerotropic leishmanias, rigidly associated with "cutaneous leishmaniasis" and "visceral leishmaniasis". One might suggest, of course, that we should look at the spleen and liver of patients with skin lesions due to sub-species of *L. braziliensis*: persons suffering from a few skin lesions are unlikely to be keen on a spleen puncture, though it would certainly be interesting to know if the parasite also lurks in their internal organs. If it does, it certainly does not produce the classical signs and symptoms of human visceral leishmaniasis due to parasites of the *L. donovani* complex.

Lord Zuckerman commented on the "mangy, sick dogs" infected with *L. donovani chagasi*, the parasite responsible for human American visceral leishmaniasis, and I have shown you a picture of an equally sick fox (*Lycalopex vetulus*) infected with the same organism, in north-east Brazil. This clearly indicates a *bad* host—parasite relationship, and we feel that neither animal is the natural host of the parasite — they are victims, like man. On the other hand, I also showed you an extremely healthy-looking specimen of another genus of fox (*Cerdocyon thous*), from which we have isolated the same parasite (or a very closely related one) on three occasions, in Pará, north Brazil (Lainson, Shaw & Lins, 1969;[2] Lainson & Shaw, 1979;[2] unpublished observations). All these foxes

---

[1] Disney, R. H. L. (1964). Visceral involvement with dermal leishmaniasis in a wild-caught rodent in British Honduras. *Trans. R. Soc. trop. Med. Hyg.* **58**: 581.
[2] See list of references, pp. 170—174.
[3] Herrer, A. & Christensen, H. A. (1975). Infrequency of gross skin lesions among Panamanian forest mammals with cutaneous leishmaniasis. *Parasitology* **71**: 87—92.
[4] Herrer, A. (1971). *Leishmania hertigi* sp. n., from the tropical porcupine, *Coendou rothschildi* Thomas. *J. Parasit.* **57**: 626—629.
[5] Lainson, R. & Shaw, J. J. (1977). Leishmanias of neotropical porcupines: *Leishmania hertigi deanei* nov. subsp. *Acta Amazônica* **7**(1): 51—57.

appeared to be in excellent health, possibly indicating this canid to be the primitive reservoir.

Bray — First of all, I'd like to say that Ralph Lainson was always immortalized for me by a remark in the *Daily Telegraph* that said he was going to South America to examine the diseases of the mouth of the Amazon. I always saw him saying "Open wide" to the Amazon. And also to my knowledge he's had leishmaniasis himself three times and once, I'm proud to say, I gave it to him. I do hope it hasn't upset his sex ratio in any way. Now, could I ask him to comment on the possibility that visceral leishmaniasis was introduced to South America by the Conquistadores?

Lainson — Well, I thought that this question would eventually turn up. Of course, other forms of New World leishmaniasis were thought to have been introduced into the Americas in this way: different forms of leishmaniasis, due to a variety of leishmanias of the *L. mexicana* and *L. braziliensis* complexes, were for some time thought to be "oriental sore" (*L. tropica tropica*) introduced from Europe in relatively recent times. In all cases, however, studies on the development of the parasites in sandflies, and their enzyme profiles, have shown the organisms to be very different. Saul Adler, perhaps the greatest authority on the leishmanias there has ever been, was very firmly of the opinion that American visceral leishmaniasis was also indigenous to the New World. He felt (Adler, 1964;[1] Adler & Theodor, 1957)[2] that introduced *L. donovani infantum*, from the Mediterranean region, was unlikely to have made an abrupt adaptation from its natural vector, *Phlebotomus* spp., of the *major* complex, to a completely new genus, *Lutzomyia longipalpis*. He pointed out that the parasite has not adapted to other sandflies outside the *major* group in the Old World, in spite of ample opportunity to do so. More recently it has been shown that *L. d. infantum*, from France, will develop well in the mid-gut of *Lu. longipalpis*, although the all-important fore-gut development was not seen and no transmission by bite achieved (Killick-Kendrick *et al.*, 1980).[3] The most convincing evidence for the introduction of *L. d. infantum* into the Americas, however, was the recent establishment of a small focus of canine visceral leishmaniasis in Oklahoma, USA, seemingly after its importation in an infected dog from Europe (Anderson *et al.*, 1980).[4] How transmission to other dogs may have been effected remains a mystery in the apparent absence of a local mammal-biting phlebotomine sandfly. Some of us (Lainson & Shaw, 1979;[4] Killick-Kendrick et al., 1980;[3] Lainson, Miles & Shaw, 1981)[5] favour the mid-line view that American visceral leishmaniasis may in part be due to imported *L. d. infantum*, but it is very likely that other, local viscero-tropic parasites may be involved. Certainly, if *all* American visceral leishmaniasis is due to *L. donovani infantum*, it would indicate a most extraordinarily rapid

---

[1] Adler, S. (1964). *Leishmania.* In *Advances in parasitology* 2: 35—96. Dawes, B. (Ed.). New York and London: Academic Press.

[2] Adler, S. & Theodor, O. (1957). Transmission of disease agents by phlebotomine sandflies. *Ann. Rev. Ent.* 2: 203—226.

[3] Killick-Kendrick, R., Molyneux, D. H., Rioux, J. A., Lanotte, G. & Leaney, A. J. (1980). Possible origins of *Leishmania chagasi. Ann. trop. Med. Parasit.* 74: 563—565.

[4] See list of references, pp. 000—000.

[5] Lainson, R., Miles, M. A. & Shaw, J. J. (1981). On identification of viscerotropic leishmaniasis. *Ann. trop. Med. Parasit.* 75: 251—253.

spread, through virtually all of Central and South America, in less than 300 years and in the face of very poor means of communication. It is true that biochemical comparison of a very limited number of stocks have so far failed to differentiate between what we call *L. donovani chagasi* of Latin America and *L. d. infantum* of the Mediterranean region, but this does not necessarily mean that they are the same organism. It is very easy to say that two parasites are very different when variations show up in the study of a few enzyme profiles: it is very difficult, however, to say how many enzymes need to be investigated before we can suggest that two organisms are the same. Thus, in spite of strong biological, clinical and epidemiological evidence for the separation of *L. b. braziliensis* and *L. b. guyanensis*, early attempts to separate them on four enzyme profiles failed (Chance, 1979);[1] they could be separated, however, when the number of enzymes was substantially enlarged (Miles *et al.*, 1981).[2] Again, although *L. b. guyanensis* was differentiated from *L. b. panamensis* on nuclear and kineto-plastic buoyant densities and enzyme mobilities of 6-phosphogluconate dehydro-genase (6PGDH) (Chance, Gardener & Peters, 1977;[3] Chance, pers. comm.), Miles *et al.* (1981)[2] were unable to distinguish the two parasites on ten other enzyme profiles.

Steck — What is actually the pathogenic basis for developing skin lesions, and what is the difference in those mammalian reservoir hosts which do not develop them?

Lainson — Again, this is a difficult question to answer fully. Regarding the first part of your question, the pathogenic basis for the development of the leish-maniotic skin lesion is, of course, that commonly observed following the inocu-lation of any foreign body. There is an initial, massive invasion of the area by monocytes and histiocytes, which ingest the organisms: susceptible micro-organisms, such as most bacteria, would disintegrate soon after ingestion, but in the case of *Leishmania* it is within these very cells of the host's defence mechanism that the organism lives and multiplies. Latterly, the lesion becomes walled off by the invasion of lymphocytes, plasma-cells and macrophages, result-ing in a diminution of the parasites. This is only partially successful, however, because the lesion does persist, although usually with a greatly reduced number of amastigotes, often for a number of years. The overlying dermis ulcerates and the epidermis shows various pathological changes which need not concern us here. After a very variable time, the process of healing starts, with the usual fibrosis and scarring.

The second part of your question is much more difficult to answer, and clearly involves a whole complex of immunological factors which are impossible to discuss adequately in the time we have available. Unfortunately, our obser-vations on the host—parasite relationship with regard to *Leishmania* are almost all on human infections or syringe-induced infections in laboratory animals — all unnatural situations for most leishmanial parasites. To my knowledge, no such

---

[1] See list of references, pp. 170–174.
[2] Miles, M. A., Lainson, R., Shaw, J. J., Póvoa, M. & da Souza, A. A. (1981). Leishmaniasis in Brazil XV. Biochemical distinction of *Leishmania mexicana amazonensis*, *Leishmania braziliensis braziliensis* and *Leishmania braziliensis guyanensis* — aetiological agents of cutaneous leishmaniasis in the Amazon basin of Brazil. *Trans. R. Soc. trop. Med. Hyg.* **75**: 524–529.
[3] Chance, M. L., Gardener, P. J. & Peters, W. (1977). Biochemical taxonomy of *Leishmania* as an ecological tool. *Colloques int. Cent. natn Rech. Scient.* No. 239: 53–62.

studies have been made using the ideal model, namely sandfly-transmitted infection in the natural mammalian host. We remain largely ignorant, therefore, as to the host—parasite interaction which determines whether infection will run a subclinical (inapparent) course or the development of a self-healing skin ulcer in the wild animal. In man, at least, it is generally considered that duration of the skin lesion reflects the time taken by the infected person to mount an adequate immunological response, leading to eventual control or elimination of the parasite. Cell-mediated immune mechanism appears to be of particular importance.

In wild animals (e.g. *L. mexicana amazonensis* in rodents) we may postulate that young, non-immune animals may, in fact, characteristically develop skin lesions, like man, but that most of these are of a sufficiently mild nature to leave no obvious scarring. Parasites eventually remain scattered in apparently normal skin on the exposed parts of the body, where the sandfly vector will feed. A minority of the rodents, we may continue to argue, occasionally develop much more conspicuous ulcers, for reasons as yet poorly understood. The other possibility is that infection in the natural host is predominantly without visible skin lesions, but that large ulcerative lesions may latterly be produced following mechanical damage of the skin and the production of localized conditions which are very suitable for multiplication of the amastigotes in the invading macrophages. In this respect, it is of particular interest to note that this situation has occasionally been recorded in man. Thus, we have examined one patient who showed amastigotes along the edges of a knife wound; on other occasions patients have insisted that their leishmanial lesion arose at the point where a tick had been removed. Finally, cases are on record in which a blow on the face has produced an abrasion, leading to a leishmanial lesion, in individuals who left an area endemic for *Leishmania* many years previously.

This is the best explanation I can offer at present: there is clearly a great need for studies on the course of the sandfly-transmitted infection in the natural mammalian host.

Thompson — You don't seem to have touched very much on prophylactics and treatment, or any attempts to deal with the disease in the host in the wild. I would have thought, with the building of the Trans-Amazon Highway, there may have been some pressure upon you or upon the authorities to protect the Highway workers who were exposed to leishmaniasis — to try to eliminate the disease in the wild animal reservoir, and to treat the human sufferers.

Lainson — Unfortunately there are no effective prophylactic measures against American silvatic leishmaniasis, other than avoiding forested areas (particularly at night) and the use of insect repellents. This is difficult for the lowly forest worker, who is lucky to secure a regular job of any sort: he cannot afford to be constantly buying insect repellents and in any case profuse sweating in the hot, humid forest reduces their effectiveness to the minimum.

If treated promptly, most forms of South American leishmaniasis respond well to pentavalent antimonials, such as "Pentostam" (sodium antimony (V) gluconate) or "Glucantime" (meglumine antimoniate), and all patients examined in our laboratory are certainly treated with one or other of these drugs. There is a great need, however, for the development of new, more effective, cheaper and less toxic drugs in the treatment of the disease.

Prevention of infection ("control") is obviously difficult in tropical rainforest which is teeming with wild mammalian hosts of the various leishmanias which infect man. Staunch protectors of wildlife will be relieved to hear that I can foresee no widespread slaughter of wild animals as a means of control: over the vast areas involved, it would be an impossible task from both the economical and practical points of view. Most of the reservoir hosts are very inaccessible to the average person, and require specialized trapping methods: sloths, for example, spend most of their time high in the forest canopy, where they are extremely difficult to locate.

Control of the sandfly vectors is more feasible, but still very difficult owing to the vast areas of forest and the enormous expense involved. Small-scale control is possible (Floch, 1957)[1] when the resting-sites of the vectors are known. Thus, *Lu. umbratilis*, the vector of "pian-bois" (*L. b. guyanensis*) in the Guyanas and north Brazil, rests on the trunks of the larger trees. Small encampments of forest-workers may to some extent be protected, therefore, by placing the camp in a cleared area and spraying the bases of nearby large trees with insecticides. Visceral leishmaniasis in Latin America is most commonly found in the drier, open areas; the disease has a peri-domestic epidemiology following the involvement of the dog as a secondary reservoir, and the adaptation of the vector, *Lu. longipalpis*, to man's dwelling places. In this case, control has been very effectively achieved by the destruction of infected dogs, treatment of diagnosed patients and repeated insecticide application in houses and farm buildings.

The ideal form of control for all forms of leishmaniasis is, of course, the production of an effective vaccine prepared from the dead organism. As most of you will be aware, however, there has been no such vaccine produced for any protozoal parasite as yet. In the case of leishmaniasis there is the added complication that there is little or no cross-immunity between the different species and sub-species of *Leishmania* (Lainson & Shaw, 1977);[2] if a vaccine is to be produced, then it would seem that it must be a polyvalent one.

---

[1] See list of references, pp. 170—174.
[2] Lainson, R. & Shaw, J. J. (1977). Leishmaniasis in Brazil XII. Observations on cross-immunity in monkeys and man infected with *Leishmania mexicana mexicana*, *L. m. amazonensis*, *L. braziliensis braziliensis*, *L. b. guyanensis*, and *L. b. panamensis*. *J. trop. Med. Hyg.* **80**: 29—35.

Symp. zool. Soc. Lond. (1982) No. 50, 181–198

Carrion-feeding Cannibalistic Carnivores and Human Disease in Africa with Special Reference to Trichinosis and Hydatid Disease in Kenya

G. S. NELSON

Department of Parasitology, Liverpool School of Tropical Medicine, Pembroke Place, Liverpool, England

SYNOPSIS

Studies in Kenya and other parts of Africa have shown that *Trichinella* is maintained in nature among cannibalistic carrion-feeding carnivores including hyaenas, jackals, leopards and lions; there is no secondary cycle in rats and domestic pigs. It is postulated that *Trichinella* may be responsible for deaths in lions and other carnivores and that the parasite may have some influence on carnivore population densities. Experimental studies on *Trichinella* from different geographical regions have shown that the Kenya parasite is of unusually low infectivity to rats and domestic pigs and that cross protection occurs between the Kenya parasite and strains from Europe and America. This could produce a natural immunological barrier preventing the introduction into Africa of strains infective to domestic pigs. Human infection in Africa results from eating the undercooked meat of the bush pig and warthog and there are strong taboos against eating wild carnivores.

Hydatid disease is a serious public health problem in Turkana but it is uncommon in man in Masailand. The main reason for the high prevalence in Turkana is thought to be the extremely close intimacy between the people and their dogs which are infected from cysts obtained from camels and other domestic animals. Silver-back and golden jackals are infected by scavenging on livestock and possibly human corpses but there is no evidence of a maintenance cycle involving wild herbivores in Turkana and spotted hyaenas are insusceptible to the local parasite.

In Masailand *Echinococcus* occurs in lions and many of their prey animals. Jackals and cape hunting dogs are also infected, so also are domestic dogs and there is a high rate of hydatid cysts in livestock. The relatively low infection rate in man compared with Turkana may be due to differences in the parasite or in human behaviour in relation to dogs, or the result of cross protection from the almost universal infection with *Taenia saginata*.

INTRODUCTION TO THE ZOONOSES

The deviation of man from the state in which he was placed by nature seems to have proven to him a prolific source of diseases. From the love of splendour, from the indulgence of luxury and from his fondness for amusement he has familiarised himself with a great number of animals. The wolf disarmed of his ferocity is now pillowed in the lady's lap. The cat, the little tiger of our island, whose natural home is the forest is equally domesticated and caressed. (Jenner, 1798.)

Edward Jenner is best known for his work on the development of the smallpox vaccine in 1796 but he was also a competent zoologist who was elected to Fellowship of the Royal Society in 1789 for his "Observations on the Natural History of the Cuckoo". As can be seen from the above quotation, he was one of the earliest physicians to appreciate the significance of the transmission of diseases from animals to man, but of more importance (because this was the origin of the first scientific vaccine) he recognized that the relatively benign cowpox could protect against the more virulent smallpox.

We now recognize the interchange of infections between man and animals as "Zoonoses" defined by a Joint WHO/FAO Expert Committee (1967) as "Those diseases and infections naturally transmitted between vertebrate animals and man". It is not generally realized that more than 80% of the infectious diseases that man is heir to are zoonoses with transmission either direct with wild animals to man or, as Jenner observed, from domestic animals or pets. The definition includes the word "infection" as well as disease, to emphasize that many of the zoonoses may be inapparent infections; they may be so benign as to go completely unrecognized in their maintenance hosts, e.g. trypanosomiasis in antelopes or yellow fever in African monkeys, but when transmitted to man they may produce a fatal disease; or some may be inapparent infections in man and result in protection against more severe disease. Because of their obsession with clinically obvious diseases, epidemiologists, immunologists and population ecologists have almost completely failed to recognize this second aspect of the zoonoses and yet this type of protection may be of fundamental significance in relation to the survival not only of man but of all mammalian species. There have been numerous laboratory studies demonstrating this phenomenon but very few epidemiological studies; to emphasize its public health importance and draw attention to this Jennerian concept the term "Zooprophylaxis" is used, defined as the "protection or amelioration of disease in man as a result of previous exposure to heterologous infections of animal origin" (Nelson, 1974).

In the following account of zoonoses which involve carnivores as maintenance hosts of the infective agent special attention is given to studies on two major problems in Kenya, namely hydatid disease and trichinosis. These have been selected for discussion because they illustrate several features which are relevant to many other zoonoses:

(a) They have a primary maintenance cycle involving many species of wild animals but man is rarely infected from a wild animal host and most human infections in the more developed areas of the world result from a secondary cycle involving domestic animals.

(b) The infective organisms exhibit considerable variation in their infectivity to different hosts in different environments and this is of fundamental importance in determining the public health significance of the infection in any particular area. There is also evidence suggesting that cross protection may affect the distribution of different strains of the parasites.

(c) Variation in the prevalence of the disease in man in different regions depends on local customs.

These are the most important zoonoses affecting man where the basic life-cycle involves meat-eating carnivores and a special study has been made of the maintenance of these infections in nature and their transmission to man in east Africa.

TRICHINOSIS

The Life-cycle

Trichinosis caused by *Trichinella spiralis* (Owen, 1835) is one of the most widespread of all parasitic infections. Although it is usually associated with pigs and rats the true maintenance hosts under natural conditions are carrion-feeding or cannibalistic carnivores, e.g. polar bears in the Arctic, foxes in Europe and hyaenas in Africa. The life-cycle is simple with infections resulting from eating the dormant larval stages which are encysted in muscles of susceptible hosts. The larvae escape from the cyst as a result of peptic digestion in the stomach and small intestine and then bury into the mucosa where within three days they become adult worms. The adults have an ephemeral existence, of about one month, and after fertilization the females produce about 1000 larvae which migrate through the vascular system and become encysted in the muscles. Here they remain viable and infective for the rest of the life of the animal so that a single infective meal by a susceptible host is all that is required to ensure continuing transmission. In any particular area the parasite is usually maintained in several different species of carnivores with secondary infection in omnivorous animals such as the wild boar (*Sus scrofa*) in Europe and the bush pig (*Potamochoerus porcus*) in Africa.

In the more developed regions of the world there is a further cycle in domestic pigs because man has made the pig a cannibalistic carnivore by feeding it with garbage containing pork products. As a consequence most of the human infections in the world result from eating undercooked pork. Under these conditions there are also secondary infections in scavenging rats, cats and dogs. The symptomalogy of the disease in man is so diverse and mimics so many other diseases that it is rarely diagnosed except during epidemics or when patients have the classical syndrome of swollen eyes, painful muscles and fever, associated with a high level of eosinophilia. The disease and the biology of the parasite have been extensively reviewed by Gould (1970).

Trichinosis in Africa with Special Reference to Kenya

Animal hosts

Although the parasite was discovered by Sir James Paget in 1835, when he was a medical student in London, and although it was subsequently found to be prevalent in many parts of the world, it was unknown in man and animals in Africa south of the Sahara until 1961 when exceptionally heavy infections were seen in 11 youths on the lower slopes of Mount Kenya (Forrester, Nelson & Sander, 1961; Nelson & Forrester, 1962). At that time the medical and veterinary authorities in Kenya were convinced that the parasite had been introduced in pork products brought in by British troops involved in operations against the Mau Mau. However, following questioning of the infected individuals we were convinced that the source of the infection was a wild pig so it was decided to see if the parasite was enzootic around Mount Kenya. Eventually the parasite was found in the bush pig (*Potamochoerus porcus*) and the leopard (*Panthera pardus*), but the medical and veterinary authorities were still unconvinced that the parasite was of local origin and they insisted that the bush pigs had eaten sausages or ham brought in by the troops and that the leopard had eaten the wild pigs! We therefore decided to examine animals in Masailand in an area where there was no likelihood of introduced infection. Here we found the most important maintenance host of *Trichinella* in Africa, namely the spotted hyaena (*Crocuta crocuta*) with more than 50% infected. We also found infection in the striped hyaena (*Hyaena hyaena*), the serval (*Felis serval*), the side-striped jackal (*Canis adustus*) and the lion (*Leo leo*). The lion was very old and was living on the Ngong Hills near Nairobi where it survived by catching and eating hyaenas. It was suffering from "rheumatism" and when we showed an interest in the animal it was immediately shot by a game warden "to put it out of its misery" and he brought the animal to our laboratory! We found that it had an active *Trichinella* infection. We also examined more than 2000 rodents and other animals which were being collected for studies on the reservoir hosts of bubonic plague, leishmaniasis, and hydatid diseases, but only the carnivores and bush pigs were found infected (Nelson, Guggisberg & Mukundi, 1963).

The studies in Kenya led to an increased awareness of the problem in Africa and further outbreaks of the disease in man were reported in Kenya by Hutcheon & Pamba (1972), in Tanzania by Bura & Willett (1977) and in Senegal by Gretillat & Vassilades (1967). *Trichinella* was also found in spotted hyaenas in the Serengeti by Sachs & Taylor (1966) and in hyaenas, jackals and a lion and leopard in the Serengeti by Sachs (1970). In South Africa Young & Kruger (1968) found that the parasite was widespread in carnivores, especially the spotted hyaena and lion, and they suggested that the parasite may have an important role in controlling the population density of large carnivores. In Kenya the densities of larvae in the muscles of the carnivores and bush pigs were almost all below 50 per gram, which is usually of no clinical significance. However, densities of larvae in the muscles of susceptible hosts are dependent on the number ingested during the first infective meal and thereafter there is a strong immunity against super-infection; it is possible that many animals die with initial infections and that all survey results under-estimate the importance of trichinosis as a disease of wildlife. It is obvious that further studies are necessary to examine for trichinosis in sick carnivores, especially in young animals.

Trichinosis is very rare in herbivores and it has not been recorded from antelopes or zebra which form the bulk of the diet of the large carnivores. The most likely source of infection in the hyaenas, jackals and lions is from other carnivores by either cannibalism or carrion feeding. The large carnivores sometimes prey on one another and on smaller carnivores, for example: leopards frequently kill and eat dogs and one notorious leopard in Masailand which relished hyaenas was known as the "Tyburn leopard" because it left its victims hanging in trees by the roadside. It is a widely held belief that "dog does not eat dog" and that cannibalism is extremely rare among carnivores. It is uncommon in the domestic dog in areas where there is plenty of alternative food but there are numerous references to cannibalism among wild carnivores. Guggisberg (1961) records instances of lions eating lions and it was our experience in Kenya that the carcasses of hyaenas that we left in the bush disappeared overnight with ample evidence that they were eaten by their fellows. If hyaenas eat the flesh of other hyaenas only once in a lifetime the parasite can be readily maintained in a community of hyaenas. The same is true of other carnivores and it was Sontgen (1939) who first suggested that the red fox was a maintenance host of the parasite in Europe because fox eats fox; this is still regarded as the basic cycle in Bulgaria (Stoimenov & Gradinarski, 1981).

Not all infections in carnivores will be due to either cannibalism or eating the infected flesh of other carnivores. Carnivores may also become infected by eating bush pig and warthogs (*Phacochoerus aethiopicus*) which may themselves become infected by eating the carcasses of other pigs or carnivores. Also scavenging carnivores and even wild pigs may sometimes become infected by eating human corpses because in many of the areas in Africa where the disease is prevalent the people rarely practise deep burial.

Characteristics of the Kenya *Trichinella*

A puzzling feature of the wildlife studies in Kenya was the total absence of *Trichinella* in rodents and the smaller carnivores such as genets and mongooses which prey on rodents. This was the first indication that the Kenya parasite might be different from *T. spiralis* in Europe or America. Subsequent comparative studies indicated that although the Kenya *Trichinella* was morphologically identical with parasites from other parts of the world it was much less infective to normal laboratory rodents and even wild Kenya rodents such as *Mastomys* and *Pattus*, but what was much more surprising, and of much more significance from the public health point of view, was the demonstration that the Kenya parasite was of very low infectivity to domestic pigs (Nelson & Mukundi, 1962). This finding was regarded as heresay because the accepted dogma was that *Trichinella spiralis* was a uniform species throughout its range and that rats and pigs were hosts par excellence. An eminent American parasitologist was so unconvinced of our findings that he advised the World Health Organization not to accept the Kenya findings! Fortunately other workers soon confirmed the observations (Kozar & Kozar, 1965) and thanks to Dr R. L. Rausch in Alaska we were able to show that a fresh isolate of an American parasite from an Alaskan bear (*Ursus arctos*) also behaved quite differently from parasites isolated from man and domestic pigs (Nelson, Blackie & Mukundi, 1966). These observations resulted in numerous detailed studies on isolates from other areas and from other animals, e.g. Schad *et al.* (1967) in India, Gretillat & Vassilades (1968) in Senegal, Read & Schiller (1969) in North America, Ozeretskowskaya *et al.* (1969) in the Soviet Union and by Siddiqi & Meerovitch (1976a,b, 1977a,b) and by Belosevic & Dick (1979, 1980) in Canada. Britov & Boev (1972) and Boev,

Britov & Orlov (1979) have fully vindicated the Kenya observations on variation of the parasite and as a result of their genetic studies they have decided to split *Trichinella spiralis* into three distinct species: *T. spiralis* for the parasite in temperate regions, *T. nativa* from the Arctic and *T. nelsoni* from southern USSR and the tropics. H. A. Flockhart (pers. comm.) has recently found that she can also separate *T. spiralis*, *T. nativa* and *T. nelsoni* using isoenzyme characters. *Trichinella pseudospiralis*, which was described by Garkavi (1974), is another species characterized by the presence of unencysted larvae in the muscles and the unusual ability to complete its development in birds. There are still arguments as to whether we are dealing with sibling species, sub-species or strains (Kim, Ruitenberg & Teppema, 1981), but in spite of the taxonomic confusion there is no doubt that the different forms with their variations in infectivity to different animal hosts have a profound effect on the epidemiology of the disease. Although the Kenya parasite is of low infectivity to rats and domestic pigs it is highly infective to man and to laboratory monkeys and baboons (Nelson & Mukundi, 1962). In the outbreak recorded by Forrester *et al.* (1961) the larval densities in the muscles averaged more than 2000 per gram which is much higher than previously recorded elsewhere in the world but the mortality in Kenya was surprisingly low. Even higher densities averaging more than 3000 per gram were seen in the outbreak in Tanzania reported by Bura & Willett (1977) and one patient with a world record of 6530 per gram recovered, whereas in Europe there have been fatalities with densities less than 100 per gram. This raises the possibility that the east African parasite is man-adapted and that it may even have evolved in the distant past owing to human cannibalism. Apart from Kuru, the slow virus disease in the New Guinea highlands, which depends on brain eating, there are no human diseases where cannibalism is a feature of the life-cycle. It is extremely unlikely that human cannibalism has had any role in the maintenance of *Trichinella* in Africa. It has always been a subject for dramatization and exaggeration and in any event we are told that whenever human flesh was eaten it was usually well cooked so this would have killed the parasite!

Although the Kenya *Trichinella* fails to develop in a variety of laboratory rats it stimulates a marked degree of protective immunity against subsequent challenge with highly infective strains of *T. spiralis* from Europe and America (Nelson, Blackie *et al.*, 1966). If this occurs in nature it could affect the establishment in the wild reservoir hosts in Africa of the domestic pig-adapted parasites, which if they were introduced from Europe or America could have serious consequences in the extensive newly developed local pig industry in Africa.

It was one of the unexpected findings of the Kenya study that although the bush pig was naturally infected with the local parasite we could not infect the domestic pig. This was certainly surprising but susceptibility to infection is still something of a mystery and we cannot even explain why one strain of laboratory rat, e.g. the hooded, is susceptible to the Kenya parasite whereas the albino rat is refractory. Unfortunately *Homo sapiens* seems to be susceptible to all the geographical forms of the parasite but it is his eating habits and not his susceptibility which determine the prevalence of the disease.

Human behaviour in relation to transmission of trichinosis

Trichinosis can only develop in man if he eats undercooked infected meat. At a temperature of $60°C$ all the larvae are killed. Trichinosis has therefore always been a rare disease in man in countries like China even when there was a high prevalence of the infection in domestic pigs. The main source of infection in developed countries is from processed pork products which are uncooked or from sausages which are inadequately cooked. Transmission of trichinosis direct from carnivore to man is relatively rare although outbreaks still occur in North America due to eating the meat of polar bears in the Arctic or black bears farther south. In Africa eating the flesh of wild carnivores is usually taboo and it is extremely rare to find tribes which permit eating the flesh of hyaenas or jackals, although some of the Nilohamites including the Masai may eat the meat of totem animals such as the lion and leopard, and there are several tribes in west Africa and the Sudan who eat specially reared dogs.

The first outbreak in Kenya was due to the breaking down of a traditional Kikuyu taboo against eating the flesh of wild animals. This occurred at the time of the Mau Mau rebellion when many of the young men were living in the forest. In other parts of Africa the bush pig is regularly hunted for food and also because it is a serious pest which destroys maize and sweet potato crops. The eating of the flesh of swine is prohibited by Mosaic law and also by Muslim law so although the warthog is one of the commonest wild animals in the savanna of west Africa, it is rarely the source of disease because the majority of the local people are followers of Mohamed. The parasite was only discovered in west Africa following an outbreak of trichinosis which incapacitated nine (presumably non-Muslim) Europeans who barbecued a warthog, and the parasite was only discovered in Tanzania following a severe outbreak which resulted from eating the meat of a warthog (Bura & Willett, 1977).

The last major outbreak of trichinosis in the British Isles was in Southern Ireland where the patients were almost all infected from eating undercooked and often raw sausages (Corridan & Gray, 1969). As in Kenya, the veterinary and medical authorities claimed that the infection had been imported and although subsequent investigations by Corridan, O'Rourke & Verling (1969) revealed the presence of trichinosis in Ireland's common carrion-feeding carnivore, the red fox (*Vulpes vulpes*), the Irish authorities claimed that the fox must have eaten infected sausages or pork products brought in by visitors!

HYDATID DISEASE (ECHINOCOCCOSIS)

The Life-cycle

Like trichinosis, hydatid disease (echinococcosis) is an extremely widespread zoonosis with a primary maintenance cycle involving wild carnivores, but in this infection there is a two-host cycle with the adult *Echinococcus* in the small intestine of the carnivore and the cystic stage in the liver or other organs of herbivores. For example, there are two species in the Arctic: *E. granulosus* with a wolf—moose cycle and *E. multilocularis* with a fox—rodent cycle; in Central America there is *E. oligarthrus* with a puma—agouti cycle and in Africa there may be a lion—warthog cycle.

The adult tapeworms live in the gut of the carnivore hosts and are very small, less than 1 cm in length, but they compensate for their small size by being present in large numbers (up to 50 000 in heavily infected dogs). The infection is transmitted through the eggs in the faeces which contaminate the pasture, food or drinking water. The most common species is *Echinococcus granulosus* (Batsch, 1786) which produces a slow growing cyst that contains many thousands of protoscolices each of which develops into an adult worm if the cyst is eaten by a susceptible host. In most parts of the world there is a secondary domestic cycle and man usually acquires infection from the domestic dog which is infected by eating the hydatid cysts which are found in domestic animals, particularly sheep, goats, cattle and camels. Although dogs and other carnivores may have large numbers of worms in the intestines these usually cause no noticeable disease in the infected animals. Also the cystic stage in sheep or wild antelopes is usually of minor clinical significance; and only in man do long-standing infections damage vital organs such as the liver, lungs or brain and eventually kill the patient. The disease is a serious public health problem in almost all areas where man uses dogs to herd his domestic livestock, and since there is no cure other than surgery, preventive measures are aimed at reducing infection in dogs and livestock. The subject has been extensively reviewed in a recent publication (Eckert, Gemmell & Soulsby, 1981).

Hydatid Disease in Africa with Special Reference to Kenya

Animal hosts

Hydatid disease was thought to be uncommon in Africa until Nelson & Rausch (1963) and Schwabe (1969) suggested that one of the most heavily infected areas in the world was Turkana in northern Kenya. The Turkana are nomadic pastoralists with large herds of camels, goats, sheep and cattle and like their distant relatives the Masai they subsist mainly on milk and blood from their domestic animals. In contrast to Masailand, Turkana is much more barren with large areas of hot stony desert and there are recurrent droughts which decimate the herds. Also, unlike the Masai, the Turkana hunt wild animals for food and as a result antelopes and large carnivores are rarely seen in Turkana whereas Masai is the world's greatest reservoir of game animals. It was known from veterinary and meat inspection records that hydatid cysts were present in a high proportion of cattle, sheep, goats and camels from all parts of Kenya with the heaviest infection in Masailand (Ginsberg, 1958) and yet the disease in man was relatively rare except in Turkana; this suggested that there a wild life-cycle with the people becoming infected from mission cycle in this area. For example, was it a different species; was there a wild life cycle with the people becoming infected from jackals or other wild carnivores; or was there some peculiar custom in Turkana which brought the people into more intimate contact with infected dogs and dog faeces? Preliminary studies by Nelson & Rausch (1963) and Rausch & Nelson (1963) indicated that the parasite from dogs in Turkana was morphologically similar to *E. granulosus* from other parts of the world and although a few *Echinococcus* adults were obtained from the silver-back jackal (*Canis mesomelas*), the spotted hyaena (*Crocuta crocuta*) and the cape hunting dogs (*Lycaon pictus*) in Masailand, they were all identified as *E. granulosus* and it was considered that the wild carnivores may have been secondarily infected from a primary cycle involving domestic dogs and domestic herbivores. Domestic dogs were particularly heavily infected in Turkana and there was no evidence of a wild life-cycle in wild herbivores in this area. It was therefore concluded that the best way of dealing with the problem in Turkana was to treat infected dogs.

Following more extensive studies on the role of wild animals in the transmission of other cestodes of medical importance in Kenya it became obvious that the picture was not so simple and that further studies were necessary to see if there was a wild life-cycle of *E. granulosus* in Africa (Nelson, Pester & Rickman, 1965). An obvious predator—prey relationship worth studying was between the lion and larger antelopes. Ortlepp (1937) described a new species *E. felidis* from the lion in South Africa but subsequent studies by Verster & Collins (1966) and by Rausch (1967) on material from the lion suggested that the parasite was synonymous with *E. granulosus*. This has been confirmed by Dinnick & Sachs (1972) who found infected lions in Masailand in Kenya and the Ruwenzori Game Reserve in Uganda and by Rodgers (1974) who reported infections in lions in the Selous Park in Tanzania. There have since been reports of hydatid cysts in a variety of wild herbivores which are the normal prey of lions. For example, Woodford & Sachs (1973) found 10% of the warthogs and 17% of buffaloes (*Syncerus caffer*) infected in the Ruwenzori Park and Eugster (1978) found 12% of wildebeest (*Connochaetes taurinus*) infected in Kenya; Young (1975) also reported hydatid cysts in 60% of zebras (*Equus burchelli*) in the Kruger Park. More recently Graber & Thal (1980) have suggested that there may be a cycle between the lion and warthog in the Central African Republic but they have produced experimental evidence to show that the local parasite is different from *E. granulosus* in morphology and infectivity to dogs.

The demonstration that lions are naturally infected with adult *Echinococcus* in east Africa and that some of their prey have hydatid cysts is not sufficient evidence to incriminate the lion as a significant maintenance host of the parasite. The prey animals of lions may be infected from other wild carnivore hosts or even from dogs because there is a very high infection rate in dogs in most of the pastoral areas. Hyaenas are far more abundant than lions and might seem ecologically well-adapted for maintenance hosts but only a few lightly infected spotted hyaenas were found by Nelson & Rausch (1963) in Kenya and by Young (1975) in the Kruger Park. More recent studies by Eugster (1978) in Masailand and by Macpherson *et al.* (in press) in Turkana have been completely negative. Of even more significance have been the experiments by Macpherson and his colleagues who have shown that the spotted hyaena is insusceptible to infection with hydatid material from Turkana although dogs and jackals are highly susceptible. Another possible host would be the cape hunting dog (*Lycaon pictus*): three out of four were found to be infected by Nelson & Rausch (1963) in Masailand, but these animals are relatively rare in Kenya. In areas where there are still large packs, for example in the Serengeti, they are well-adapted for acquiring and disseminating the parasite but because of their scarcity elsewhere they are unlikely to be important hosts. Jackals are much more likely to be of significance as natural maintenance hosts, not because of a predator—prey cycle as in the case of the lion but because they are widespread scavengers likely to be infected from herbivores killed by other predators and also from animals that have died as a result of disease or drought. The first record in Africa was by Nelson & Rausch (1963) who found one out of nine silver-back jackals (*Canis mesomelas*) infected in Masailand. Verster & Collins (1966) found nearly 10% infected out of more than 200 examined in South Africa and Eugster (1978) found five out of 13 infected in Masailand. Macpherson & Karstad (1980) and Macpherson *et al.* (in press) found 11 out of 38 silver-back jackals and six out of 22 golden jackals (*Canis aureus*) infected in Turkana. They have shown that the silver-back species of jackal is even more susceptible to the Turkana human parasite than the domestic dog but they found no hydatid cysts in 152 wild herbivores in Turkana and suggest that here, in contrast to Masailand, where infected antelopes are abundant, the main source of infection for jackals are the carcasses of domestic livestock and possibly human corpses.

Characterization of the Kenya *Echinococcus*

Morphological data from specimens of *Echinococcus* from wild and domestic animals in Kenya are all consistent with measurements of *E. granulosus* from other parts of the world. But *E. granulosus* is not a uniform species and there are striking variations in the infectivity and pathogenicity of the parasites in different geographical areas. For example, the parasites isolated from the wolf—moose cycle in North America differ from the parasites from the dog—sheep cycle and even in Britain there are two distinct forms of *E. granulosus*. In Wales the parasite maintained in the sheepdog—sheep cycle is a common cause of human hydatid disease, whereas the foxhound—horse parasite which is more prevalent in southern England seems to be non-infective to man (Nelson, 1972; Smyth, 1977). Although they are morphologically indistinguishable they can be separated on the basis of their behaviour *in vitro* and their infectivity to laboratory animals (Smyth, 1979; Thompson, 1979). They also have different bio-chemical and isoenzymatic characteristics. These techniques have been applied to material from Kenya and it is already clear that the parasites in Turkana differ not only from *E. granulosus* in other parts of the world but also from the parasites isolated from wild animals in

Masailand (McManus & Macpherson, 1980; McManus, 1981). It has been suggested by French & Nelson (in press) and French, Nelson & Wood (in press) that the high prevalence of the disease in Turkana might be because the local parasite has increased infectivity to man because dogs and jackals have access to infected human corpses and the parasite has become man-adapted. This is more plausible than the similar hypothesis related to trichinosis and it would fit with the observation by Smyth & Smyth (1964) that polyembryony in the cystic stage allows for the development and dispersal of new mutants — but in considering the unusually high prevalence, we still have to explain how the people become infected from their dogs.

Human behaviour in relation to the transmission of hydatid diseases
Infection can only arise if man ingests eggs from dog faeces and, as Schwabe (1969) has emphasized, in areas where there is an exceptionally high prevalence of the disease this can usually be traced to some peculiar custom such as the use of dog faeces to tan leather. The Turkana have an unusually close intimacy with their dogs and Nelson & Rausch (1963) suggested that the high incidence of hydatid disease in the area might be due to the custom of the Turkana mothers of using dogs as "nurses" to guard their infants and lick them clean when they vomit or defaecate. Another plausible explanation of the high prevalence is that the Turkana might eat the heavily infected intestines of infected dogs or wild carnivores. They certainly relish the partially cooked gut of their domestic animals but usually they deny eating carnivores or their intestines. Macpherson & Karstad (1980) believe that this is not always true and they noted during their studies in Turkana that the women were pleased to receive the carcasses of jackals after they had examined them.

In other parts of the world, including mediaeval England, dog faeces have been incorporated in medicaments and C. M. French (pers. comm.) claims that the women in Turkana use "cooked" dog faeces as part of a concoction for treating wounds; women are more frequently infected than men and they have a much closer relationship with the dogs; they also incorporate dog faeces in the lubricants which they use to protect their skins from abrasion by the massive weight of necklaces that they wear. The primitive "manyatta" homestead must be heavily contaminated with *Echinococcus* eggs from dogs and there is plenty of opportunity for dogs and jackals to contaminate the scarce water holes. This could adequately explain the infection of domestic animals with cysts but it is the very close intimacy with a large number of dogs which probably explains the high prevalence of cysts in man. The relatively low prevalence of hydatid disease in man in Masailand in the presence of a high prevalence in animals including dogs, is thought to be due to the less affectionate relationship that the Masai have with their dogs but it is possible that the parasites are different in their infectivity or it may be that this is another example of the Jennerian principle of zoo-prophylaxis, and that the universal presence in the Masai of infection with the benign *Taenia saginata* protects them from the much more pathogenic *Echinococcus*.

ACKNOWLEDGEMENTS

I am greatly indebted to the staff of the Kenya Medical Department and especially the Division of Insect Borne Diseases for their support in carrying out the field and laboratory studies on trichinosis and hydatids. I am particularly grateful to Dr Timothy Arap Siongok for his continuing interest in the problem and to Dr Michael Wood and Dr Marcus French of the African Medical and Research Foundation, Nairobi, for their help and enthusiasm in trying to solve the problem of hydatid disease in Turkana. I am also indebted to Dr Calum Macpherson for information on current studies on *Echinococcus* in wildlife in Kenya.

REFERENCES

Belosevic, M. & Dick, T. A. (1979). *Trichinella spiralis*. Comparison of stages in host intestine with those of an Arctic *Trichinella spiralis*. *Expl Parasit.* 48: 432–446.

Belosevic, M. & Dick, T. A. (1980). *Trichinella spiralis*. Comparison with an Arctic isolate. *Expl Parasit.* 49: 266–276.

Boev, S. N., Britov, V. A. & Orlov, I. V. (1979). Species composition of Trichinellae. *Wiad. Parazyt.* 25: 495–503.

Britov, V. A. & Boev, S. N. (1972). Taxonomic status of different strains of *Trichinella* and the nature of their focal occurrence. *Vsesoyuznaya Konferentsiya (VIII) po prirodnoi ochagovosti boleznei zhivetnkich i ohrane ikh chrislennosti* 1: 83–84.

Bura, M. W. T. & Willett, W. C. (1977). An outbreak of trichinosis in Tanzania. *E. Afr. Med. J.* 54: 185–193.

Corridan, J. P., O'Rourke, F. J. & Verling, M. (1969). *Trichinella spiralis* in the red fox (*Vulpes vulpes*) in Ireland. *Nature, Lond.* 222: 1191.

Corridan, J. P. & Gray, J. (1969). Trichinosis in South West Ireland. *Br. Med. J.* 1969(2): 727–730.

Dinnik, J. A. & Sachs, R. (1972). Taeniidae of lions in East Africa. *Z. Tropenmed. Parasit.* 23: 197–210.

Eckert, J., Gemmell, M. A. & Soulsby, E. J. L. (1981). *FAO/UNEP/WHO guidelines for surveillance, prevention and control of echinococcosis/hydatidosis.* Geneva: WHO.

Eugster, R. O. (1978). *A contribution to the epidemiology of* Echinococcus *hydatidosis in Kenya (East Africa) with special reference to Kajiado District*. Thesis for Degree of Doctor of Veterinary Medicine: University of Zurich.

Forrester, A. T. T., Nelson, G. S. & Sander, G. (1961). The first record of an outbreak of trichinosis in Africa south of the Sahara. *Trans. R. Soc. trop. Med. Hyg.* **55**: 503–513.

French, C. M. & Nelson, G. S. (In press). Hydatid disease in the Turkana District of Kenya (II). A study in medical geography. *Ann. trop. Med. Parasit.*

French, C. M., Nelson, G. S. & Wood, M. (In press). Hydatid disease in the Turkana District of Kenya (I). The background to the problem with hypotheses to account for the remarkably high prevalence of the disease in man. *Ann. trop. Med. Parasit.*

Garkavi, B. L. (1974). Potential hosts of *Trichinella pseudospiralis. Parazitologiya* **8**: 489–493.

Ginsberg, A. (1958). Helminthic zoonoses in meat inspection. *Bull. Epizootic Dis. Afr.* **6**: 141–149.

Gould, S. E. (1970). *Trichinosis in man and animals*. Springfield, Illinois: Charles C. Thomas.

Graber, M. & Thal, J. (1980). L'échinococcose des artiodactyles sauvages de la République Centrafricaine: existence probable d' un cycle lion-phacochère. *Revue Élev. Méd. vét. Pays trop.* **33**: 51–59.

Gretillat, S. & Vassilades, G. (1967). Présence de *Trichinella spiralis* (Owen, 1835) chez les carnivores et suidés sauvages de la région du delta du fleuve Sénégal. *C.r. hebd. Séanc. Acad. Sci. Paris* **264**: 1297–1300.

Gretillat, S. & Vassilades, G. (1968). Particularités biologiques de la souche ouest-africaine de *Trichinella spiralis* (Owen, 1835). Receptivité et sensibilité de quelques mammifères domestiques et sauvages. *Revue Élev. Méd. vét. Pays trop.* **21**: 85–99.

Guggisberg, C. W. A. (1961). *Simba, the life of a lion*. London: Bailey & Swinfen.

Hutcheon, R. A. & Pamba, H. O. (1972). Report of a family outbreak of trichinosis in Kajiado District, Kenya. *E. Afr. Med. J.* **49**: 663–666.

Kim, C. W., Ruitenberg, E. J. & Teppema, S. (1981). *Trichinellosis*. (Proceedings of the 5th International Conference on Trichinellosis, September 1980). Surrey: Reedbooks Ltd.

Kozar, Z. & Kozar, M. (1965). A comparison of the infectivity and pathogenicity of *Trichinella spiralis* strains from Poland and Kenya. *J. Helminth.* **39**: 19–34.

McManus, D. P. (1981). A biochemical study of adult and cystic stages of *E. granulosus* of human and animal origin from Kenya. *J. Helminth.* **55**: 21–27.

McManus, D. P. & Macpherson, C. N. L. (1980). A biochemical and isoenzymatic study of adult and cystic stages of *E. granulosus* of human and animal origin from Kenya. *Parasitology* **81**: xxxi–xxxii.

Macpherson, C. N. L. & Karstad, L. (1980). The role of jackals in the transmission of *Echinococcus granulosus* in the Turkana District of Kenya. In *Wildlife disease research and economic development*: 53–56. Karstad, L., Nestel, B. & Graham, M. (Eds). Ottawa: International Development Research Centre.

Macpherson, C. N. L., Karstad, L., Stevenson, P. & Arundel, J. H. (In press). Hydatid disease in the Turkana District of Kenya (III). The significance of wild animals in the transmission of *Echinococcus granulosus*, with particular reference to Turkana and Masailand in Kenya. *Ann. trop. Med. Parasit.*

Nelson, G. S. (1972). Human behaviour in the transmission of parasitic diseases. In *Behavioural aspects of parasite transmission. Zool. J. Linn. Soc.* (Suppl. I) 51: 109—122. Canning, E. U. & Wright, C. A. (Eds). London and New York: Academic Press.

Nelson, G. S. (1974). Zooprophylaxis with special reference to schistosomiasis and filariasis. In *Parasitic zoonoses. Clinical and experimental studies*: 273—285. Soulsby, E. J. L. (Ed.). London and New York: Academic Press.

Nelson, G. S., Blackie, E. J. & Mukundi, J. (1966). Comparative studies on geographical strains of *Trichinella spiralis. Trans. R. Soc. trop. Med. Hyg.* 60: 471—480.

Nelson, G. S. & Forrester, A. T. T. (1962). Trichinosis in Kenya. *Wiad. Parazytol.* 8: 17—28.

Nelson, G. S., Guggisberg, C. W. A. & Mukundi, J. (1963). Animal hosts of *Trichinella spiralis* in East Africa. *Ann. trop. Med. Parasit.* 57: 332—346.

Nelson, G. S. & Mukundi, J. (1962). A strain of *Trichinella spiralis* from Kenya of low infectivity to rats and domestic pigs. *J. Helminth.* 37: 329—338.

Nelson, G. S., Pester, F. R. N. & Rickman, R. (1965). The significance of wild animals in the transmission of cestodes of medical importance in Kenya. *Trans. R. Soc. trop. Med. Hyg.* 59: 507—525.

Nelson, G. S. & Rausch, R. L. (1963). *Echinococcus* infections in man and animals in Kenya. *Ann. trop. Med. Parasit.* 57: 136—149.

Ortlepp, R. J. (1937). South African helminths, part I. *Onderstepoort J. Vet. Sci.* 9: 311—336.

Ozeretskowskaya, N. N., Tumolskaya, N. I., Tchernyaeva, A. I., Pereverzeva, E. V., Uspenskiy, S. M., Romanova, V. I. & Bronshiein, A. M. (1969). Characteristics of human trichinellosis caused by natural (Arctic) and several sinanthropic strains of *Trichinella spiralis* in the USSR. *Int. Conf. Trichinellosis* 2: 18—22.

Rausch, R. L. (1967). A consideration of infraspecific categories in the genus *Echinococcus* Rudolphi 1801 (Cestoda, Taeniidae). *J. Parasit.* 53: 484—491.

Rausch, R. L. & Nelson, G. S. (1963). A review of the genus *Echinococcus* Rudolphi 1801. *Ann. trop. Med. Parasit.* 57: 127—135.

Read, C. P. & Schiller, E. L. (1969). Infectivity of *Trichinella* from the temperate and Arctic zones of North America. *J. Parasit.* 55: 72—73.

Rodgers, W. A. (1974). Weights, measurements and parasitic infestation of six lions from Southern Tanzania. *E. Afr. Wildl. J.* 12: 157—158.

Sachs, R. (1970). Zur Epidemiologie der Trichinellose in Africa. *Z. Tropenmed. Parasit.* 20: 117—126.

Sachs, R. & Taylor, A. S. (1966). Trichinosis in a spotted hyaena (*Crocuta crocuta*) of the Serengeti. *Vet. Rec.* 78: 704.

Schad, G. A., Nundy, S., Chowdhury, A. B. & Bandyopadhyay, A. K. (1967). *Trichinella spiralis* in India, II. Characteristics of a strain isolated from a civet cat in Calcutta. *Trans. R. Soc. trop. Med. Hyg.* 61: 249—258.

Schwabe, C. W. (1969). *Veterinary medicine and human health.* 2nd Edition. London: Bailliere, Tindall & Cassel.

Siddiqi, M. N. & Meerovitch, E. (1976a). Host-parasite relationship in trichiniasis — I. Infectivity of various strains of *Trichinella spiralis* in rats. *Pakistan J. Zool.* 8: 183—189.

Siddiqi, M. N. & Meerovitch, E. (1976b). Host-parasite relationship in trichiniasis — II. Infectivity of various strains of *Trichinella spiralis* in mice, guinea pigs and two strains of rats. *Pakistan J. Zool.* 8: 191—197.

Siddiqi, M. N. & Meerovitch, E. (1977a). Host-parasite relationship in trichiniasis — III. Stability of various strains of *Trichinella spiralis*. *Pakistan J. Zool.* **9**: 47—50.

Siddiqi, M. N. & Meerovitch, E. (1977b). Host-parasite relationship in trichiniasis — IV. Intestinal phase of *Trichinella spiralis*. *Pakistan J. Zool.* **9**: 51—57.

Smyth, J. D. (1977). Strain differences in *Echinococcus granulosus*, with special reference to the status of equine hydatidosis in the United Kingdom. *Trans. R. Soc. trop. Med. Hyg.* **71**: 93—100.

Smyth, J. D. (1979). An *in vitro* approach to taxonomic problems in trematodes and cestodes especially *Echinococcus*. *Symp. Br. Soc. Parasit.* **17**: 75—101.

Smyth, J. D. & Smyth, M. M. (1964). Natural and experimental hosts of *E. granulosus* and *E. multilocularis* with comments on the genetics of speciation in the genus *Echinococcus*. *Parasitology* **54**: 493—514.

Sontgen, K. (1939). Zur Frage der Fuchstrichinose. *Z. Fleisch-Milchyg.* **49**: 334—336.

Stoimenov, K. & Gradinarski, I. (1981). On the epizootiology of *Trichinella spiralis*. In *Trichinellosis:* 373—376. (Proceedings of the Fifth International Conference on Trichinellosis). Kim, D. W., Ruitenberg, E. J. & Teppema, J. S. (Eds). Surrey: Reedbooks Ltd.

Thompson, R. C. A. (1979). Biology and speciation of *Echinococcus* granulosus. *Austr. vet. J.* **55**: 93—98.

Verster, A. & Collins, M. (1966). The incidence of hydatidosis in the Republic of South Africa. *Onderstepoort J. vet. Sci.* **33**: 49—72.

Woodford, M. H. & Sachs, R. (1973). The incidence of cysticercosis, hydatidosis and sparganosis in wild herbivores of Queen Elizabeth Park, Uganda (now Ruwenzori Park). *Bull. Epizoot. Dis. Afr.* **21**: 265—271.

World Health Organisation (1967). Zoonoses. Third report of the Joint FAO/WHO Expert Committee. *WHO Tech. Rep. Series* No. 378.

Young, E. (1975). Echinococcosis (hydatidosis) in wild animals of the Kruger National Park. *Jl S. Afr. vet. Ass.* **46**: 285—286.

Young, E. & Kruger, S. P. (1968). *Trichinella spiralis* (Owen, 1835) Railliet, 1895. Infestation of wild carnivores and rodents in South Africa. *S. Afr. med. Ass.* **38**: 441—443.

DISCUSSION

Walsh — I would like to make the comment that in northern Ghana, near Bolgatanga, the local people eat dogs, in fact there is a dog market, and dog meat is the best of the available meats.

Nelson — This is very interesting because domestic dogs are often infected with *Trichinella*. We found them infected on Mount Kenya, but people in Kenya don't eat dogs. In the southern Sudan they do, and also I believe in parts of Nigeria where large numbers of dogs are consumed at wedding feasts.

Clarke — There was an outbreak of trichiniasis in Liverpool about 20 years ago. I expect it was brought over from Ireland. I was interested in the end result of the disease — calcification of the lesions. These, I think, do not give rise to symptoms unless they are located in the brain when they may produce epilepsy. Is that right?

Nelson — It is true that they calcify but it is tapeworm cysts that cause epilepsy, not *Trichinella*. The patients in Kenya had the heaviest infection ever recorded in man but the disease was relatively mild. For example, in the Liverpool epidemic people died with larval densities of less than 1000 larvae per gram whereas the Kenyan boys survived with densities of over 2000 larvae per gram. Only one died, and none of them was treated with specific anthelminthics. They were treated with corticosteroids or aspirin and six months later they were all back at work. Two years later when I re-examined them, they still had over 2000 living worms per gram in their muscles, but there was no reaction around the worms and the inflammation had gone.

Rees — Professor Nelson mentioned that they had found *Trichinella* in foxes, in Ireland; can he tell us if they have examined any wildlife in Great Britain? We examine, regularly, around 120 000 pig samples each year; we haven't found any signs of a *Trichinella* infestation for over four years. What is the position in wildlife in Great Britain?

Nelson — There hasn't been a wildlife study in the UK for a long time, but there is one record of *Trichinella* from a fox in Cornwall, and there have been several records of the parasite in rats. The big difference between the Kenya parasite and the Alaska parasite and the one in the UK is that they have no secondary cycle in rats, whereas here and in Europe rats can be found infected around abattoirs and in the sewers, but only as a secondary cycle to the artificial cycle in domestic pigs. It is now a very rare human disease in England and, with the introduction of deep freezes and the regulations about cooking garbage fed to pigs, it has practically disappeared as a domestic zoonosis even in America, where at one time 11% of the population was infected.

Macpherson — In view of the largely nebulous international borders, with people coming and going, the incidence in jackals, the suggestion you made that there is a human strain of *Echinococcus* in Turkana, all compounded with the fact that the people have a very close relationship with their dogs, do you think there is any real chance of a control programme in the area?

Nelson — Dr Macpherson has worked in Turkana, so he knows that this is an extremely difficult problem. But if the wildlife cycle is of only secondary importance, and if the dog is the primary host, then I think that a mass treatment campaign using "Praziquantel" could be successful, especially if the Turkana get rid of all stray dogs and only keep registered dogs which look after the children and the animals.

Khalil — I have examined a large number of wildebeest in the Kajiado district of Kenya, and found a number of hydatid cysts, but most of them are sterile: there are no daughter cysts or hydatid sand. What part do game animals and carnivores play in Masailand in the transmission of hydatid disease to man?

Nelson — There is still no proof that there is a wildlife cycle of *Echinococcus* in Kenya. I have no doubt that there is a true wildlife cycle of *Trichinella*, but there is not sufficient evidence to be sure that there is a genuine maintenance cycle of *Echinococcus* in the same way as in northern Canada where there is a cycle of *Echinococcus* between the wolf and the moose. Dr de Savigny who has been studying this recently told me that a wolf can recognize a moose with hydatid, and that all the moose that are killed by the wolves have hydatid cysts and the rest of the animals are uninfected; that the wolves are actually specifically taking out of the population of moose those animals with hydatid cysts, even though they look normal, they don't look ill.

GENERAL DISCUSSION

Clarke (Chairman) — I'm going to open this general session by asking Prince Philip whether he has anything to say about conservation and disease, or anything else that he cares to comment on.

HRH The Prince Philip — Well, I didn't actually come here to say anything, and you've rather taken me by surprise. I'd just like to point out that the fact that I am President of the World Wildlife Fund doesn't mean to say that I am a scientist, and I think I ought to make it clear to everybody here that the whole technical side of conservation is in the hands of the IUCN, and not of the World Wildlife Fund which, as its name implies, is a fund, and the purpose of the exercise is to raise money for projects provided or promoted by IUCN. But I think that the reason that I welcome this sort of discussion very much indeed is because there is a tremendous area of misunderstanding and misconception about the inter-relationship of disease between domestic animals and man and wild animals. The problem is more complicated than it appears. On the other hand there are conservationists who believe that there are a number of scientists or agriculturalists who would like to control this cycle of disease by trying to take out either the host or the secondary host, and it has in fact, I think, been tried with trypanosomiasis in Africa, where they tried to kill out the wildlife and of course it has failed. And so there is a suspicion amongst animal conservationists that scientists and agriculturalists are trying to control this disease cycle by attacking wildlife, and you know the emotional attitude to any kind of control of wildlife is a very difficult one. So I think that the fact that this sort of seminar takes place — and I hope that the record of it will find its way to the headquarters of IUCN and the World Wildlife Fund and a lot of other conservation bodies — is most useful, because I think it's vitally important that people should understand what actually is going on and what the problems are, and also so that they can take some intelligent part in the conversation about — at least, the discussion about — ways in which we control diseases and conserve wildlife. So I really am most grateful to the Zoo and to the organizers of this seminar for putting it on today.

Clarke — We are most grateful; thank you very much indeed.

Zuckerman — May I put a question to Professor Nelson, who referred in his talk to the fact that the people in charge of wildlife in the Kruger National Park regard animal numbers, for example that of the lion, as being determined by their parasitic diseases. Could you explain the reason behind this?

Nelson — Yes. The parasite I was talking about was *Trichinella*, and the severity of the disease produced by this parasite is proportional to the number of parasites ingested during the first meal. Thereafter, if you recover, you are almost completely immune; and this would be true of animals as well as man. What they were suggesting in the Kruger National Park was that when young lions have their first meal on a heavily infected carcass of another carnivore or on a wild pig this would result in a considerable mortality at the early age.

Clarke — Could I ask the panel a general question? In the north of China, there is an area where cancer of the gullet — cancer of the oesophagus — is about 50 times more common than it is in this country. Nobody quite knows the cause, but the chickens in these high incidence areas get it too, probably because they are fed on the scraps which humans eat; but the exact component isn't known. Does the panel think there might be a similar occurrence between us and our pets? After all, dogs get exactly like their owners, or *vice versa*; and I wonder whether some types of malignant disease might be transmitted from man to, particularly, dogs?

Kaplan — There are quite a number of diseases that we study in comparative medicine — if you are speaking particularly about the dog — that are of interest. There are a large number of viruses of man and dogs that are identical. The flu virus is one that I mentioned before. You can isolate the flu from dogs during epidemics, but there doesn't seem to be a transmitting chain that occurs with other animals. There are paramyxoviruses related to the flu virus. The common environment that dogs experience with human beings was studied by the World Health Organization with respect to the incidence of lung tumours in dogs in urban populations to determine whether it would reveal any information concerning air pollutants to which the general population as a whole is exposed. We found lung tumours in dogs, but the dog doesn't seem to be very prone to lung tumours of the type found in human beings, so that we could not make a correlation. We studied breast tumours and all sorts of tumours of dogs, for comparison with human disease. The closest relationships are the microbes — bacteria, mycoplasmas and viruses — of which there are large numbers that man and his pet animals share.

Molyneux — I'd like Professor Nelson to come back to the wolf—moose cycle, which I find fascinating, because it implies that the parasite is in some way altering the behaviour of the moose. Now I would like to reflect very briefly on the work that Vale (1980)[1] has done in Zimbabwe on the attractiveness of cattle to tsetse through the odour which is emanating from ox breath. It is possible — and this is purely hypothetical — that the trypanosome actually changes the metabolism of the host so that the host is more attractive to tsetse and thereby increases the likelihood of completion of the cycle. We know, moreover, that the parasite affects the behaviour of the fly to increase the chances of transmission (Jenni, Molyneux, Liversey & Galun, 1980).[2] I would suspect that in some way hydatid is affecting the behaviour of the moose in such a way that the wolf recognizes it. But what advantage does the wolf get?

Nelson — The information I have on the wolf—moose cycle was obtained from Dr de Savigny, who told me that the wolf will preferentially select animals infected with hydatids. The hydatids in the moose are mainly in the lungs, and it may be that the excretions or exhalations from the lungs of an infected animal are recognizable by the wolf. This is an excellent example of a predator—prey relationship favouring transmission of a parasite. Another example would be *Echinococcus multilocularis*, which has a life-cycle between the Arctic fox and the microtine rodents. In this case there is a definite behaviour change in the rodents, because the parasite is very unlike the one in the moose or in the cow or the sheep, it is a very rapidly growing parasite which destroys the liver and it makes the heavily infected animal wobble and actually incapable of moving. So the fox can then detect that infected animal preferentially; it looks big and juicy, but in fact this is due to an enormous growth of the *Echinococcus* in the liver of the rodent.

Scott — I would like to make a brief comment about hydatid disease in moose and wolf in Canada. Current research at McGill University (by Rau, published in

---

[1] Vale, G. (1980). Field studies of the responses of tsetse flies (Glossinidae) and other Diptera to carbon dioxide, acetone and other chemicals. *Bull. ent. Res.* **70**: 563—570.
[2] See list of references, pp. 47—52.

Can. J. Zool.) has shown that the first moose to be shot during the hunting season are in fact the most heavily infected with hydatid, and there is an indication that there may be some behavioural influence of the hydatid on the moose.

HRH The Prince Philip — Could I just ask one question about the vectors, the sandflies and the mosquitoes and the fleas and the things that transmit the disease. Are they affected by it in any way? I mean, does it have a positive or negative effect, or is it totally neutral?

Lainson — First of all, as far as sandflies are concerned, I must point out that not all species of these insects are capable of transmitting, or even of becoming infected. Although vector species do often have a great deal of their intestine packed with the promastigotes of *Leishmania* there is no concrete evidence to show that they suffer any pathological effects from the infection. That is to say, they do not, to our knowledge, die of the infection. On the contrary, we have been able to re-feed infected sandflies (*Lu. flaviscutellata*) four times in some experiments, during which time the insects appear to be healthy and lay their eggs normally. Conclusions are difficult, however, for we have few guidelines on the longevity of American sandflies in their natural, forest habitat and, under laboratory conditions at least, many sandflies die after their first oviposition, whether they are infected or not.

HRH The Prince Philip — But the sandfly wouldn't be *un*healthy if she didn't have the special parasite in her?

Lainson — No, I don't think so. There is no evidence to show that the presence of *Leishmania* in the sandfly gut is necessary for the insect's survival. Vector species continue to thrive in areas where there appears to be no transmission going on, and laboratory-bred colonies of these vectors survive and multiply normally. It is quite possible that under good laboratory conditions the survival of a given female sandfly is longer than that of the individual facing predation and other hazards of the natural habitat.

Zuckerman — That means you are able to keep them alive much longer when they are uninfected?

Lainson — Well, let us say this: in the laboratory there is no conclusive evidence to show that infected flies die earlier than non-infected ones. As I have said, the issue is complicated by the fact that a high proportion of sandflies of some species die when they lay their eggs — whether or not they may be infected with *Leishmania*.

Kaplan — But that's not true, is it, of the louse? For example, in typhus fever the louse transmitter of typhus gets infected with the *Rickettsia* which packs its intestinal tract and passes in its faeces. This in turn infects the wound. The louse is fatally infected by the parasite that it is dealing with. That is one exception.

Nelson — Another exception is the mosquito which transmits elephantiasis. If it ingests large numbers of the larval forms which cause the elephantiasis, so many may migrate to the thoracic muscles that the mosquito has difficulty in flying.

Molyneux — I tend to disagree with Dr Lainson. We have in fact shown that if you infect tsetse with trypanosomes, the probing and feeding behaviour of the infected tsetse is different to that of the uninfected fly. The capacity to transmit is increased substantially, up to five times in fact, in *brucei*-infected flies (Jenni, Molyneux, Liversey & Galun, 1980)[1] and a recent publication by Roberts

---

[1] See list of references, pp. 47–52.

$(1981)^1$ has confirmed the same observations in *congolense*-infected *Glossina*. This work is partially stimulated by Dr Lainson's own work in Belize, the work of his colleague and my colleague Dr Killick-Kendrick at Ascot, which shows that *Leishmania*-infected sandflies have more difficulty obtaining a blood-meal than uninfected sandflies do (Killick-Kendrick, Leaney, Ready & Molyneux, 1977):^{2} multiple probing has been observed in infected flies. We believe this is due to the interaction between parasites and receptors in the anterior part of the gut. And I would remind the audience, of course, of the well-known observations on the transmission of plague, where that is largely due to the blocked, or partially blocked, proventriculus which really does have an important effect on the probing behaviour of the flea (Bacot & Martin, 1914).^{3} And as Professor Nelson said, the effects of the filarial worms on mosquitoes are well defined. There is evidence that with trypanosomes in sandflies — where the sandflies were transmitting trypanosomes amongst bats — the gonadotrophic concordance is disrupted by the presence of very heavy trypanosome infections in the sandfly (Williams, 1976).^{4} Similarly, there is a pathogenic effect of *Trypanosoma rangeli* in *Rhodnius prolixus* (Watkins, 1971).^{5} *Trypanosoma rangeli* has been suggested as a biological control agent for bugs which also transmit *T. cruzi*. So there are many examples of the effects of parasites on insect vectors. In many cases the parasite's presence enhances its chances of transmission, by one mechanism or another.

Lainson — I can't speak for your trypanosomes, but with regard to sandflies I will go along with you that multiple probing (and transmission without the taking up of new blood-meal) can take place. Whether or not this is always a result of the infection is debatable: I have frequently seen them make this repeated probing in the field — perhaps these were infected flies, too! All I can say is that although the infection may result in the necessity for the sandfly to probe more than once to obtain her blood-meal, she does seem eventually to achieve this aim. I would certainly agree with you, however, that this multiple probing will greatly increase the chances of transmission of the *Leishmania*. This might explain the frequent development of multiple lesions on the arms of persons collecting *Lu. umbratilis*, the vector of *L. b. guyanensis*, from tree-trunks; although we also have to consider the fact that the infection rate in this species may sometimes be exceptionally high. It is more difficult to explain the fact that multiple lesions in infections with *L. b. braziliensis* are much less common. Perhaps the situation varies in different vectors. Finally, unless I am mistaken, the studies on the effects of leishmanial infection on the feeding of sandflies were done with experimental infections in sandflies which were not the

^{1} Roberts, L. W. (1981). Probing by *Glossina morsitans morsitans* and transmission of *Try-panosoma (Nannomonas) congolense*. *Am. J. trop. Med. Hyg.* **30**: 948—951.
^{2} Killick-Kendrick, R., Leaney, A. J., Ready, P. D. & Molyneux, D. H. (1977). *Leishmania* in phlebotomid sandflies. IV. The transmission of *Leishmania mexicana amazonensis* by the bite of experimentally infected *Lutzomyia longipalpis*. *Proc. R. Soc. Lond* (B) **196**: 105—115.
^{3} Bacot, A. W. & Martin, L. J. (1914). Observations on the mechanism of the transmission of plague by fleas. *J. Hyg.* (Plague Suppl.) **3**: 423—439.
^{4} Williams, P. (1976). Flagellate infections in cave-dwelling sandflies (Diptera: Psychodidae) in Belize, Central America. *Bull. Ent. Res.* **65**: 615—629.
^{5} Watkins, R. (1971). Histology of *Rhodnius prolixus* infected with *Tryanosoma rangeli*. *J. invert. Path.* **17**: 59—66.

natural vectors (e.g. *L. mexicana amazonensis* in *Lutzomyia longipalpis*). Perhaps we should extend these observations to naturally infected vectors.

Clarke — Dr Lainson, is your information about China up to date? When I was there, the idea that the Chinese had *any* endemic diseases was looked upon with horror — they'd got rid of all that, parasites, flies, the lot, and I just wondered whether this is really so.

Lainson — I am basing my information on the latest WHO meetings on leishmaniasis, at which we had the personal experience of scientists from China. A very interesting document was presented on the situation regarding visceral leishmaniasis in different parts of China; it was dated 1981, so I assume that it was up to date. It would seem that although the disease has been cut drastically, there was no doubt that cases of visceral leishmaniasis do still occur in some regions.

Bray — By chance last week I spent half a day with Professor Zhong who is the head of the Peking Institute of Tropical Medicine, and he told me that the old endemic of visceral leishmaniasis in the hills outside Peking is still prevalent, but new endemics, or at least new to them, are occurring in Inner Mongolia and right out in the Gobi Desert, which is bothering the Chinese army to some extent.

Kaplan — I have a comment on animals acting as sentinels for disease. You brought up the Chinese situation with respect to gullet cancer in chickens, and the very high incidence of oesophageal cancer in human beings. What they are looking for, as you said, is a common irritant in the feed of the chickens that might be shared by man. You may remember the aflatoxin incident, where groundnut meal was contaminated with aflatoxin from a fungus. This was first shown here in Britain in turkeys or chickens with liver tumours, and it is now associated in Africa with African hepatomas, one of the most common of all cancers in the African population, a very disturbing situation which is considered also to be caused by ingestion of aflatoxin, of which there are several types.

Clarke — Could I change the subject, and ask you a question about an occasional complication of inoculation against influenza, namely the Guillain-Barré syndrome? For those of you who are non-medical, the Guillain-Barré syndrome is a very unpleasant disease in which you get a creeping paralysis and you never know quite where it is going to stop, but it usually gets better. It has been in the news recently because Tony Benn had it — and he is fortunately now improving rapidly. Now I knew it was sometimes caused by inoculation against influenza, and I met Mrs Benn and I asked her whether he had had such an inoculation. She put me firmly in my place because she had read the subject up much more than I had, and she said "Oh, no, not at all; you should know it only occurs after inoculation against the swine form of the disease"; and I just wonder what the panel thinks about this?

Kaplan — Actually this was observed after the almost hysterical reaction in the United States that I mentioned before, when swine flu occurred in the soldiers of Fort Dix in 1976, and the authorities proceeded with a mass campaign of vaccinations with the swine strain. But the Guillain-Barré disease syndrome occurred, I think, in the ratio of about one to 300 000 people inoculated. What is known about the syndrome is that it may follow not only swine influenza vaccine, but also many types of vaccinal interventions. When you do vaccinal interventions on a mass scale you are going to run into trouble, and Guillain-Barré is one of them. Thus I would not say that it is only swine influenza vaccine that's involved, but other types of insults, immunological insults. We really don't know why this syndrome occurs.

Clarke — So I can tell Mrs Benn she is wrong, can I?

Zuckerman — Now that swine fever has been mentioned, may I ask whether the quality of lability that you have demonstrated in the flu virus also applies to any other virus which affects the animal world?

Kaplan — Yes. You will recall that I tried to explain the "drift" syndrome where a virus is mutating gradually, and finally you have to make a new type of vaccine to deal with an epidemic or pandemic. The foot-and-mouth disease virus is another example of a drifting mechanism where the antigenic change is very slight, but it keeps changing, and suddenly you have a strain that causes an epizootic. There are a number of viruses in which this occurs. But I do want to recall what I didn't have time perhaps to stress before. The flu virus has been found in whales, striped whales, in the Pacific; we have seen it in seals — 150 died out of 500 in Boston Harbour. Virus was isolated from the brain and the lung. I explained the tern outbreak, in which there was extensive mortality in terns in South Africa from an avian flu virus. These strains can interchange among species, and I think in wildlife we are going to find much more evidence of natural harm occurring in outbreaks of influenza, not to speak of its possible contribution to human influenza.

Mullaney — I would like to ask Dr Kaplan to comment on flu vaccines. A few years ago we were quite happily using a two-injection primary course of inactivated influenza vaccine in horses, followed by a booster a few months later and subsequently annual doses. Then horses didn't win races and everybody was worried about respiratory diseases. When sera were checked, very low titres of HI antibodies were found in many cases. A number of trainers then changed to six-monthly boosters, and later following American systems used 90 days, then 60 days, and now some few establishments are using 30 days injections, and still some horses show low humoral antibody levels especially to the A equi-2 component. If this latter antigen is treated with detergent and ether we get much higher HI titres, but we don't know how this relates to immunogenicity. Certainly, with more frequent boosters we seem to get better maintained plateau levels, but we don't know if the horses are any more resistant than they were before these procedures, since there have been no serious challenge outbreaks in the last year or so. We have not checked on neuraminidase antibodies because it's a bit more difficult for routine examinations.

Kaplan — Well, it's hard to answer that quickly. The equine flu vaccines have not been very satisfactory, it is true. I think it is a function of the antigen content of the vaccine, which must be improved. About the splitting of the haemagglutinen, I did not have time to mention in my talk that you must have cleavage, that is, enzymatic splitting of the haemagglutinen, before you get the antigenic and infective properties revealed. It makes no sense to me to be vaccinating at six-month intervals, three months, two months and one month. This is just asking for trouble. I think that one should really go to the vaccine manufacturers and give them a little prod in the back.

Mullaney — Do you think a split vaccine or a disrupted vaccine would tend to give better results?

Kaplan — I would think so, yes.

Ormerod — I would like to comment on the ancient history of two of the topics discussed today. First to draw Professor Nelson's attention to some pictorial evidence (Zeuner, 1963)[1] that the Egyptians in 2750—2500 B.C. kept striped

---

[1] Zeuner, F. E. (1963). *A history of domesticated animals*. London: Hutchinson.

hyaenas in captivity and used them as food animals. Secondly, there is another interpretation of the pre-Columbian pottery figures of mutilated faces, to which Dr Lainson has referred. Salaman (1949)[1] suggests that they represent the sacrificial victims of a potato cult who have had their noses removed to represent potato "eyes".

Clarke — I think that's a very good closing remark to this session, and I would like to thank the speakers very much indeed for their contributions.

---

[1] Salaman, R. N. (1949). *The history and social influence of the potato*. Cambridge: University Press. (Reprinted 1970.)

Major Factors in the Spread of Infections

C. E. GORDON SMITH

London School of Hygiene and Tropical Medicine, Keppel Street (Gower Street), London WC1, England

SYNOPSIS

Disease is an indicator of infection but, depending on virulence of the pathogen and on species or individual susceptibility, the infections are usually much more widespread than evident disease and their epidemiology cannot be understood without identifying the full distribution of infections. Certain types of disease contribute to the spread of their causative agents, others play no such role and some reduce the likelihood of transmission. Aspects of species, individual and genetically determined susceptibility to infections and of the infectiousness of various types of disease, are reviewed, as are the counterpart characteristics of the pathogens: infectivity, virulence and stability outside the host. The epidemiology of the spread of infections is then considered in terms of their modes of transmission: ingestion, skin and mucosal infections, arthropod transmission, biting between vertebrates, respiratory, congenital and neonatal infections. Factors in the longer range transport of infections are then considered — many of these attributable to man's activities.

INTRODUCTION

The title originally suggested for this paper was "the spread of diseases". It has been changed to "the spread of infections" because, in epidemiological terms, disease is no more than an indicator of infection. Some types of disease are of course important factors in transmission of their causative agents, but some are irrelevant and some actually diminish the likelihood of transmission. A cough or a desquamating skin infection can play an active role in promoting transmission from a diseased animal to an uninfected one; but, for example, encephalitis plays no role in transmission of its causative agents and often occurs after the animal has ceased to be infective. An infection spread by contact in a herd will cease to be transmitted by a sick animal which segregates itself from the herd, or is too sick to keep up with its movements. Many infections are asymptomatic or cause only mild disease even though the affected animals are infectious in some instances for long periods. These considerations are particularly important in relation to wildlife in which, in the absence of specific and sophisticated studies, disease is generally recognizable only if it is of relatively high incidence and has some easily observable sign — usually death: i.e. a fatal epizootic. Even in domesticated animals, the inability to elicit symptoms severely limits the recognition of mild illnesses.

Our knowledge of the infections of wildlife is very scanty and this can perhaps be illustrated by considering only one small sector of the range of pathogens: the large heterogeneous group of arboviruses (Smith, 1977). Of the 103 first isolated in Africa (Berge, 1967) two-thirds were isolated in three countries and all of them in six; these were isolated almost entirely by specialized units at Entebbe, Johannesburg, Cairo, Yaba, Ibadan, Dakar and Bangui. Of 112 which have been isolated in Africa 67 have been isolated from mosquitoes, 13 from ixodid ticks, nine from argasid ticks, two from phlebotomines (sandflies) and nine from *Culicoides* midges; 37 have been isolated from man, four from lower primates, 19 from rodents, 23 from birds, seven from bats and 15 from other vertebrate species. For the majority of these viruses (i.e. other than those associated with a few diseases of man or domestic livestock) little is known about the species they infect or the diseases (if any) they may cause. Many large areas and a multitude of species have not even been examined.

To understand the spread of infections it is necessary to consider susceptibility to infection and to disease in the hosts; and those characteristics of the infective agents which determine their mode of transmission.

CHARACTERISTICS OF HOSTS

Susceptibility

Although many infective agents can be adapted by laboratory experiment to infect other species such as laboratory mice, in nature susceptibility to infection with a particular organism may be limited to a single species or group of species. There is for example no evidence that rubella virus naturally infects any other species than man; and certain infections otherwise specific to man (e.g. viral hepatitis) seem to be capable only of infecting his closest relations, the higher apes. Only Equidae seem to be susceptible, in nature, to African horse sickness virus, and only animals with cloven hooves to foot and mouth disease virus. Only rabbits are susceptible to infection with myxomatosis virus: the disease was discovered as a lethal enzootic among laboratory (*Oryctolagus*) rabbits in Uruguay in 1896 but it was not until 1942 that it was realized that myxomatosis is a natural and relatively mild disease of *Sylvilagus* rabbits in the New World (Fenner & Ratcliffe, 1965). Several virus infections, notably Herpesvirus infections, show markedly enhanced virulence when transferred from their natural host to another species. At the other end of the spectrum of pathogens, the hosts of helminths (e.g. of the tapeworm *Taenia solium*, pig and man) are often not at all closely related. The precise reasons for species susceptibility or insusceptibility are far from clear but presumably depend on biochemical relationships between the host and the parasite. In the case of viral infections the presence or absence of receptors or inhibitors at the host cell surface or the presence or absence of a particular enzyme are probably critical. In bacterial infections growth factors are of similar importance: for example, Keppie *et al.* (1965) elegantly demonstrated the importance of the sugar, erythritol, in promoting the growth of the Brucellae in the foetal and placental tissues of cattle. This is a neglected area of research now technically susceptible to attack.

Within a host species, susceptibility to disease may vary quite widely in degree. The clearest demonstration of the natural selection of insusceptibility in vertebrates has been provided by Fenner & Ratcliffe (1965) in their studies of a rabbit (*Oryctolagus*) population in Australia exposed to enzootic, and repeated epizootics of, myxomatosis over seven years. Susceptibility to a standard dose of the virus by a standard route changed over this period so that, in experimental infections, the incidence of severe disease fell from 93 to 54% and the mortality from 90 to 25%.

Susceptibility to infection has been studied in several species of arthropod, most notably the genetic basis of susceptibility in the mosquito, *Aedes aegypti*, to infection with the filarial worms, *Wuchereria bancrofti*, *Brugia malayi* and *Brugia pahangi*. A single gene was found to control the development of these parasites in the mosquito (see W. W. Macdonald, 1967, who reviews genetic factors in the susceptibility of vectors to parasites). Gubler & Rosen (1976) compared 13 strains of *Aedes albopictus* of differing geographical origin for susceptibility to infection by feeding with dengue viruses. They found differences as large as 100-fold and that susceptibility was heritable. In similar experiments Tesh, Gubler & Rosen (1976) showed differences as large as 1000-fold in susceptibility to chikungunya virus between *A. albopictus* strains and, again, indications of genetic control of susceptibility.

In individual animals, there is increasing evidence of a genetic basis for susceptibility to disease (and perhaps to infection): people with blood group A seem to be more susceptible to influenza than those of group 0; HLA-linked specific immune response genes seem likely to be important predisposing factors in a variety of infectious diseases of man; in mice, single major genetic factors confer resistance to virus infections and single gene control of susceptibility to leishmaniasis has been described (Bradley, 1974; and see Bradley, 1980, for a review of genetic factors in susceptibility to protozoal infections). Susceptibility is also subject to a variety of acquired and environmental influences: in higher vertebrates it may be greatly increased by immunodepressive factors (cytotoxic drugs, irradiation, malnutrition, "stress" or certain associated viral or protozoal infections). It is, for example, well known that if sheep are transported from one pasturage to another during the viraemic phase of louping ill they are much more likely to develop encephalomyelitis. On the other hand susceptibility to both infection and disease may be reduced (even eliminated) by cell- or antibody-mediated immunity following previous infection by the same or an antigenically related organism, or vaccination. In lower animals such as arthropods, which have no comparable immunological defences, overcrowding and malnutrition may increase susceptibility, for instance of the cuticle to fungal invasion.

Chemical poisons (e.g. insecticides) may increase susceptibility to infection in insects, for example, healthy larvae of the codling moth, *Cydra pomonella*, and of the tent caterpillar, *Malacostoma neustria*, were insusceptible to infection with spores of *Beauvaria bassiana* unless first treated with a weak dose of DDT or HCH (Benz, 1971). Degree of susceptibility is also influenced by the route of infection. For instance, many more species of mosquitoes are susceptible to infection with certain arboviruses if the latter are injected directly into their tissues, than are susceptible if the virus is ingested in a blood-meal — the natural route (Smith, 1964a). Mammals, including man, are more susceptible to infection with quite a range of viruses and bacteria when they are inhaled in a small particle aerosol which impacts in the small bronchioles, than in a large partical aerosol which enters no farther than the upper respiratory tract. For example, Semliki Forest virus (an alphavirus) is lethal to rabbits by the respiratory route but not by a massive intravenous dose (Boulter, Zlotnik & Maber, 1971).

Infectiousness

Infectiousness defines both the duration of the period when the animal is infective (i.e. capable of making effective contact) and the relative amount of infection it is capable of transmitting (i.e. the relative dose which an animal in effective contact might receive). One form of a disease can be much more infectious than another, e.g. pulmonary compared with other forms of tuberculosis; lepromatous cases of leprosy compared with tuberculoid cases.

CHARACTERISTICS OF PATHOGENS

Infectivity, virulence and stability are the most important epidemiological characteristics of infective agents. Infectivity and virulence are the obverses of susceptibility and cannot be separated. Our knowledge of either side of this coin is largely limited by the availability of animals in which experimental infections can be studied. While this information is supplemented, sometimes anecdotally, by observations of naturally occurring infections and diseases, not all species are naturally exposed to all infections even if they occur simultaneously in the same geographical area — ecological and behavioural factors may isolate a particular species from infection with a particular pathogen. An extraordinary (if serendipitous) example from recent years was the discovery that armadillos are highly susceptible to infection with *Bacillus leprae*, previously thought to be non-infective for all species except man. Thus all statements about the range of species susceptible to a particular infection or of infectivity of a particular pathogen for a range of host species hangs on a rather limited range of evidence and often on inferences which may or may not be correct.

Infectivity

Infectivity can be defined in terms of the dose of an organism required for a high probability of establishing an infection in its host species by a particular route. Different strains of the same organism may differ in infectivity for a certain species and be even more widely dependent on the route of infection, e.g. in mice, between subcutaneous and intracerebral inoculation. Kramer & Scherer (1976) showed that two epizootic strains of Venezuelan encephalitis virus were transmitted more frequently by *Aedes taeniorhynchus* than two enzootic strains and that the level of viraemia necessary to infect this mosquito was slightly lower for epizootic than enzootic strains.

Virluence

Virulence describes the capacity of an organism (or strain of organism) to cause disease in its hosts, in terms of both frequency and severity. Virulence is subject to natural selection, readily manifest as adaptation in laboratory experiments. Adaptation is probably an expression of the heterogeneity of so-called "strains" (unless they have been rigorously cloned) and also of the mutability of populations of micro-organisms. Virulence tends to increase with serial passage in a single species and to be "buffered" by alternate passage in vertebrate and arthropod hosts. Taylor & Marshall (1975) showed that serial passage of Ross River virus (an alphavirus) in suckling mice rapidly enhanced its virulence; alternate passage in mice and *Aedes aegypti* gave no change in virulence for mice.

The best studied natural example of selection is in myxomatosis where interesting contrasts can be drawn between the effects of the differing ecological circumstances in Australia and in Europe (Fenner & Ratcliffe, 1965). In Australia the infection is transmitted by mosquitoes which probe the infected lesions; the virus persists on their probosces and, when they bite another rabbit, the virus is inoculated. In these circumstances, the longer a diseased rabbit lives (i.e. a rabbit with a less virulent infection), the more mosquitoes it will tend to infect. This has led to a progressive decrease in the virulence of the prevalent virus strains: at the outset all strains isolated were 100% lethal for rabbits; but 13 years later, only a quarter of strains caused death in more than 90% of rabbits inoculated. In Europe, the infection is mainly flea-transmitted and, after nine years, half the virus strains still killed more than 90% of inoculated rabbits. Fleas tend to stay with their individual host and only to seek another when it dies; additionally, a sick rabbit is less able to divest itself of fleas, for example, by scratching. The duration of disease thus has a smaller selective effect, the same number of fleas tending to become infected however long the host survives.

Stability

Stability of an organism — here defined as how long it can remain infective outside its hosts — is an important epidemiological characteristic. The greatest stability is exhibited by the spores of organisms such as the bacilli of anthrax or tetanus which may persistently infect soil for tens or even hundreds of years. At the other extreme there are very labile organisms such as many of those which are sexually transmitted, which hardly survive at all outside their hosts, and which depend for continued transmission on being transferred directly from a moist mucous surface of one to a similar surface of another.

EFFECTIVE CONTACT

The circumstance or circumstances in which, for a particular host—parasite relationship, there is a high probability of the infection being successfully transmitted from one host to another can conveniently be called effective contact. Its nature is dependent mainly on the routes by which infection leaves an infective host, the stability of the organisms and the routes by which a susceptible host can be infected. As regards vertebrates, organisms are released through one of their orifices (e.g. respiratory, enteric or urinary infections), from skin lesions (e.g. poxviruses) or mucous membranes (e.g. venereal infections), or in their blood or tissue fluids when these are released by injury or surgical procedure (e.g. viral hepatitis) or by arthropods which feed on their tissues (e.g. mites in scrub typhus) or blood (e.g. mosquitoes in malaria). Infection can of course also spread by predation, or from the dead by scavengers; many infections are also dangerous during autopsies. In numerous diseases of insects, the organisms are released only by dissolution of their cadavers.

Correspondingly, routes of infection include ingestion, inhalation, direct contact, or indirect contamination of mucous membranes or skin (particularly broken skin) and the bite of arthropods or other vertebrates (e.g. rabies). Effective contact can vary widely in space and time, ranging from immediate and direct physical contact (e.g. sexually transmitted infections), to contamination of a wound with tetanus-infected soil many years after the organism was excreted on it, or to a tick-borne infection (such as louping ill) in which the interval between a tick taking a blood-meal may be as long as a year or more.

The number of effective contacts an individual infective animal makes is dependent on a number of behavioural and population density factors but perhaps most on the duration of its infectivity. This varies widely between different types of infection as well as between individuals. Infectivity in some viral respiratory infections or in viraemic infections transmitted by arthropods may be measured in hours or days, whereas in some bacterial infections (e.g. tuberculosis, leptospirosis) infectivity may continue for the remainder of the life of the animal, and in helminth infections such as onchocerciasis infectivity may last for a decade or more. The duration of infectivity has of course profound implications for the design of disease control measures. Hosts which remain infective for long periods without overt evidence of disease are called carriers and prevention of transmission from them depends on their recognition by laboratory tests.

CLASSES OF HOST

It is helpful to distinguish certain epidemiological classes of host. These classifications are particularly important where more than one species harbours the infection, whether or not they exhibit disease. The one or more species on which the infecting organism depends for its long-term continued existence in nature are defined as its maintenance hosts — these may be the same species world-wide or the maintenance species may vary with the climatic, geographical and ecological features of each infected area. Thus, although monkeys are the vertebrate maintenance hosts of yellow fever both in Africa and in South America, the species involved are different; and while the infection causes no apparent disease in African monkeys, some New World monkeys die of it. In both continents, the arthropod maintenance hosts are mosquitoes; the species differ but are ecological counterparts, biting mainly in the forest canopy.

Animals of other species may be susceptible and become infected but if they do not contribute importantly to the maintenance of the infection in the area, they are described as incidental (or dead-end) hosts. Colorado tick fever (an unpleasant febrile illness which occurs in the Rocky Mountains area of North America) is transmitted by the tick *Dermacentor andersoni*, which is unlikely to be infected by blood containing less than a *threshold* concentration. The vertebrate hosts are the golden mantled and Columbian ground squirrels. In the latter the viraemia level never exceeds the tick threshold while in the former the viraemia exceeds the threshold for several days (Burgdorfer, 1960). Thus the golden mantled squirrel is a maintenance host and the Columbian squirrel an incidental one.

However, in some circumstances, incidental hosts can play different but very important roles in the epidemiology of an infection. The first known outbreak of Kyasanur Forest disease (a serious systematic virus disease of man and monkeys) in Mysore, India (Work, 1958), appears to have arisen because of a substantial increase in cattle grazing in the forest which, in turn, built up a large increase in the population of the tick species (the more mature stages of which feed mainly on large mammals), which were responsible for infecting man and monkeys. Although cattle become infected with this virus they do not have a viraemia sufficient to infect ticks and are therefore incidental hosts of the infection.

Zoonoses are infections which are maintained in a species other than man but which infect man and cause disease — brucellosis (usually maintained in cattle or goats) and leptospirosis (often maintained in rodents) are examples. By definition, man is always an incidental host of a zoonosis. He may, however, be a temporary maintenance host of an infection which is maintained in the longer term as a zoonosis: e.g. urban yellow fever or pneumonic plague.

In some zoonoses, the maintenance hosts are wild species but infection is also frequent in another species which lives in closer juxtaposition to man. Such a species is called an amplifier host because its presence increases the risk of infection to man. Good examples are provided by the mosquito-borne virus diseases, Japanese and St Louis encephalitis, in which pigs and house-sparrows respectively are amplifier hosts in certain areas. In Japan, Japanese encephalitis virus is maintained throughout the spring and summer by the mosquito, *Culex tritaeniorhynchus* and nestling herons, mainly black-crowned night herons (*Nycticorax*). As the mosquito population increases seasonally and as the intensity of infection rises, the mosquitoes infect pigs. These live closer to man and transmission among them is by *Culex tritaeniorhynchus* breeding in ricefields (also close to man) (Scherer & Buescher, 1959). Infections in pigs tend to occur about two weeks earlier in the season than infections in man (Takahashi *et al.*, 1966).

SPREAD OF INFECTIONS

This must be considered firstly within communities or localities and secondly on a wider geographical basis. Transmission within communities or localized areas can best be considered in terms of routes of infection.

Ingestion

Effective contact depends on ingestion of infected water or food (including other animals) or by oral contact with infected tissue or skin. Among wild species the risk of water-borne infections is probably highest at drinking places, especially at water holes in arid areas where animals and their excreta are most concentrated. Animals licking each other, for example the peri-anal skin, are likely to acquire intestinal helminth and other such infections. In certain infections (e.g. brucellosis, bovine tuberculosis) the pathogen is secreted in the milk. Scavengers or predators can become infected by eating an infected carcass or animal, e.g. anthrax or trichinosis. Predation can also disseminate infective organisms without actual infection of the predator: bird and rodent predators in virus-infected sawfly larvae and pupae (*Gilipinia* spp.) excrete large amounts of virus in their faeces over a considerable area (Entwistle, 1973).

Skin and Mucous Membranes

Infections maintained in rodents and shed in their urine or faeces are generally transmitted to man and other animals by contact between skin and/or mucous membranes and infected water; or the oral mucous membrane may be infected by ingestion of infected water or food. Direct urinary contamination of skin also occurs and may be particularly important in transmission in the nest or burrow between an animal and its progeny. Leptospirosis, a good example, does not normally cause detectable disease in its rodent maintenance hosts, whose kidneys remain infected, shedding leptospires into the urine for prolonged periods, probably often for life. Rodents are characteristically territorial animals each living mainly within its "home range": when a population is sparse, home ranges overlap little, effective contact between animals (through water or wet soil) is relatively infrequent, and the infection rate is low — for example, in primary tropical forest. When populations are dense — for example in secondary forest or more so in food crops such as sugar or rice — the home ranges greatly overlap and infection rates are high — as much as 50% of a ricefield rat population (*Rattus argentiventer*) may be excreting leptospires.

In a mixed population of rat species, the incidence of infection in any species tends to be proportional to the proportion that species makes of the total population. In forest, it was estimated that the minimum population density for the maintenance of leptospirosis was about two rats per hectare. Species behaviourally associated with water tend to be more heavily infected. A study in Malaya showed that, in the same area, 32% of *R. norvegicus* (living on river banks, in drains and sewers) were infected compared with only 2% of *R. diardi* (living in roofs and walls). In forest, ground living rat species, particularly those living in the bottoms of forested valleys (*R. bowersi, R. muelleri*), were more heavily infected than species which live scrambling on vegetation or are arboreal (Smith *et al.*, 1961a, b, c).

A further major factor is the acidity or alkalinity of the soil and surface waters — leptospires survive for days in neutral or alkaline conditions but die very quickly if the pH is at all acid (Smith & Turner, 1961): thus in Sumatra leptospirosis is common in ricefield workers only where the water is neutral or alkaline (Sardjito & Zuelzer, 1929). The nature of the soil also probably plays a part — certain clays (e.g. Montmorillonite clays) in fine suspension appear to adhere to and inactivate leptospires (Smith & Turner, 1961). Salinity is also lethal to leptospires and accounted for the rapid decline in the incidence of leptospirosis in Amsterdam among persons falling into the canals in the blackout during World War II, when German action admitted sea water to the canals (Wolff & Ruys, 1953).

Leptospirosis is an economically important disease of domestic animals, particularly cattle and pigs, in many parts of the world, and it causes "yellows" in dogs, which can infect their owners. Recreational factors are well illustrated by the history of a river pool suitable for bathing near Kuala Lumpur in Malaya. It was a popular picnic site at which food tended to be left behind causing a local increase in rodents. In due course a small outbreak of leptospirosis occurred among bathers; use of the pool was then forbidden and the rodent population declined; after an interval, bathing (and picnics) started again without adverse incident, but after a period a further outbreak of leptospirosis occurred. Leptospirosis was the commonest disease among British troops operating in Malaya during the late 1950s — they were shown to acquire this in the jungle, not in standing camps, but particularly in forest valleys and most of all in temporary camps in forest clearings where aborigines had grown food and where rodent populations were dense (McCrumb *et al.*, 1957).

The arenaviruses are similarly maintained in rodents and excreted in their urine for long periods.[*] When viruses in this group are transmitted to immunologically immature mice a tolerant state with prolonged infection and infectivity but little or no disease is induced. Such infections probably occur in the nest from mother to progeny by urinary contamination. Lymphocytic choriomeningitis virus is widely enzootic in *Mus musculus*. Argentinian haemorrhagic fever is a severe and important occupational disease of agricultural workers with much the same epidemiology as leptospirosis; it is maintained by rodents. A spectacular outbreak of the related Bolivian haemorrhagic fever occurred in a small town in Bolivia with a population of about 2500 in which there were 470 cases and 142 deaths

---

[*] Lassa fever, causing increasing concern among travellers from parts of west Africa, is caused by an arenavirus and is readily transmitted by blood or urine to doctors and nurses attending cases.

(Mackenzie *et al.*, 1964). A rodent, *Calomys callosus*, which in this area occupies the niche elsewhere filled by the house mouse, *M. musculus*, had built up a very high population, contaminating food, water, and perhaps skin, with infected urine and faeces. This rodent, once infected, excretes the virus in its urine for several months and probably often for life. The fact that the cats in the town had recently died of DDT poisoning, as a result of spraying for malaria control, may have been a contributory factor.

A particularly complex set of events is required to maintain the life-cycle of certain pathogenic worm infections which have obligatory developmental stages in both their vertebrate and invertebrate hosts. These infections differ from all others in that the infecting organisms are sexually dimorphic and the stages which infect the vertebrate reproduce in it sexually.

Schistosomiasis (an important disease of cattle and sheep in some parts of Africa) is a good example. The schistosome eggs are shed (depending on species) in either faeces or urine. In water, they hatch miracidia which infect snails and multiply. Cercaria are released and invade through the skin (or, if infected water is drunk, through the upper gastrointestinal tract), pass through the lymph- and bloodstreams to the lungs, and thence to the general circulation where larvae can be found in tissues in all parts of the body. The larvae mature and mate; worms then migrate to the colon and rectum, or to the bladder, depending on species, and lay eggs which are shed.

The larval schistosome which invades skin is already sexually determined. One such larva, if it survives the considerable hazards of growing up to maturity in its host, can give rise to only one adult worm, either male or female; the continuation of the transmission chain depends upon the adult worm finding and mating with a member of the opposite sex. Clearly, these spatial complications require that a fairly difficult set of problems in probability mathematics must be solved if full epidemiological understanding of these infections is to result. The first efforts to solve the mathematical difficulties involved were made by Hairston (1962, 1965a, b) and G. Macdonald (1965) who produced interestingly contrasting approaches to an analysis of the problem in schistosomiasis. Later, Hairston & Jackowski (1968) and Hairston & de Meillon (1968) analysed the problem in Bancroftian filariasis of which the invertebrate host is a mosquito, and in which the spatial complications of mating are perhaps more complex.

Filarial infections are widespread in animals in Africa and other parts of the tropics.

Arthropod Transmission

A wide range of infections are transmitted by arthropods, not always by bite, sometimes by causing irritation and inoculation of the arthropod's infected excreta by scratching.

Various virus infections are transmitted by mosquitoes, sandflies, midges, soft or hard ticks; rickettsial infections (e.g. typhus) by lice, ticks or mites; bacterial infections by fleas (e.g. plague) or soft ticks (e.g. African swine fever); protozoal infections by mosquitoes, tsetse flies, sandflies, ticks or bugs; helminths by mosquitoes or simuliid flies. Epidemiologically the main contrasts are between infections transmitted by free-flying Diptera (mosquitoes, sandflies, midges, etc.), those transmitted by acarines (ticks, mites) and those transmitted by such ectoparasites as lice and fleas. It is also important to distinguish between those cases where the arthropod transmits merely mechanically, and those where it becomes infected and is an obligate (or at least an ecologically obligate) host of the infection.

In mechanical transmission, the mouthparts of the biting arthropod become infected on biting an infective host, and inoculate infection by biting a subsequent host. Relatively stable organisms such as pox-viruses are most readily transmitted mechanically by arthropods: myxomatosis is a very good example. Less stable agents (e.g. dengue viruses) may be transmitted mechanically where the interval between bites is very short, for example, in interrupted feeding by mosquitoes which, when disturbed on one host, may immediately complete their feed on another.

In biological transmission, a process of infection in the arthropod intervenes: the simplest case is an arbovirus (e.g. yellow fever) in a mosquito: virus is ingested in a blood-meal, invades the tissues from the gut, multiplies, spreads and infects the salivary glands and hence the saliva which is then injected when a later blood-meal is taken. In protozoal and helminth infections, there is an essential phase of development of the parasite in the arthropod. For example, in malaria, gametocytes in blood release gametes which mate in the mid gut of the mosquito; the resultant zygotes invade the tissues and multiply within an oocyst; this bursts to release sporozooites which migrate to the salivary gland and are injected.

In simplistic terms (Smith, 1975), the intensity of transmission of a mosquito-borne infection among its maintenance hosts varies directly with:

(a) the number of mosquito hosts feeding on each vertebrate host;
(b) the square of the proportion (h) of mosquito hosts feeding on maintenance hosts;

(c) the blood-feeding frequency of the mosquito hosts (b, the interval in days);

(d) the effective duration of viraemia in the vertebrate hosts;

(e) the proportion of vertebrate hosts susceptible to infection.

It varies as $p^i/-\log_e p$ in terms of mosquito longevity (p, the proportion surviving each 24 h) and the extrinsic incubation period (i, the interval between ingestion of the infection and the arthropod becoming infective) which depends on the temperature of its environment.

Thus, with yellow fever in *Haemagogus* mosquitoes, Bates & Roca-Garcia (1946) found i to be 28 days at 25°C and 10 days at 30°C. As mosquitoes are relatively shortlived this sort of difference has a profound effect on the number of transmissions achieved.

Other factors being equal, a change from 28 days to 10 days would represent about a 50-fold increase in transmission rate for $p = 0.8$ and about 70-fold for $p = 0.9$.

These factors (p, i) together with the biting habits (h, b) are particularly powerful in influencing transmission rates (Smith, 1975); all of them, except h, are clearly temperature-dependent and therefore vary with season. At least with some species in some areas, h also varies seasonally.

In almost any situation our present knowledge of the numerical values in this process is fragmentary and probably often misleading. Few serious attempts have ever been made to assemble information on all the factors. Reeves & Hammon (1962) after a long and detailed study of St Louis (SLE) and western encephalitis (WE) in the southern valley of California found that WE virus was detected in or following two-week periods in which mean daily temperatures exceeded 80°F: but SLE virus, with one exception, only when mean daily temperatures exceeded 85°F at least once and usually over a number of days or weeks. Reeves (1970) found that SLE transmission continued while *Culex tarsalis* light traps indices (LTI) remained averaging over 10 females/night; but WE transmission continued until the average LTI fell to one. In early summer WE virus was first detected when the average LTI reached four, SLE virus not until the average LTI had been as high as 16 for at least two weeks. Human and equine cases of western encephalitis occurred only at LTIs of five or more.

Dow, Reeves & Bellamy (1957) suggested that excessively high *C. tarsalis* populations may maintain transmission relatively inefficiently because of their deflection at feeding by their hosts, especially birds, and Tempelis *et al.* (1965) demonstrated an increased proportion of non-avian blood-meals in *C. tarsalis* when populations were unusually dense. Reeves (1971) suggested that the seasonal shifts in feeding patterns of *C. tarsalis* and *C. nigripalpus* (Edman & Taylor, 1968) might result from (1) increased mosquito population causing increased interference and disturbance during feeding on their hosts and (2) increased biting activity on the predominant (avian) hosts causing increased host intolerance and diversion to other species. Such a shift in feeding habit from maintenance to incidental hosts of a virus infection may be particularly important in enhancing the risk to man and his livestock.

Of the numerical factors involved, mosquito populations are perhaps the easiest to monitor, but without sufficient knowledge of their biting habits, may be difficult to interpret usefully.

Maintenance depends on an infective mosquito biting a host capable of infecting further mosquitoes. Few if any mosquito species feed solely on one vertebrate species and as a mosquito must bite a susceptible host of a maintenance species at least twice, the transmission rate is, in this respect, determined by the square of the probability (proportion) of the mosquito species biting such species. For example, in some areas, the yellow fever mosquito *Aedes aegypti* mainly bites primates; however, McClelland & Weitz (1963) found an area of east Africa where only about a quarter had bitten primates and the probability of a mosquito in such a population twice biting a primate would be one-thirteenth that if, say 90% were biting primates ($0.25^2 = 0.06$ compared $0.9^2 = 0.81$). This and other behavioural factors in both arthropod and vertebrate hosts determine which species (often of a number which are susceptible to infection) are the maintenance hosts in any one local ecosystem (Smith, 1964b).

Blood-feeding patterns vary widely between mosquito species — some feeding predominantly on only a few vertebrate species, others very catholic in their tastes — and also geographically and seasonally (see Smith, 1975). What is not at all clear is whether, within a mosquito species in a particular area, there are sub-populations which consistently bite particular vertebrate species, or have a higher probability of doing so, or whether all individuals within a population of a species have the same probability of biting a particular vertebrate species, the event being controlled wholly or mainly by availability of the vertebrate species concerned. If sub-populations could be identified this would constitute an important advance in the interpretation of transmission rates, and it could improve control measures by directing attention to the relevant sub-population. It is a particularly difficult problem for research — however Rutledge *et al.* (1975) have presented evidence of a heritable biting habit in *A. aegypti*.

Information on blood-feeding patterns and on changes in them cannot be extrapolated from one place to another and must be locally measured and monitored — fortunately, techniques have been much improved in recent years (Boreham, 1975; Tempelis, 1975) and this whole area could be greatly clarified by well-designed experiments directed at interpreting important arbovirus disease situations.

Longevity in mosquitoes appears to be influenced by temperature and humidity: broadly, within moderate ranges, it is proportional to relative humidity and inversely to temperature (see Clements, 1963; Lewis, 1966). Longevity is, however, probably predominantly determined by the microclimatic characteristics of their resting places — most species rest for a high proportion of their life wherever the environment provides a high humidity and lower temperature (in vegetation, under culverts, etc.). However, such factors as availability of blood and nectar feeds are also important. Longevity deserves more detailed study, particularly of important species in the field, not only because of its importance in modelling transmission rates, but also because of its important effects on the rates of mosquito population changes. In studies of *A. aegypti*, Crovello & Hacker (1972) found for instance that the forest (*formosus*) form had a lower reproductive potential than the urban form — presumably because adaptation to fluctuating urban climatic conditions had selected for a higher reproductive potential than the stable conditions of the forest. Similarly, Walter & Hacker (1974) found that a Bangkok strain of *Culex pipiens quinquefasciatus* had a lower reproductive potential than American strains and attributed the difference to the more stable climate of Thailand. On the other hand, Lansdowne & Hacker (1975) failed to show any differences in adult life-table characteristics between five strains of *A. aegypti* kept experimentally at fluctuating or constant temperatures and relative humidities.

Estimates of longevity have been few: e.g. p 0.91—0.97 *Anopheles gambiae* (Davidson, 1954), three *Anopheles* spp. in New Guinea p 0.69, 0.68, 0.95, one species varying from 0.91—0.69 during a year (van den Assem, 1959); *A. aegypti* Bangkok p 0.88 (Sheppard *et al.*, 1969). Despite the wide variances of these estimates, p probably varies widely and seasonally: to illustrate the meaning of these estimates, the implications are p 0.67: average survival 2.5 days; p 0.97: average survival 33 days. With more carefully measured data on temperature effects on i and p (for both of which surprisingly few data exist) for important mosquito species (using wild-caught mosquitoes and recently isolated, minimally passaged virus strains) much better predictions of seasonal changes in populations and infection rates would become possible.

These factors operate differently in acarine-transmitted infections, particularly when the vertebrate maintenance host is a rodent or other territorial animal. The home ranges of such a rodent population may be wholly or partially suitable for survival of acarines which drop off their hosts, e.g. the soil may be wet enough only in certain parts, or only part of the vegetation may provide a suitable micro-climate. When the vertebrate population spends more time outside than inside the habitat favourable for their survival, the tick or mite population of the habitat is depleted, and the probability of trans-mission reduced. If the host ranges are small relative to the area of suitable habitat, depletion is unlikely; if the ranges are relatively large, depletion is probable (Smith, 1964b).

An additional factor in viral, rickettsial and perhaps some other infections maintained by arthropod transmission, is transovarial transmission (see below). In cold climates (e.g. Canada, Finland) transovarial transmission of arboviruses in mosquitoes probably plays an important role in enabling the infection to over-winter where mosquito populations bite actively during a relatively short season — it has been shown to be a common phenomenon in California group arboviruses transmitted in these regions. Transovarial transmission has now been shown to occur at least experimentally in infections maintained in warm climates where its epidemiological role, if any, is less clear. Other possible mechanisms for the overwintering of Diptera-transmitted infection include hibernating insects, and hiber-nating vertebrates in which the infection process is suspended at lowered body temperature and recommences when hibernation ceases.

For a systemic infection which is capable of causing lasting immunity in its vertebrate host(s) to persist in a restricted area, there must be continuous replenishment of the population of susceptible vertebrates at a rate compatible with the frequency of successful transmission, i.e. with depletion of the susceptibles by resulting death, or, more often, immunity. Replenishment depends on vertebrate reproduction unless new susceptible hosts migrate into the area from uninfected or less infected areas. Because ticks and mites have very restricted powers of movement (except attached to their hosts) and feed only at relatively long intervals, and because their hosts are often small mammals or birds with relatively rapid repro-duction rates, tick- and mite-borne infections occur characteristically in fixed foci in areas of vegetation and climate mutually suitable for the tick and its vertebrate hosts. These tend to remain infected indefinitely so long as no major ecological change occurs. For instance, the Malayan tick-borne Langat virus (Smith, 1956) was isolated each year for seven years from rodent ticks in an area of forest only a few yards across. In contrast, mosquitoes have a short life and bite frequently so that, in infections in which lasting immunity follows (e.g. virus infections), they rapidly exhaust the susceptibles in a maintenance population. Such infections must, therefore, either migrate to areas which have not recently been infected or die out. Yellow fever, for instance, persists permanently in the Congo and Amazon Basins where, despite the slow reproduction rates of their primate maintenance hosts, the areas of suitable habitat are so large that there is always room for migration of the infection, as a wave through the primate populations. On the other hand, when sylvan yellow fever crossed the Panama Canal in 1949 and spread through the narrow forest on the Atlantic side of the mountains it proceeded at a fairly steady rate (fluctuating with rainfall) until it reached the Mexican border in 1956 and petered out. Because of the narrowness of the habitat, the infection was unable to turn back and could migrate in one direction only (Smith, 1971).

Population-waves in small mammals such as voles, which may occur at fairly regular intervals, can greatly increase the risk to man and domestic animals of infections they carry and those transmitted by their ectoparasites such as ticks. Outbreaks of central European tick-borne encephalitis have followed "mouse-years" (Havlik, 1954). In Russia, in 1958–1959, an outbreak of about 1000 cases of haemorrhagic fever with renal syndrome — a disease probably identical with Korean haemorrhagic fever — was preceded by an unprecedented peak in bank vole (*Clethrionomys glareolus*) populations (one at least of its maintenance hosts) due apparently to an unusually large seed crop from trees in 1957 (Myasnikov, Levacheva & Yegiazaryan, 1961).

Infections transmitted by soft ticks, fleas and lice are probably mainly transmitted among animals within burrows or nests. The remarkable maintenance cycle of African swine fever in east Africa uncovered by Plowright, Perry & Greig (1974) may represent the ultimate (or perhaps the original) maintenance cycle for an arbovirus. The African swine fever virus is moreover the vertebrate virus most closely related to an insect virus — it is an iridovirus. Although the *Ornithodorus* ticks infect warthogs in their burrows, the warthogs do not have a sufficient level of viraemia to infect the ticks and the infection appears to be maintained by transmission from tick to tick, probably sexually.

Biting between Vertebrates

Some infections are transmitted by bite from one vertebrate to another, by far the most important being rabies, an almost universally fatal disease of the nervous system with an unusually long incubation period. The virus infection spreads from the infecting bite to the peripheral nerves up which it travels to the central nervous system, causing either the well-known and much feared "furious" disease or a less dangerous progressive paralysis. Meanwhile the virus has spread, again along nerves, to infect the salivary glands from which it is secreted in the saliva. Effective contact is obviously enhanced in the furious form of the disease.

Other infections transmitted by vertebrate bite include *Herpesvirus simiae* which causes ulcers on the skin or oral mucous membranes of rhesus monkeys (Keeble, Christofinis & Wood, 1958) but encephalitis, usually fatal, in persons bitten by infected monkeys. There are two kinds of rat-bite fever in man, one caused by *Streptobacillus moniliformis*, the other by *Spirillum minus*. There are also several virus infections of bats (such as Rio Bravo virus) which are probably maintained at least partially by biting in bat colonies.

Respiratory

The spread of respiratory infections is enhanced by crowding, especially indoors and presumably in burrows and lairs. Effective contact depends on the inhalation of infective particles (each of which may contain a few or many organisms). An infective aerosol, generated by coughing, sneezing or barking, lowing, and so on, very rapidly dries down into particles, the sizes of which depend on the content of solids in the suspending fluid (mucus, saliva, etc.). If the particle diameter is over about 5 μm it is likely to impact in the upper respiratory tract (nose and throat); if smaller it may penetrate to the bronchioles. Not all particles infective by inhalation originate, however, from the respiratory tract: poxvirus infections may be acquired by inhalation of material shed from skin lesion; Q fever — a widespread rickettsial disease — is caused by a very stable organism and is usually acquired by inhalation of infected particles aerosolized from the urine, faeces, and especially the placentas of infected sheep, goats or cattle which are often healthy in appearance. Infections can occur several miles from their aerosol source (Welsh *et al.*, 1958). Stability of the infective organism in the aerosol influences the nature of effective contact: very stable organisms like those of poxvirus infections, foot-and-mouth disease and Q fever can be transmitted over considerable periods of time or over some distance (e.g.

in dust, fodder, etc.); many organisms are only moderately stable and relatively close contact within an enclosed space is necessary for a high probability of transmission. Relative humidity has a complex effect on stability in the aerosol: some organisms survive better in aerosols at a low and some at a high relative humidity.

Most dramatic of the examples of long-distance aerosol spread was the airborne spread in the epidemic of foot-and-mouth disease in Britain in 1968; and analysis of some previous outbreaks in Britain suggests strongly that infection has spread on the wind from Holland across the North Sea (Hugh-Jones, 1973; Sellers *et al.*, 1973).

Many infections not naturally transmitted by the respiratory route have been shown to be infective in this way experimentally, and some infections, transmitted normally in other ways, are sometimes transmitted by the respiratory route under special circumstances. For example, in certain bat-caves in the south-west USA and in Mexico where the bats have a high prevalence of rabies, animals (including man) exposed to the aerosol of their excreta in the caves become infected and suffer from rabies (Constantine, 1962).

Congenital and Neonatal

These mechanisms are most important in epidemiological terms where they play a role in the maintenance of infections. For instance, in arthropods, transovarial infection occurs in certain arbovirus infections; for example, tick-borne encephalitis in its *Ixodes* tick hosts or vesicular stomatitis virus in phlebotomine sandflies (Tesh, Chaniotis & Johnson, 1972). In mosquitoes, transovarial infection appears to be important in the over-wintering of California group arboviruses (Watts & Eldridge, 1975).

In most, if not all, of these cases, transovarial transmission is probably only supplementary to the main maintenance cycle which is based on the alternation of vertebrate and arthropod hosts. However, in infections maintained by arthropods which each feed only on one vertebrate host (for instance, mites transmitting scrub typhus — Audy, 1968) transovarial infection is a *sine qua non*. In another mechanism, common among the insect nuclear polyhedrosis viruses (NPV), the surface of the egg is contaminated with virus and infects the hatching larva (for instance the NPVs of *Gilpinia* sawflies).

There are also some vertebrate-maintained infections in which congenital infection may play a significant role in maintenance, notably the rodent-maintained arenaviruses (e.g. lymphocytic choriomeningitis and the South American haemorrhagic fever viruses) where animals infected *in utero* (or, probably much more important, neonatally by the mother) become chronically infected and excrete infective urine for a large proportion (if not the whole) of their lives without notable disease.

Congenital and neonatal infections are important causes of disease, but their epidemiology is largely that of the spread of infection to, and the duration and timing of infection in, the mother. Congenital (i.e. *in utero*) infections usually derive from blood-borne infection in the mother and the type and severity of disease in the foetus depends largely on its stage of development when infected (e.g. Richards & Cordy, 1967). A substantial number of arthropod-borne viruses have been associated with the occurrence of congenital defects in domestic animals and man (Parsonson, Della Porta & Snowdon, 1981).

Longer Range Movement of Infections

The commonest factor in larger scale geographical movements of infections is probably the movement of infected vertebrates — one has to think only of the periodic pandemics of influenza in man, accelerated by modern air transport. Another example is provided by the appearance of an epizootic and epidemic of Rift Valley fever in Egypt in 1977–1978, causing abortions and mortality in sheep, cattle, buffaloes and camels, and 20 000 to 100 000 cases of illness in man with somewhere between 60 and 600 deaths (Meegan, 1979; Laughlin *et al.*, 1979). The infection was not known north of Kenya until an outbreak in the Sudan in 1975. The infection was probably introduced to Egypt by camels of which 50 000–100 000 enter from the Sudan annually (Hoogstraal, Meegan & Khalil, 1979).

Migrating birds may play a role in the movement of certain infections which they maintain (Hoogstraal, 1973), but are probably more important in transporting ectoparasites which may be infected. The distribution of the Congo–Crimean haemorrhagic fever suggests transport of infected ticks between the Crimean region and Egypt and other parts of the Middle East (Hoogstraal, 1979).

Migrating deer, antelopes and the like transfer ectoparasites (particularly ixodid ticks) from one area to another and probably infections. Bluetongue virus which is transmitted by *Culicoides* midges was until recent years thought to infect only sheep, cattle and goats. In recent years it has been clear that a much wider range of species become infected: deer, big-horn sheep, and most species of African antelope. In African game species the infection is usually clinically inapparent (Verwoerd, Huismans & Erasmus, 1979). The risks of transporting infections and ectoparasites by domestic livestock are well known and most countries have regulations designed to minimize the risk.

That such regulations cannot prevent the introduction of wind-borne foot-and-mouth disease virus was realized after demonstration of air-borne spread as the main transmission mechanism of the British epizootic in 1968 (Hugh-Jones, 1973; Sellers *et al.*, 1973). Sellers *et al.* (1973) and Sellers (1980) extended their studies and concluded that a wide range of infected insect populations (mosquitoes, midges) are transported on winds (particularly those associated with movements of the Intertropical Convergence Zone) and that this is an important mechanism for the introduction of infections into new areas — especially the midge transmitted viruses: African horse sickness, bluetongue, Ibaraki disease and bovine ephemeral fever. African horse sickness seems normally to be enzootic in sub-Saharan Africa but epizootics arise elsewhere from time to time: Egypt in 1928, 1943—1944, 1953, 1971; Palestine, 1944; Cape Verde Islands, 1944; north Africa and southern Spain, 1965—1966; and most widespread of all, the Middle East and the Indian sub-continent 1959—1960. Sellers (1980) found evidence for the wind carriage of infected midges to account for spread from Morocco to Spain (1966), Turkey to Cyprus (1960), Senegal to the Cape Verde Islands (1943), from the Sahara to north Africa (1965) through the Middle East and from Pakistan to India (1960).

The control of onchocerciasis in the Volta River basin by a large-scale programme of larviciding to kill the *Simulium* flies which transmit the infection is impaired on its western boundaries by wind-borne invasion of infected flies from neighbouring endemic areas (Garms, Walsh & Davies, 1979).

CONCLUSIONS

This chapter has necessarily skated somewhat lightly over a large and complex subject area apparently rich with information and ideas but, on closer examination, riddled with ignorance — particularly about infections of animals other than man and his domesticated species.

The rate of spread of infection in any community of animals depends on the average number of effective contacts each infective animal makes with susceptible individuals. If this average is less than one the infection will tend to die out over time; if it is more than one and this level is sustained an epizootic will result. An enzootic is a situation where the average remains close to one. The frequency of effective contacts depends largely on population density (which may be general or localized, e.g. around waterholes) of the relevant species, vertebrate or arthropod, and this is influenced not only by many natural and behavioural factors, only some of which are understood, but also by man's activities. Irrigation schemes in arid areas may engender dense populations of snails or mosquitoes, and, if food crops are grown, dense populations of birds and rodents — with obvious implications for the infections they can carry and transmit. Deforestation can lead to simplification of the species structure of, for example, rodents; and if the forest is replaced say by oil palms then there is likely to be a dense population of virtually a single rodent species. Transportation can move infected animals into uninfected areas, or arthropods to colonize new areas. Species introduced for biological control purposes can be the source of new or enhanced infection risks, e.g. mongoose rabies in Grenada.

REFERENCES

Audy, J. R. (1968). *Red mites and typhus*. London: Athlone Press.

Bates, M. & Roca-Garcia, M. (1946). The development of the virus of yellow fever in *Haemagogus* mosquitoes. *Am. J. trop. Med.* 26: 585—612.

Benz, G. (1971). Synergism of micro-organisms and chemical insecticides. In *Microbial control of insects and mites*: 327—355. Burges, H. D. & Hassey, N. W. (Eds). New York and London: Academic Press.

Berge, T. O. (Ed.) (1967). *Catalogue of arthropod-borne viruses of the world*. U.S. Dept of Health, Education and Welfare. Pub. No. 1760; and supplements to 1975.

Boreham, P. F. L. (1975). Some applications of bloodmeal identifications in relation to the epidemiology of vector-borne tropical diseases. *J. trop. Med. Hyg.* 78: 83—91.

Boulter, E. A., Zlotnik, I. & Maber, H. B. (1971). A lethal respiratory infection of rabbits with a strain of Semliki Forest virus (SFV). *Br. J. exp. Path.* 52: 638—645.

Bradley, D. J. (1974). Genetic control of natural resistance to *Leishmania donovani*. *Nature, Lond.* 250: 353.

Bradley, D. J. (1980). Genetic control of resistance to protozoal infections. In *Genetic control of natural resistance to infection and malignancy*: 9—28. Skamene, E., Kongshaun, P. A. L. & Landy, M. (Eds). New York and London: Academic Press.

Burgdorfer, W. (1960). Colorado tick fever: II. The behaviour of Colorado tick fever virus in rodents. *J. Infect. Dis.* 107: 384—388.

Clements, A. N. (1963). *The physiology of mosquitoes*. Oxford: Pergamon Press.

Constantine, D. G. (1962). Rabies transmission by nonbite route. *Publ. Hlth Rep. Wash.* 77: 287—289.

Crovello, T. J. & Hacker, C. S. (1972). Evolutionary strategies in life table characteristics among feral and urban strains of *Aedes aegypti* (L). *Evolution* 26: 185—196.

Davidson, G. (1954). Estimation of the survival-rate of anopheline mosquitoes in nature. *Nature, Lond.* 174: 792—793.

Dow, D. P., Reeves, W. C. & Bellamy, R. E. (1957). Field tests of avian host preference of *Culex tarsalis* Coq. *Am. J. trop. Med. Hyg.* **6**: 294—303.

Edman, J. D. & Taylor, D. J. (1968). *Culex nigripalpus*: seasonal shift in the bird-mammal feeding ratio in a mosquito vector of human encephalitis. *Science, N.Y.* **161**: 67—68.

Entwistle, P. F. (1973). The primary epizootiology of a nuclear polyhedrosis virus disease in populations of spruce sawfly, *Gilpinia hercyniae*. *Abst. 5th International Colloqium on Insect Pathology and Microbial Control, Oxford*. Paper 64.

Fenner, F. & Ratcliffe, F. N. (1965). *Myxomatosis*. Cambridge: Cambridge University Press.

Garms, R., Walsh, J. F. & Davies, J. B. (1979). Studies on the re-invasion of the Onchocerciasis Control Programme in the Volta River Basin by *Simulium damnosum* s.l, with emphasis on the South-western areas. *Tropenmed. Parasit.* **30**: 345—362.

Gubler, D. J. & Rosen, L. (1976). Variation among geographic strains of *Aedes albopictus* in susceptibility to infection with dengue viruses. *Am. J. trop. Med. Hyg.* **25**: 318—325.

Hairston, N. G. (1962). Population ecology and epidemiological problems. In *Bilharziasis*: 36—62. Wolstenholme, G. E. W. & O'Connor, M. (Eds). (Ciba Foundation Symposium). London: Churchill.

Hairston, N. G. (1965a). On the mathematical analysis of schistosome populations. *Bull. Wld Hlth Org.* **33**: 45—62.

Hairston, N. G. (1965b). An analysis of age-prevalence data by catalytic models: a contribution to the study of bilharziasis. *Bull. Wld Hlth Org.* **33**: 163—175.

Hairston, N. G. & Jackowski, L. A. (1968). Analysis of the *Wuchereria bancrofti* population in the people of American Samoa. *Bull. Wld Hlth Org.* **38**: 29—59.

Hairston, N. G. & de Meillon, B. (1968). On the inefficiency of transmission of *Wuchereria bancrofti* from mosquito to human host. *Bull. Wld Hlth Org.* **38**: 935—941.

Havlik, O. (1954). Význam myšich kalamit pro epidemiologii Čs. klišťové encefalitidy. *Czech. Hyg. Epidem. Mikrobiol. Immunol., Prague* **3**: 300—303.

Hoogstraal, H. (1973). Viruses and ticks. In *Viruses and invertebrates:* 384—392. Gibbs, A. (Ed.). Amsterdam: North Holland Publishing Co.

Hoogstraal, H. (1979). The role of birds in CCHF virus dissemination. In *The epidemiology of tick-borne Crimean-Congo haemorrhagic fever in Asia, Europe and Africa. J. med. Ent.* **15**: 369—373.

Hoogstraal, H., Meegan, J. M. & Khalil, G. M. (1979). The Rift Valley fever epizootic in Egypt 1977—78. 2. Ecological and entomological studies. *Trans. R. Soc. trop. Med. Hyg.* **73**: 624—629.

Hugh-Jones, M. E. (1973). The epidemiology of airborne animal diseases. In *Airborne transmission and airborne infection*: 399—404. Hers, J. F. Th. & Winkler, K. C. (Eds). Utrecht: Dosthoch Publishing Co.

Keeble, S. A., Christofinis, G. J. & Wood, W. (1958). Natural virus-B infection in rhesus monkeys. *J. Path. Bact.* **76**: 189—199.

Keppie, J., Williams, A. E., Witt, K. & Smith, H. (1965). The role of erythritol in the tissue localisation of the Brucellae. *Br. J. exp. Path.* **46**: 104—108.

Kramer, L. D. & Scherer, W. F. (1976). Vector competence of mosquitoes as a marker to distinguish Central American and Mexican epizootic from enzootic strains of Venezuelan encephalitis virus. *Am. J. trop. Med. Hyg.* 25: 336–346.

Lansdowne, C. & Hacker, C. S. (1975). The effect of fluctuating temperature and humidity on the adult life table characteristics of five strains of *Aedes aegypti. J. med. Ent.* 11: 723–733.

Laughlin, L. W., Meegan, J. M., Strausbaugh, L. J., Morens, D. M. & Watten, R. H. (1979). Epidemic Rift Valley fever in Egypt: observations of the spectrum of human illness. *Trans. R. Soc. trop. Med. Hyg.* 73: 630–633.

Lewis, D. J. (1966). Nile control and its effects on insects of medical importance. In *Man-made lakes*: 43–46. Lowe-McConnell, R. H. (Ed.). London and New York: Academic Press.

Macdonald, G. (1965). The dynamics of helminth infections, with special reference to schistosomes. *Trans. R. Soc. trop. Med. Hyg.* 59: 489–506.

Macdonald, W. W. (1967). The influence of genetic and other factors on vector susceptibility to parasites. In *Genetics of insect vectors of disease*: 567–584. Wright, J. W. & Pal, R. (Eds). Amsterdam: Elsevier.

Mackenzie, R. B., Beye, H. K., Valverde, Ch. L. & Garrón. (1964). Epidemic hemorrhagic fever in Bolivia. I. Preliminary report of the epidemiologic and clinical findings in a new epidemic area in South America. *Am. J. trop. Med. Hyg.* 13: 620–625.

McClelland, G. A. H. & Weitz, B. (1963). Serological identification of the natural hosts of *Aedes aegypti* (L.) and some other mosquitoes (Diptera, Culicidae) caught resting in vegetation in Kenya and Uganda. *Ann. trop. Med. Parasit.* 57: 214–224.

McCrumb, F. R. jr., Stockard, J. L., Robinson, C. R., Turner, L. H., Levis, D. G., Maisey, C. W., Kelleher, M. F., Gleiser, C. A. & Smadel, J. E. (1957). Leptospirosis in Malaya. I. Sporadic cases among military and civilian personnel. *Am. J. trop. Med. Hyg.* 6: 238–256.

Meegan, J. M. (1979). The Rift Valley fever epizootic in Egypt 1977–78. I. Description of the epizootic and virological studies. *Trans. R. Soc. trop. Med. Hyg.* 73: 618–623.

Myasnikov, Y. A., Levacheva, Z. A. & Yegiazaryan, K. K. (1961). Epidemiological peculiarities of an outbreak of haemorrhagic fever with renal syndrome. *J. Microbiol. Epidem. Immunobiol.* 32: 815–824.

Parsonson, I. M., Della-Porta, A. J. & Snowdon, W. A. (1981). Developmental disorders of the fetus in some arthropod-borne virus infections. *Am. J. trop. Med. Hyg.* 30: 660–673.

Plowright, W., Perry, C. T. & Greig, A. (1974). Sexual transmission of African swine fever virus in the tick *Ornithodoros moubata procinus*, Walton. *Res. vet. Sci.* 17: 106–113.

Reeves, W. C. (1970). Evolving concepts of encephalitis prevention in California. *Proc. Ann. Conf. Calif. Mosquito Control Assoc. Inc.* 37: 3–6.

Reeves, W. C. (1971). Mosquito vector and host interaction: the key to maintenance of certain arboviruses. In *The ecology and physiology of parasites*: 223–230. Falles, A. M. (Ed.). Toronto: University of Toronto Press.

Reeves, W. C. & Hammon, W. McD. (1962). Epidemiology of the arthropod-borne viral encephalitides in Kern County, California, 1943–1952. *Univ. Calif. Publs publ. Hlth* 4: 1–257.

Richards, W. P. C. & Cordy, D. R. (1967). Bluetongue virus infection: pathologic response of nervous systems in sheep and mice. *Science, N.Y.* **156**: 530–531.

Rutledge, L. C., Khan, A. A., Skidmore, D. L. & Maibach, H. I. (1975). Genetic transmission of host-seeking behaviour in colonized *Aedes aegypti* (L.). *Mosquito News* **35**: 189–194.

Sardjito, M. & Zuelzer, M. (1929). Weiterer Beitrag zur Biologie der *Spirochaeta biflexa* syn. *Leptospira icterohaemorrhagiae* syn. *Spirochaeta icterogenes* in den Tropen. *Zbl. Bakt.* (Abt. I. Orig.) **110**: 180–187.

Scherer, W. F. & Buescher, E. L. (1959). Ecological studies of Japanese encephalitis virus in Japan. I. Introduction. *Am. J. trop. Med. Hyg.* **8**: 644–650.

Sellers, R. F. (1980). Weather, host and vector — their interplay in the spread of insect-borne animal virus diseases. *J. Hyg., Camb.* **85**: 65–102.

Sellers, R. F., Barlow, D. F., Donaldson, A. J., Herniman, K. A. J. & Parker, J. (1973). Foot-and-mouth disease, a case study of airborne disease. In *Airborne transmission and airborne infection*: 405–412. Hers, J. F. Th. & Winkler, J. C. (Eds). Utrecht: Oosthoch Publishing Co.

Sheppard, P. M., Macdonald, W. W., Tonn, R. J. & Grat, B. (1969). The dynamics of an adult population of *Aedes aegypti* in relation to dengue haemorrhagic fever in Bangkok. *J. Anim. Ecol.* **38**: 661–697.

Smith, C. E. G. (1956). A virus resembling Russian spring-summer encephalitis virus from an ixodid tick in Malaya. *Nature, Lond.* **178**: 581–582.

Smith, C. E. G. (1964a). Factors influencing the behaviour of viruses in their arthropodan hosts. *Symp. Br. Soc. Parasit.* **2**: 1–32.

Smith, C. E. G. (1964b). Factors in the transmission of virus infections from animals to man. Chapter VIII. *Scient. Basis Med. A. Rev.* **1964**: 125–150.

Smith, C. E. G. (1971). Human and animal ecological concepts behind the distribution, behaviour and control of yellow fever. *Bull. Soc. Path. Exot.* **64**: 683–694.

Smith, C. E. G. (1975). The significance of mosquito longevity and blood-feeding behaviour in the dynamics of arbovirus infections. *Med. Biol.* **53**: 288–294.

Smith, C. E. G. (1977). Research priorities for the control of arbovirus infections. In *Medicine in a tropical environment*: 327–349. Gear, J. H. S. (Ed.). Cape Town, Rotterdam: Balkema.

Smith, C. E. G. & Turner, L. H. (1961). The effect of pH on the survival of leptospirosis in water. *Bull. Wld Hlth Org.* **24**: 35–43.

Smith, C. E. G., Turner, L. H., Harrison, J. L. & Broom, J. C. (1961a). Animal leptospirosis in Malaya. I. Methods, zoogeographical background and broad analysis of results. *Bull. Wld Hlth Org.* **24**: 5–21.

Smith, C. E. G., Turner, L. H., Harrison, J. L. & Broom, J. C. (1961b). Animal leptospirosis in Malaya. II. Localities sampled. *Bull. Wld Hlth Org.* **24**: 23–34.

Smith, C. E. G., Turner, L. H., Harrison, J. L. & Broom, J. C. (1961c). Animal leptospirosis in Malaya. III. Incidence in rats by sex, weight and age. *Bull. Wld Hlth Org.* **24**: 807–816.

Takahashi, K., Matsuo, R., Kuma, M., Noguchi, H., Higashi, F. & Fujiwara, O. (1966). Studies on epidemic of Japanese encephalitis virus in Nagasaki Prefecture in the 1965 season. *Endem. Dis. Bull., Nagasaki Univ.* **8**: 1–28.

Taylor, W. P. & Marshall, I. E. (1975). Adaptation studies with Ross River virus: retention of field level virulence. *J. gen. Virol.* **28**: 73–83.

Tempelis, C. H. (1975). Host feeding patterns of mosquitoes with a review of advances in analysis of blood meals by serology. *J. Med. Ent.* **11**: 635—653.

Tempelis, C. H., Reeves, W. C., Bellamy, R. E. & Lofy, M. F. (1965). A three-year study of the feeding habits of *Culex tarsalis* in Kern County, California. *Am. J. trop. Med. Hyg.* **14**: 170—177.

Tesh, R. B., Chaniotis, B. N. & Johnson, K. M. (1972). Vesicular stomatitis virus (Indiana serotype): transovarial transmission by phleobtomine sandflies. *Science, N.Y.* **175**: 1477—1478.

Tesh, R. B., Gubler, D. J. & Rosen, L. (1976). Variation among geographic strains of *Aedes albopictus* in susceptibility to infection with chikungunya virus. *Am. J. trop. Med. Hyg.* **25**: 326—335.

van den Assem, J. (1959). Daily mortality in four species of New Guinea anophelines. *Trop. geogr. Med.* **11**: 223—236.

Verwoerd, D. W., Huismans, H. & Erasmus, B. J. (1979). Orbiviruses. In *Comprehensive virology* **14**: 285—345. Fraenkel-Conrat, H. & Wagner, R. R. (Eds). New York and London: Plenum Press.

Walter, N. M. & Hacker, C. S. (1974). Variation in life table characteristics among three geographic strains of *Culex pipiens quinquefasciatus*. *J. med. Ent.* **11**: 541—550.

Watts, D. M. & Eldridge, B. F. (1975). Transovarial transmission of arboviruses by mosquitoes: a review. *Med. biol.* **53**: 271—278.

Welsh, H. H., Lennette, E. H., Abinanti, F. R. & Winn, J. F. (1958). Air-borne transmission of Q fever: the role of parturition in the generation of infective aerosols. *Ann. N.Y. Acad. Sci.* **70**: 528—540.

Wolff, J. W. & Ruys, A. C. (1953). Changes in the frequency of Weil's disease in Amsterdam during the last 18 years. *Docum. med. Geogr. Trop. (Amst.)* **5**: 247—252.

Work, T. H. (1958). Russian spring-summer virus in India. Kyasanur Forest disease. *Progr. med. Virol., N.Y.* **1**: 248—279.

DISCUSSION

Tyrrell (Chairman) — Do you think there is any way, Dr Gordon Smith, of quantitating some of these things that you talk about? I know it must be incredibly difficult to manage the multiple different systems which may impinge on one particular question, but has any work of that sort been done?

Gordon Smith — Well, I think you are talking about infections which involve species other than man; and the answer is yes, great attempts have been made, but not enough is ever known. The slide I showed of the work on western encephalitis in California — Bill Hammon and later Bill Reeves have been studying the mosquitoes, the viruses, the hosts, the weather, the water, the effects of insecticide resistance, and all sorts of aspects of this since 1930; and they now have what looks like a reasonably good prediction system based on light traps catching mosquitoes (the numbers of mosquitoes indicating the risk) and they have sentinel bleedings of horses and birds; nevertheless, in what according to all the rules of the game was the most favourable year for an outbreak, it didn't happen. So clearly there is still something that is not known. But I think it is only by assiduous study of rather restricted areas that we might eventually begin to understand some of these questions.

Tyrrell — Yes, you dangled this possibility before us with your "balance", suggesting what we could do if we could predict what the reproduction rate would be — if it were over one then we'd have to look out for an epidemic, and we could choose, out of the different measures available, one which might pull it down below one and so we might control the infection. I'm baiting you a bit because I think that though it's a nice idea it hasn't actually ever been successfully turned into practical administration.

Gordon Smith — Well, it's being turned all the time. If you vaccinate, and if you vaccinate enough, you will cut down the reproduction rate, or the effective contact rate, and the disease will disappear. If you are dealing with very much more complex systems, of course, it's more difficult, but if you can reduce the longevity of mosquitoes adequately, you will in fact have a very profound effect because it works in a logarithmic fashion. A mosquito has to live long enough in the first place to become infective, which may be 10 or 14 days; if something like 67% of them only survive every 24 h none of them will actually live long enough to transmit. If on the other hand 95% of them live 24 h then you will get perhaps 15 days of transmission afterwards.

Rees — To pursue this question from Dr Tyrrell, this has been done in certain instances, such as liver fluke forecasting by the Ministry of Agriculture; but your point is relevant, that if we could derive some form of risk index, this would help very much in predicting what would happen and what policies should be pursued. I don't know if somebody could describe the liver fluke forecasting service put out by the Ministry laboratory?

Wilesmith — It is not possible to describe in detail Dr Ollerenshaw's methods for forecasting the incidence of liver fluke, nematodiriasis and swayback, but they are based on evaluating data collected over a number of years on both disease incidence and meteorological conditions. In addition some account is taken of the disease incidence in the previous year before formulating these forecasts which have proved to be very successful.

Gordon Smith — There are a number of situations where if you get a rodent population peak you should be on the look-out for an increase in the transmission of plague or tick-borne encephalitis or whatever is being maintained in the species.

Tyrrell — Yes, I know some studies have also been done through the Ministry of Agriculture and the Meteorological Office looking at aphid populations and things like that as predicting risks of crop disease as well. This really is a successful application of the broad idea you've been floating.

Gordon Smith — They're much better at it than we are.

Ormerod — Dr Gordon Smith has pointed out the advantages of considering the spread of "infection" from one animal to another rather than the spread of "disease". I would like to emphasize this point by stating that "disease" is essentially an anthropocentric concept. Disease rarely occurs in wild populations under balanced ecological conditions. The simplest example of the concept is illustrated by the antelope which is rapidly removed by a predator as soon as its performance is reduced. Pathogenesis in an infecting organism puts it at a severe disadvantage: thus in the presence of active predators, a pathogenic organism will rapidly have to come to terms with its host or risk extinction. Although pathogenic organisms can often be isolated from wild animals, disease as understood in medical or veterinary practice is a rare phenomenon indicating ecological imbalance (Ormerod, 1979).[1]

---

[1] Ormerod, W. E. (1979). Human and animal trypanosomiases as world public health problems. *Pharmac. Ther.* 6: 1–40.

Gordon Smith — I think you would expect such a phenomeon to occur, for example, when a species barrier is crossed. After all, myxomatosis is a relatively mild disease in *Sylvilagus* rabbits; it was only when it was artificially transferred into *Oryctolagus* rabbits that it caused very dramatic effects. If you go back to Marburg fever and how it came to Europe in monkeys from Africa, it was shown that if you put one Marburg virus into a monkey it will die. It is extremely unlikely, therefore, that monkeys are where Marburg virus lives when it is not causing an epizootic — we have not the foggiest idea where it does live. I think if you interfere with the environment of a zoonosis, if you change it, if you increase populations of its hosts, an epidemic or epizootic may result. For example, in the Camargue in the 1960s there was an outbreak in horses and man of West Nile fever. This was because they had started to grow rice in the Camargue. The salinity of the surface water was such that the *Culex* mosquito concerned could not breed there in large numbers until they started to grow rice. This introduced large amounts of fresh water and washed the salt out of the soil; hence large amounts of mosquitoes, and an epidemic. These sorts of processes are going on all the time but you cannot always work out the reasons for them.

Tyrrell — I was trained by Sir Christopher Andrews, who wrote a fascinating book called *The natural history of viruses*, and he insisted that I should take a virocentric view and that I should only regard mankind as important because it provided me with rhinoviruses on which I could earn my living and educate my family; but I felt there was something wrong with the loop of the argument somewhere!

Symp. zool. Soc. Lond. (1982) No. 50, 237–270

Surveillance and Control of Bubonic Plague in the United States

ALLAN M. BARNES

Chief, Plague Laboratory, Center for Infectious Diseases, Department of Health and Human Services, PO Box 2087, Fort Collins, Colorado 80522, USA

SYNOPSIS

Bubonic plague is a flea-transmitted disease of rodents caused by *Yersinia pestis* and is transmissible to humans by vector fleas or by direct contact contamination from infected animals. The disease is characterized by quiescent or interepizootic periods during which it is restricted to highly discrete and geographically limited foci of infection, followed by sporadic or periodic epizootics among susceptible wild rodent/flea complexes which expand its geographic distribution and expose other animal populations, including humans who live in affected areas. Human cases are not numerous in the United States, but have increased to an average of 16 per year since 1974. Mortality has remained high (18%), despite the availability of effective antibiotic therapy. Human cases are most frequent among persons under 20 years of age and are concentrated in two regions, the southwest (Arizona, southern Colorado, New Mexico) and the Pacific States (California, southern Oregon).

At our present state of knowledge, plague outbreaks among wild rodents are not predictable as to time and location. Our primary objective is to develop data that will enable us to detect and, ultimately, to predict epizootic outbreaks and geographic expansions that may lead to human exposure. A collaborative surveillance program involving longitudinal serologic surveys of wild carnivore and domestic dog populations is a key program element. Another important program element is consistent liaison with state and local health departments and an informal surveillance network of biologically-oriented public agencies and individuals. Rodent surveys and control actions are undertaken on the basis of carnivore data indicating epizootic plague in areas of human risk.

Data from these studies have reconfirmed and somewhat expanded the known geographic distribution of *Y. pestis* activity and identified rodent/flea complexes important in epizootic amplification and transmission of plague to humans. Carnivore data are presented and human risk factors associated with plague in major epizootic foci are discussed along with strategies for control to prevent human infection.

INTRODUCTION

Plague is a flea-transmitted disease of rodents caused by the bacterium *Yersinia pestis*, known to exist in widespread, discontinous foci in parts of Africa, Asia, and the Americas (Poland & Barnes, 1979; Velimirovic, 1979). In its natural foci, the organism circulates among moderately to highly resistant rodent species with little or no overt disease (Baltazard, 1953; Baltazard *et al.*, 1963; Kartman, Prince, Quan & Stark, 1958); its persistence is dependent upon a complex set of interrelationships among reservoir hosts, flea vectors, the organism itself, and external environmental factors that influence its circulation. Each plague focus is discrete and unique in terms of its rodent reservoirs, flea vectors, the external environmental factors that influence transmission of the pathogen, and its epizootic manifestations (Macchiavello, 1954; Pollitzer & Meyer, 1961; WHO, 1980). In these respects plague is typical of zoonoses as characterized by Audy (1958) and Pavlovsky (1966).

The disease in animals is characterized by explosive and often devastating sporadic or periodic epizootics among susceptible and receptive rodent and flea populations. During epizootic amplification, it tends to expand its distribution geographically, often for great distances, involving chance victims among ecologically associated animal species including humans who enter or live in the affected areas. Epizootics among both wild and commensal rodents are self-limiting (Meyer, 1942) and ultimately move on or recede to focal areas, at times seeding new pockets of infection that may persist for varying periods of time, or occasionally, remain as new nidi of infection. One of our principal concerns involves its possible transfer to urban or rural commensal rat populations with their potential for causing epidemic plague transmission as described by Miles, Kinney & Stark (1957) and by Hudson & Quan (1975). Another is the potential for development of secondary pneumonia among human victims of bubonic plague followed by its transmission via infectious airborne droplets to other persons, thus creating a pneumonic plague epidemic.

Historical Background

Bubonic plague has been known to exist in North America for 81 years. Its introduction to Pacific and Gulf Coast ports, the commensal rodent-borne epidemics that followed, and its ultimate discovery among native wild rodent populations near San Francisco in 1908 (McCoy, 1908, 1911; Wherry, 1908) are well-documented episodes of the latest world pandemic (Meyer, 1942; Link, 1955; Pollitzer, 1954). The last urban epidemic to take place in North America occurred in Los Angeles, California, in 1924—1925 (California Board of Health, unpublished) and involved both commensal rats and California ground squirrels and their fleas as sources of human infection (Link, 1955). The Los Angeles episode also represents the last occasion on which person-to-person spread of pneumonic plague is known to have occurred in the United States. Since 1925, all human cases reported from North America have had their zootic origins from native wild rodents and their fleas, although both wild and domestic animals associated with an epizootic source have at times played an intermediary role as sources of human infection.

After futile attempts to eradicate plague from California ground squirrels in California, the disease eventually spread across much of the State, not only among California ground squirrels but also in other rodent species (Meyer, 1942). The first reported human plague case of wild rodent/flea origin in southern Oregon occurred near the California border in 1934. The Oregon case was eventually followed by first cases in Utah (1936), Nevada (1937), Idaho (1940), New Mexico (1949), Arizona (1950), Colorado (1957) and ultimately Wyoming (1978). Extensive animal plague surveys were conducted by the US Public Health Service through the 1930s and 1940s to determine the distribution of plague and the possibility of its expansion into the more populous eastern states (Eskey & Haas, 1940; Meyer, 1942; Link, 1955; Kartman, Prince *et al.*, 1958; Miles, Wilcomb & Irons, 1952; Ecke & Johnson, 1952). These surveys, dependent entirely on bacterial isolations, found animal plague in 15 western states extending eastward to about the 101st Meridian in western Texas, Oklahoma, Kansas, and North Dakota. No human plague cases of wild rodent origin have been reported from Kansas, Montana, Oklahoma, North Dakota, Texas,* or Washington.

From 1925–1964 human plague cases in the USA averaged about two per year as depicted in Fig. 1. Most of the cases reported were from California. In 1965 an outbreak of six cases occurred among Navajo Indian residents of New Mexico (Collins *et al.*, 1967). Since that time, the annual number of human cases has consistently increased, with the majority of cases in the south-west, particularly New Mexico. With this geographic shift, the occurrence of human plague initially appeared to peak each five years, suggesting a definite periodicity. Since 1975, however, cases per year have shown little significant change but have ranged from 12–20 with an average of 16 cases per year.

General Epidemiology

Mortality among North American plague cases has remained high; in the United States, 19 of the 105 cases (18%) that occurred from

---

* Since the present paper was completed, one human case has been reported from Ector County Texas.

Allan M. Barnes

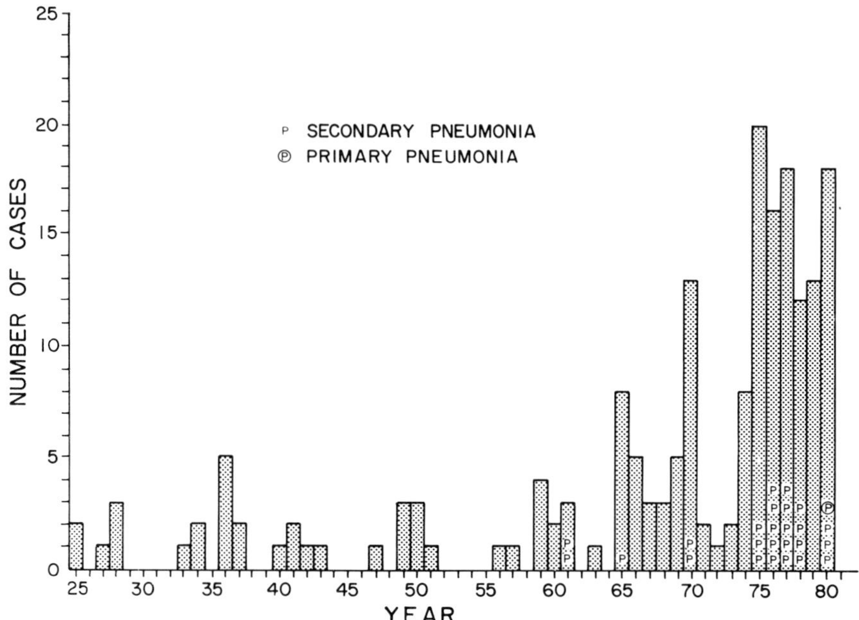

FIG. 1. Incidence of human plague and plague pneumonia cases in the United States, 1925–1980.

TABLE I

Number of human plague cases and fatalities in the United States according to age group, 1974–1980

Age group (years)	No. of cases	No. of fatalities
0–9	28	6
10–19	34	5
20–29	8	0
30–39	10	0
40–49	9	2
50–59	10	2
> 60	6	4
TOTALS	105	19

1974 to 1980 (Table I) died despite the availability of specific anti-biotic therapy (White, Gordon, Poland & Barnes, 1980). Untreated bubonic plague cases have a case fatality rate of 50–60% (Pollitzer, 1954). The disease occurs predominantly in young people; the case fatality rate is highest among the young and the elderly. The youngest person reported to have had plague in North America was a one-week-old child from Colorado who recovered from her infection acquired from a flea presumably transported into her house by a pet dog (White, Rosenbaum, Canfield & Poland, 1981). The eldest was a 79-year-old woman who died of plague septicemia (CDC Records). In the past, cases among males outnumbered those in females by about a 2:1 margin; since 1975, 58 of the 105 reported cases have been in females. Most of the cases investigated (78%) are acquired at home, rather than in a work or recreational setting, and most occur during summer months from May to October, coinciding with the peak period of activity among moderately to highly susceptible rodents and their fleas (Table II).

TABLE II

Number of plague infections acquired in home environment[a] versus those acquired in work or recreational environments, 1974–1978

State	At home	Away from home	Total
Arizona	10	2	12
California	3	4	7
Colorado	3	3	6
Nevada	1	0	1
New Mexico	38	5	43
Oregon	1	1	2
Utah	2	0	2
Wyoming	0	1	1
TOTALS	58	16	74

[a] Within 1 mile radius of home.

The development of secondary plague pneumonia among bubonic and/or septicemic plague patients has increased in the United States in recent years. As shown in Fig. 1, 21 of 97 (22%) confirmed cases in the United States during the years 1975–1980 developed secondary plague pneumonia. An analysis of 91 available and sufficiently complete case records from wild rodent associated plague cases in the United States during the years 1925–1974 disclosed that only five (5%) had plague pneumonia secondary to more typical forms of the disease. No obvious reasons for the increase are evident. Analysis of case records has not revealed significant differences in clinical approaches which might have resulted in the increase in recent years nor have preliminary studies on *Y. pestis* strains from pneumonic cases revealed differences between them and isolates from previous years. Of the 21 recent pneumonic cases, six (35%)

died. Case contacts deemed sufficiently exposed to require observation and/or prophylaxis have averaged about 75 per case. A number of pneumonic cases and case contacts have traveled to distant places during incubation, thus potentially exposing persons living in non-plague areas where the medical community most likely would be naive to the disease. No secondary cases are known to have resulted among contacts.

No one expects that plague will ever again cause the mortality it once did in the United States. Epidemics are preventable by appropriate management of commensal rodent populations. The effective use of insecticides against vector fleas can control outbreaks should a lapse in rodent management occur. The disease in man is successfully treatable with readily available antibiotics if diagnosed in time (White, Gordon *et al.*, 1980), and reporting of human cases and epidemiological reaction are effective in detecting human cases to prevent epidemic spread. Nevertheless, human plague cases from wild rodents and their fleas have increased substantially during recent years. From 1970 through 1980, 123 human cases occurred in the western United States, the only developed country in the world to report human plague (WHO, 1981). If we are to reduce these figures it is essential that we gain a far greater understanding of plague in animal and flea populations than we presently possess, particularly with regard to the factors involved in epizootic outbreaks and consequent human hazard.

PLAGUE SURVEILLANCE

Human Case and Epizootic Investigations

As required by the International Health Regulations of the WHO (WHO, 1971), all human cases reported in the United States are investigated epidemiologically by state and local health agencies, and CDC, to: (1) confirm the infection as plague; (2) determine the locality in which the infection was acquired; (3) determine possible animal (or human) source of infection; (4) locate, observe, and if necessary, treat case contacts or others exposed to the suspected source of infection; (5) inform the medical community of possible additional cases and provide information on diagnostic and therapeutic methods; and (6) inform the public of the existence of plague and means of avoiding infection, and seek their cooperation with health authorities in reporting events among animal populations in which plague is suspected.

Epizootiological investigations are conducted by local and state health agencies in consultation or collaboration with the Plague Laboratory, CDC, to: (1) locate the site at which infection was acquired; (2) determine animal and vector species involved, using standardized survey techniques and observations supported by laboratory diagnostic tests; (3) determine the geographic scope of epizootic plague among animal populations; (4) determine areas of human risk; and (5) implement control measures adapted to the characteristics of the focus and animal species involved.

Plague surveillance network
Liaison is maintained not only with health agencies but also with an informal surveillance network of biologically trained persons in other fields, e.g. park managers, game managers, agricultural agents, foresters, and others, who report unusual events among rodent and rabbit populations such as animals found sick or dead, unusual disappearances among animal populations, or unusually high rodent populations. Information from such observers, followed by submission of specimens for laboratory studies, have been instrumental in the early detection of epizootic plague events and implementation of control to prevent human exposure.

Carnivore Serosurveys

Background and methods
Studies by Rust, Cavanaugh, O'Shita & Marshall (1971), Rust, Miller *et al.* (1971) and Archibald & Kunitz (1971) suggested the possibility that domestic dogs might serve as sentinels of plague activity. Dogs were shown to produce detectable antibody to specific Fraction 1 of *Y. pestis* following the ingestion of plague-infected meals, but to suffer little or no morbidity. The potential of dogs as sentinels has particular significance on south-western Indian Reservations where they usually are allowed to run freely and to forage for themselves. Serosurveys among domestic dogs also have been used to elucidate the geographic and temporal distribution of plague in Zimbabwe during and following an outbreak (Cruickshank, Gordon, Taylor & Naim, 1976; Taylor, Gordon & Isaacson, 1981). Early surveys on Indian reservations, begun in 1970, indicated that serologic conversions in dog populations frequently appear before rodent plague is observed and human cases occur. Sera were collected in the field by the Indian Health Service during annual rabies vaccination campaigns conducted from March through June, timed to precede the peak of human plague exposure during the summer months. Sera were tested

by the passive hemagglutination test (PHA) for antibody to Fraction 1 of *Y. pestis* as described by WHO (1970). Antibody positive specimens were subjected to the passive hemagglutination inhibition (PHI) test as a control to eliminate cross-reactions with antibodies to related organisms. Survey data, integrated with observation on the distribution and estimated population densities of susceptible rodent populations, were used to evaluate risk of human exposure, make preventive control decisions, and to define the distribution and temporal fluctuations of the disease in rodents.

Serosurveys among wild carnivores were begun in 1971 following pilot studies on field and laboratory techniques undertaken in collaboration with the US Fish and Wildlife Service, Predator Damage Control Program. Wild carnivore blood samples, unlike those from dogs, were collected on filter paper strips and tested by PHA and PHI as described by Wolf & Hudson (1974). Strip samples were collected in the field by the US Fish and Wildlife Service, state predator control programs, researchers, and occasionally by professional fur trappers. After air drying, specimens were placed individually in envelopes, on which the collection data were inscribed, and sent to state health agency co-ordinators who collated the data and forwarded data and specimens to the Plague Laboratory, CDC. The program was positively reinforced by feedback of information to each collector.

In limited laboratory studies in conjunction with early surveys we found that responses to oral challenge among coyotes (*Canis latrans*), skunks (*Mephitis mephitis*) and raccoons (*Procyon lotor*) were similar to those described by Rust, Cavanaugh *et al.* (1971) for domestic dogs. In our studies, challenge did not produce a change in temperature or other signs of morbidity and PHA antibody became detectable in eight to 14 days and usually reached a peak in 20 to 30 days, then subsided to low or undetectable levels in six to eight months. Our data on duration of titers in wild carnivores are similar to those of Taylor *et al.* (1981), who, working with dogs in Zimbabwe, noted that dogs initially found antibody-positive usually were negative when tested four to five months later (although one dog remained positive for eight months).

Results by state and region

Table III presents the percentages by state of sera found positive among a total of 12 405 wild and 7037 domestic carnivore specimens tested from 1976–1980. Arizona, Colorado and New Mexico stand out as high plague incidence areas, an observation supported by identified rodent epizootics and the occurrence of human cases; how-

TABLE III

Carnivore sera tested for antibody to Yersinia pestis, *15 Western States, USA, 1976–1980*

	Wild				Domestic			
State	No. tested	% of total	No. pos.	% pos.	No. tested	% of total	No. pos.	% pos.
New Mexico	2257	18.2	564	25.0	4136	58.8	663	16.0
Oregon	3380	27.2	362	10.7	43	0.6	1	2.3
California	2129	17.2	243	11.4	650	9.2	64	9.8
Arizona	484	3.9	135	27.9	1536	21.8	278	18.1
Washington	1259	10.2	29	2.3	26	0.4	0	0
Colorado	749	6.0	168	22.4	236	3.3	42	17.8
Other states[a]	2147	17.3	167	7.7	410	5.8	57	13.9
TOTAL	12 405	100.0	1668	13.4	7037	100.0	1105	15.7

[a]Other states include, in descending order of numbers submitted: Texas, Idaho, Nevada, South Dakota, Wyoming, Utah, Oklahoma, and Kansas.

ever the data contain built-in biases which limit the accuracy of comparisons on a state-by-state basis. For example, only the hyper-endemic northern plateau area of Arizona was surveyed during the 1976–1980 survey period. Conversely, the entire state of Oregon has been represented in surveys since the early 1970s. Plague in Oregon is all but confined to the semi-arid eastern part of the state except for its occasional incursions into the cool wet temperate region extending from the Cascade Mountains west to the Pacific Ocean. Such incursions are important epidemiologically; one caused a human case in the coastal town of Coos Bay in 1978. However, they appear to be ephemeral and, based on our data, are not as widespread or intensive as epizootic plague east of the Cascades where animal plague is as active as it is in adjoining areas in California and in the south-western states.

Wild carnivores

Wild carnivore data are shown by species, percentage of each species in the total sample, and percentage of the total with antibodies for each species in Table IV. Coyotes (*C. latrans*) predominate among serum specimens tested with 83% of the total, reflecting the objectives of the federal and state predator damage control programs that provided most of the specimens. Bobcats (*Lynx rufus*), badgers (*Taxidea taxus*), foxes (*Vulpes* spp. and *Urocyon* sp.), striped and spotted skunks (*M. mephitis* and *Spilogale gracilis*), and raccoons (*P. lotor*) provided significant but smaller proportions of the total, ranging from 4.9 down to 2.3% respectively. The 104 bears (*Ursus americanus*), weasels (*Mustela frenata*) and ringtailed cats (*Bassariscus astutus*) were captured in low numbers and in only one or a few states.

TABLE IV

Wild carnivore sera tested for antibody to Yersinia pestis, *15 Western States, USA, 1976–1980*

Carnivore	No. tested	Percent of total tested	No. positive	Percent positive
Coyote	10 296	83.0	1403	13.6
Bobcat	603	4.9	73	12.1
Badger	383	3.1	84	21.9
Fox	342	2.7	24	7.0
Skunk	308	2.5	12	3.9
Raccoon	289	2.3	49	17.0
Bear	104	0.8	24	23.0
Marten	49	0.4	15	30.6
Weasel	19	0.2	5	26.3
Ringtailed cat	12	0.1	6	50.0
TOTAL	12 405	100.0	1695	13.7

Antibody positive sera were found among all the species listed: 13.6% of coyotes tested had antibodies; others ranged from 3.9% in skunks to 50% in the few ringtailed cats represented. The higher incidence of antibody found among badgers, marten, weasel, and ringtailed cats may indicate their greater avidity for rodent prey as compared to coyotes which feed on animals other than rodents and also consume vegetable material when in need. The antibody prevalence indicated among bobcats may understate their association with plague since field evidence suggests that bobcats often succumb to plague and human cases are known to have been acquired from them (Poland, Barnes & Herman, 1973).

Domestic dogs and cats

Domestic dog and cat data are shown in Table V. Cats are not routinely bled during surveys and only 183 (2.6%) are in the 7037 animals included. Only 17 (9.3%) cats had antibody, probably because many that become infected do not survive. Of the 6854 dogs tested from 1976 to 1980, 1088 (15.9%) had significant antibody titers.

The primary purpose of dog serologic surveys on south-western Indian reservations since 1971 has been the early detection of rodent plague activity on a regional and local basis. Evidence of a rise in frequency and in the means of titers found in dog populations are followed up by observations on rodent populations and, if deemed necessary for human case prevention, by insecticidal and/or rodenticidal control actions.

Table VI shows the number of collection locations (pueblos,

TABLE V

Domestic dog and cat sera tested for antibody to Yersinia pestis, *15 Western States, USA, 1976–1980*

Carnivore	No. tested	Percent of total tested	No. positive	Percent positive
Cat	183	2.6	17	9.3
Dog	6854	97.4	1088	15.9
TOTAL	7037	100.0	1105	15.7

settlements, etc.) sampled, the number with serologically positive dogs, and the number of epizootics identified by isolation of *Y. pestis* from rodents and fleas. The incidence of epizootics discovered was relatively low in the early 1970s, a period with few human cases and little identified rodent plague in the western United States. The incidence of locations with seropositive dogs peaked in 1975–1977. The discovery of associated rodent epizootics was most frequent in 1975–1976, coincident with a sharp increase of human cases in the south-west (Fig. 1) and an increase in animal plague throughout the western United States (Vector-Borne Diseases Division, 1976, 1977). From 1977 to 1980, the percentage of dogs with titers remained high, affected in part by residual titers from challenges in previous years. However, the number of epizootics identified declined, in part because of control measures carried out in previous years in the locations sampled.

TABLE VI

Number of collection locations with plague seropositive dogs and rodent epizootics identified on Arizona and New Mexico Indian Lands, 1971–1980.

Year	No. of locations sampled	No. of locations with serologically positive dogs (%)	No. of separate epizootics identified[a]
1971	54	7 (13%)	3
1972	52	7 (13%)	2
1973	74	7 (9%)	5
1974	59	9 (15%)	2
1975	45	22 (49%)	13
1976	74	54 (73%)	18
1977	93	45 (48%)	6
1978	71	31 (44%)	5
1979	97	60 (62%)	7
1980	97	35 (36%)	3

[a] Epizootics identified by isolation of *Yersinia pestis* from animal tissues and/or fleas as a result of follow-up investigation of serologically positive dog sera.

 Allan M. Barnes

Geographic Distribution of Plague in the USA, 1970–1980

Human cases

As shown in Fig. 2, the majority of human plague cases in the United States from 1970 to 1980 occurred in the Rocky Mountains subregion. The focus includes the northern half of New Mexico, north-eastern Arizona, southern Colorado and a small portion of south-eastern Utah. A second and smaller grouping of cases is seen in California, southern Oregon and a small segment of western Nevada. It is interesting to note here that all of the few human cases to occur outside these two focal areas resulted from direct contact with plague-

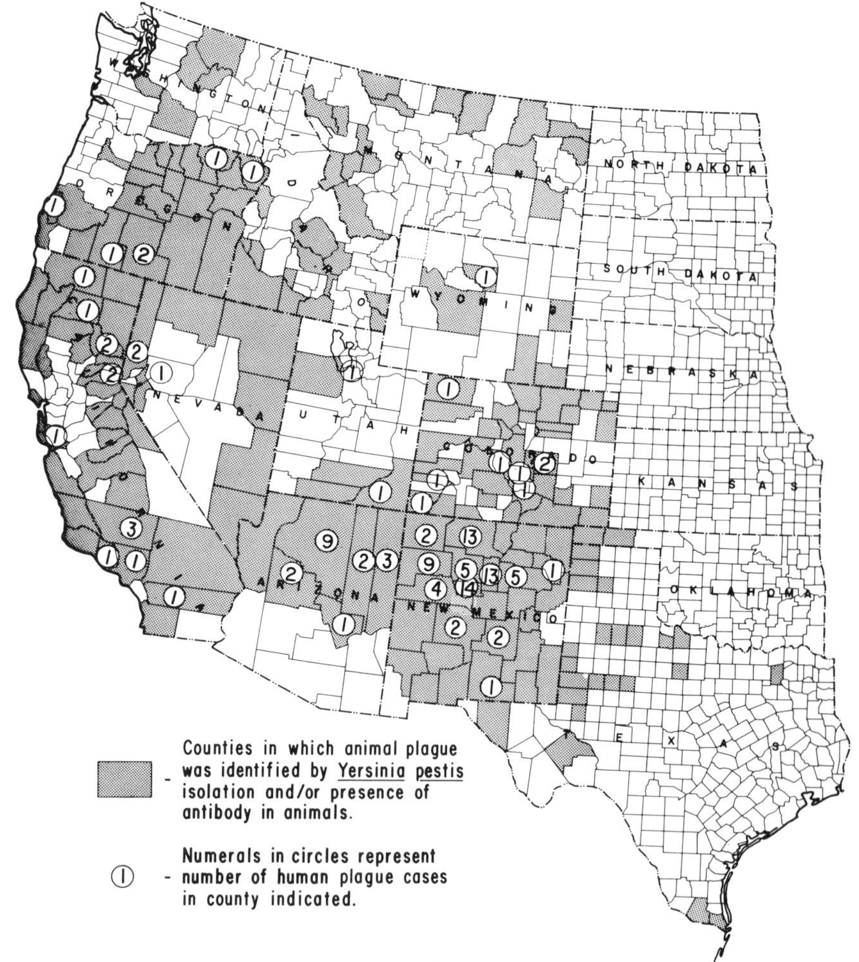

FIG. 2. Geographic distribution of human and animal plague in the United States, 1970–1980.

infected rabbits rather than from flea-bite inoculation of the organism (von Reyn, Barnes, Weber & Hodgin, 1976). It is also interesting to observe that no human plague cases have occurred at all in Montana, despite serological evidence of its presence, nor have cases occurred in the plains of eastern Colorado, west Texas, or Oklahoma, despite extensive epizootics among the plains prairie dog (*C. ludovicianus*), observed periodically in the plains areas from the Rocky Mountains eastward, and concurrent epizootics among the wood rats, *Neotoma micropus* and *N. albigula* in west Texas, as described by Miles, Wilcomb & Irons (1952). The absence of reported human cases in much of Utah and Nevada may be due, at least in part, to the sparse human population in the region.

Animal plague

Animal plague activity detected in the United States during the period 1971–1980 also is shown in Fig. 2. *Y. pestis* activity was identified by isolation of the organism from flea vectors and animal tissues and also by the presence of plague antibodies in the sera of animals, primarily rodents and carnivores.

The 1971–1980 data reconfirm the geographic range of plague activity in the United States as described by Eskey & Haas (1940), extending it somewhat eastward from the 101st to about the 97th Meridian. Serologic positives from border counties in Montana, Washington, California and Texas strongly suggest its presence, at least on occasion, in Canada and Mexico. Recent findings have demonstrated the presence of plague antibodies in two carnivores collected in Hidalgo and Cameron Counties at the southernmost tip of Texas. Other new records include areas in Pacific north-west temperate rain forest in north-western California and south-western Oregon. Although plague activity is well-known in the inland areas of both states, it had been assumed that the plague organism did not occur in the cool, wet environment typical of the coastal area.

Plague was not discovered by earlier surveyors in hot desert areas of the south-west, nor were we able to discover evidence of plague activity in extensive surveys for antibodies among wild carnivores in the deserts of southern Arizona in 1975–1980.

The data as represented in Fig. 2 indicate that the plague organism remains widespread and strongly entrenched among wild rodent populations in the western United States as shown by earlier workers (Eskey & Haas, 1940; Kartman, Prince *et al.*, 1958). However, quantitative changes in antibody prevalence observed from year to year among wild carnivores indicate that in much of its distribution it is relatively short-lived and demonstrates a pattern of geographic amplification followed by regression to focal areas. For example, in longitudinal studies at Fort Collins, we have observed plague activity among carnivore and rodent populations in the Colorado Front Range every year since 1971. Conversely, carnivore populations on the bordering eastern plains from the Rocky Mountains eastward to Kansas have had antibody only during and the year or so following identified epizootics among the plains prairie dog (*C. ludovicianus*).

As a second example of fluctuant distribution, a bubonic plague epizootic was identified in 1978 among two of the smaller ground squirrel species, *Spermophilus richardsoni* and *S. townsendi*, in southern Idaho when high plague antibody titers appeared among coyote (*Canis latrans*) populations. In 1976, only seven of 160 coyote sera collected in the entire state were antibody positive and in 1977, only eight of 211. In each year the titers of positive animals did not exceed 1:256. In 1978, 91 of 300 carnivores collected in Idaho had antibody titers ranging from 1:32 to 1:4096. The epizootic centered in the southern part of the state where ground squirrels are abundant, particularly in Twin Falls County where 68 of 136 coyotes had significant titers. By 1979, plague had invaded ground squirrel populations in adjacent counties in northern Nevada where widespread and devastating epizootics among *S. beldingi* occurred. Plague subsided in southern Idaho in 1979 and by 1980 only four of 327 coyotes captured had antibody and the highest titer was 1:256. An increased incidence of plague titers among carnivores in eastern Oregon led to discovery of epizootic plague there among *S. beldingi* in 1981, suggesting that the epizootic first discovered in southern Idaho in 1978 had moved through populations of these small ground squirrels distributed over much of the northern Great Basin region within about three years, leaving little residual evidence of infection where it was first noted.

Hosts and Vectors of Plague Infection

Plague infection among animals

Evidence of plague infection has been found in a multitude of mammalian species. Poland & Barnes (1979) note that on a world basis, members of eight Orders, 95 genera, and well over 200 species have been found plague-positive or to have antibodies indicating prior infection. For the United States during the 1970—1980 surveillance period, we found evidence of plague infection in 76 species of five mammalian Orders (Rodentia, Lagomorpha, Insectivora, Artiodactyla (deer), and Primates (man)). While some of the species represented may function as maintenance hosts and others as epizootic hosts of infection, many others (e.g. humans and deer) represent dead-end hosts from which no further transmission might be expected. Infections in some species may be of transient importance and are passed on by a few individuals even though the host—flea population complex of itself is incapable of sustaining a transmission cycle. Such individuals may at times play a key intermediary role between an epizootic source of infection and man, either by transport of infective rodent fleas, or by serving as a direct source of infection.

TABLE VII

Animal species associated epidemiologically with human plague cases in the United States, 1974—1980.

Mammal groups	No. of cases	Percent of cases
Sciurid Rodents		
Spermophilus variegatus	44	41.9
S. beecheyi	11	10.5
Cynomys gunnisoni	4	3.8
Eutamias spp.	1	0.9
Undetermined sciurids[a]	3	2.9
Cricetid Rodents		
Neotoma cinerea	2	1.9
Rodents		
Undetermined species	11	10.5
Lagomorphs		
Sylvilagus spp.	7	6.5
Carnivores		
Canis latrans	1	0.9
Felis catus	4	3.8
Association unknown	17	16.2
TOTAL	105	100.0

[a] Either *Eutamias* spp. or *S. lateralis.*

Animal sources of human infection

Table VII shows animal species found to be epidemiologically associated with 105 human cases from 1974 through 1980. Epizootics among sciurid rodents and their fleas were responsible for 63 (60%) of the 105 cases; 44 (41.9%) originated from *S. variegatus* (rock squirrels) and 11 from *S. beecheyi* (California ground squirrels), while eight came from other squirrels and their fleas. Two co-primary cases originated in southern Oregon from *N. cinerea*, a wood rat. Although we have found wood rats of this and other species associated with plague cases in the past, they have always before been found to play an intermediary role between squirrel epizootics and the human victim by the transport of infective fleas. The one human case associated with *C. latrans* (coyote) resulted when an 11-year-old boy attempted to remove the pelt from an animal he found dead near his home (von Reyn, Barnes, Weber, Quan & Dean, 1976).

Seven cases from 1974 to 1980 were directly attributable to plague infected rabbits or hares. Von Reyn, Barnes, Weber & Hodgin (1976) reviewed eight human cases associated epidemiologically with wild lagomorphs from 1950 to 1974, including three of the seven cases counted in Table VII. Of the total 12 cases, 11 occurred during winter months from November through January; one case occurred in July. All were associated with the killing and skinning of animals or their preparation as food, and inoculation of *Y. pestis* through a break in the skin on the victim's hand or arm.

Von Reyn, Barnes, Weber & Hodgin (1976) suggest the possibility of a winter epizootic cycle among rabbits and their fleas. Our observations, however, show no evidence of epizootics among them but, rather, suggest that rabbits are individual victims of plague acquired from ecologically associated rodents.

Four cases were contracted directly from domestic cats (*Felis catus*). In three cases, infection resulted when owners attempted to care for their plague-infected pets (CDC Records). In a California case, the victim died of primary pneumonic plague, the first such case in the United States since 1925 (California Department of Health, 1980). Her cat, having died shortly before, was exhumed and shown to have expired from secondary plague pneumonia acquired from an epizootic among chipmunks in the neighborhood. In another case, a child became infected after being bitten and clawed by his cat and in a 1981 case (not counted in Table VIII) a veterinarian acquired plague after being bitten by a feline patient. Four of the cats died and *Y. pestis* was isolated from their tissues. The veterinarian's patient recovered and was later found to have an antibody titer of 1:8192 to plague.

Epizootic host/flea complexes

Plague epizootics at any particular point in time usually involve one or, sometimes, two rodent/flea complexes that play principal roles in the dynamics of plague amplification and without which the epizootic would collapse. Arguments in the literature concerning "monohostal" versus "polyhostal" plague foci (Rall & Federov, 1960; Kalabukhov, 1965) would seem to be without general relevance; if two susceptible rodent populations coexist or overlap in place, plague infection in one will almost inevitably transfer to individuals of the other. If the second population is receptive in terms of its population density, flea vector abundance and other factors that encourage epizootic amplification, it will become centrally involved in its own epizootic. If it is not receptive in terms of one or more key factors, infection may be confined to individuals that acquire it directly from the first population. To further complicate the situation, time and season of receptivity vary among species and populations. In this complex and shifting milieu, it is often difficult to separate the principals from the bit players, especially since their roles may change with time and in space. The roles of some, however, are all too obvious in terms of flea abundance and infection rates, animal mortality, ultimate effect on the rodent herd, and human exposure potential. Principal rodent/flea complexes among which epizootics were observed during the 1970–1980 surveillance periods are listed in Table VIII, along with key landscape indicators and the regions in which they were found to occur. Further discussion will focus on those epizootic complexes best known for their human exposure potential or their widespread and prominent occurrence.

Regional Epizootic Centers

The south-west plague focus

As shown earlier (p. 244), by far the greatest number (83%) of wild-rodent-associated human plague cases in the United States during the past 10 years have originated in the south-west within an area which includes most of New Mexico, north-eastern Arizona, southern Colorado, and southernmost Utah. Two major amplifying host/flea complexes exist in the area, overlapping in ecotonal situations: one composed of prairie dogs (*C. gunnisoni*) and its fleas, *O. hirsutus* and *O. tuberculatus*; the other of the rock squirrel (*S. variegatus*) with its fleas *Diamanus montanus* and *Hoplopsyllus anomalus*. The prairie dog complex occurs in semi-arid grasslands throughout the south-west; the rock squirrel complex, also widespread, occurs in uplands characterized by open to closed pinon-juniper and Gambel oak woodland at elevations of about 1500 to 2500 m.

Epizootics among *C. gunnisoni* described by Kartman, Quan & Lechleitner (1962), Lechleitner, Tileston & Kartman (1962) and Lechleitner, Kartman, Goldenberg & Hudson (1968), involved

TABLE VIII

List of host–flea complexes found involved in epizootic plague amplification in western North America, by geographic region

	Principal rodent hosts and flea vectors involved	
State and regions	Rodent species	Flea vectors
Arizona, New Mexico, southern Colorado, southern Utah	*Spermophilus variegatus*	*Diamanus montanus, Hoplopsyllus anomalus*
Arizona, New Mexico, Colorado, Utah (Rocky Mountains and west)	*Cynomys gunnisoni*	*Opisocrostis hirsutus, O. tuberculatus cynomuris*
Colorado (east of Rocky Mountains), western Texas, Oklahoma, Kansas	*C. ludovicianus*	*O. hirsutus, O. tuberculatus cynomuris*
Wyoming, north-western Colorado, north-eastern Utah (high plains)	*C. leucurus*	*O. tuberculatus cynomuris, O. hirsutus*
Colorado, Idaho, Montana, Wyoming (mountain parks, high plains grassland)	*S. richardsoni*	*O. labis, Oropsylla idahoensis,* (Rocky Mountains) *O. t. tuberculatus, Thrassis bacchi*
California, Oregon, northern Nevada, south-eastern Idaho (montane meadows, Great Basin sagebrush-grasslands)	*S. beldingi*	*Thrassis francisi, T. pandorae, T. petiolatus, Opisocrostis t. tuberculatus*
Southern Idaho, eastern Oregon, Nevada, Utah (Great Basin sagebrush)	*S. townsendi*	*T. francisi*
Idaho, Utah, Wyoming (Great Basin and montane, 4000–8000′)	*S. armatus*	*T. pandorae, T. francisi*
California, Oregon, western Nevada (valleys, foothill savanna, open pine forest, to temperate rain forest edge)	*S. beecheyi*	*D. montanus, H. anomalus*
Arizona, California, Colorado, Idaho, Montana, Nevada, New Mexico, Oregon (montane areas, open pine forest)	*S. lateralis*	*Oropsylla idahoensis, D. montanus* (Sierra-Cascade), *O. labis* (Rocky Mountains)
Western United States from Rocky Mountains westward	*Eutamias* spp.[a] 16 species	*Monopsyllus eumolpi M. ciliatus, M. eutamiadis, M. fornacis* (last three from the Pacific States only)
Western United States from Texas to the Pacific States (desert to high montane shrubby habitats)	*Neotoma* spp.[b] 8 species	*Orchopeas sexdentatus, O. neotomae, Anomiopsyllus* spp.

TABLE VIII (*continued*)

State and regions	Principal rodent hosts and flea vectors involved	
	Rodent species	Flea vectors
Colorado, Wyoming, California (urban residential and rural environments)	*Sciurus niger* [c]	*Orchopeas howardi*

[a] Individuals of nine species were found to have been plague-infected or carried plague-positive fleas.
[b] Individuals of five species were found to have been plague-infected or carried plague-positive fleas.
[c] This peridomestic species has been introduced to western cities as a park squirrel along with its flea, *O. howardi.*

populations resident in high mountain parklands in Colorado and the flea *Opisocrostis tuberculatus cynomuris* as the predominant transmitter. Subsequent observations by Maupin (1970) have shown that at lower elevations the flea *O. hirsutus* is dominant in summer and is replaced by *O. t. cynomuris* in winter. Both species appear to transmit the disease equally well among prairie dogs; an epizootic involving *O. tuberculatus* in the spring continues at the same pace as *O. hirsutus* becomes abundant and *O. tuberculatus* numbers diminish with the onset of summer.

Devastating plague epizootics are common among *C. gunnisoni*, having been reported from one locality or another in the south-west in each of the past 15 years. Mortality in a colony during a plague epizootic typically exceeds 99%. Epizootics may be sporadic and localized in small colonies, but in large continuous colonies may sweep across hundreds of square kilometers. Colonies almost eradicated by the disease require approximately four to five years to regenerate to their former levels and once more become receptive to a plague epizootic. Many colonies kept under observation as a surveillance measure have regenerated only to be attacked again and nearly eradicated. Similar epizootics have been observed among *C. ludovicianus*, *C. leucurus*, and *C. parvidens*.

Although mortality is great and flea infection rates may reach tremendous levels, human cases resulting from prairie dog plague, regardless of the species, are relatively few (Table VII). Those cases that are acquired from prairie dog sources usually result from direct contact with an infected animal rather than by flea-bite, suggesting that *Opisocrostis* spp., although excellent vectors among prairie dogs, may be reluctant to bite humans.

Epizootics among rock squirrels (*S. variegatus*), have not been recognized previously in the literature as a major source of human plague infection, although a human case was attributed to the species in 1938 and *Y. pestis* was isolated from them and their fleas as early as 1936 in Utah (Eskey & Haas, 1940). *S. variegatus* typically live in small aggregations of low to moderate density, digging burrow systems into sharply cut dry stream banks and under rocky outcrops. They quickly respond to added harborage provided by man in the form of refuse dumps, masonry walls, and similar artifacts. Given the harborage, rock squirrels tend to live in peridomestic association with people, thus increasing human exposure potential during epizootics. Although rock squirrels suffer heavy mortality from plague, survivors with antibodies frequently are found after an epizootic, indicating that the population response to plague among rock squirrels, unlike that among prairie dogs, is heterogeneous.

Rock squirrels are heavily parasitized by two flea species: *D. montanus* and *Hoplopsyllus anomalus*. Both species also parasitize the California ground squirrel (*Spermophilus beecheyi*), a near relative of the rock squirrel in the Pacific States. *Diamanus* typically is most abundant during the moister seasons (Ryckman, 1971), thus in California's Mediterranean climate it is a winter flea, but in the south-west we have found it to be most abundant during the summer rainy season. *H. anomalus* replaces *D. montanus* during the dry season. Of the two species, *Diamanus* has been found to be an efficient plague vector in laboratory studies; *H. anomalus* a poor one (Eskey & Haas, 1940; Wheeler & Douglas, 1945; Burroughs, 1947; Kartman & Prince, 1956). In one field study in New Mexico (Vector-Borne Diseases Division, 1979) we found *D. montanus* to comprise approximately 85% of the flea biomass removed from the total of all rodents captured in saturation trapping studies conducted in favorable *S. variegatus* habitat. We have observed *Diamanus* to be an extremely aggressive biter that readily feeds on alternative hosts, including man, when separated from its normal host during and after an epizootic. In plague case investigations it frequently is found to be the predominant species on other host species in the environment, including domestic pets and humans.

Rapid human population growth and new developments in the pinon-juniper uplands of the south-west appear to have increased the *S. variegatus* population rather than decreased it, by providing additional harborage. In a recent case control study, it was shown that residences with refuse piles, dilapidated outbuildings and other evidence of poor general sanitation were more likely to produce plague cases than better-tended residences of their nearest neighbors (Mann *et al.*, 1979). Thus, it appears that south-western plague cases are a premises problem as well as a "wild" rodent disease.

The Pacific coast focus

Plague cases from California and southern Oregon in recent years have continued to originate from two primary epizootic complexes: (1) *S. beecheyi* and its fleas, *D. montanus* and *H. anomalus*, and (2) a somewhat diffuse and not entirely defined group of montane sciurids which includes several species of chipmunks (*Eutamias* spp.) and golden-mantle squirrels (*S. lateralis*). Future analysis may prove the latter entity to consist of more than one complex.

S. beecheyi, the first North American wild rodent species from which plague was isolated (Wherry, 1908), remains the most important sylvatic species in plague epidemiology in the Pacific Focus (Nelson, 1980). Most of the human cases that occur in California are associated with this rodent and its flea *D. montanus*. The species is widely distributed in California and southern Oregon, where it occurs from sea level up to about 2500 m elevation. In valley grasslands and foothill savanna in California, *S. beecheyi* often lives in dense and widespread aggregations. In montane forested areas, the squirrel exploits open or disturbed areas where it lives in small, dispersed family groups; however, aggregations are quickly formed in areas opened by construction, particularly those that provide cutbanks, road embankments, rock piles, or stumps left by the removal of trees. The construction of electrical transmission lines through coniferous forest areas, which requires tree removal, provides habitat for dense populations of *S. beecheyi* in otherwise less hospitable areas of California and southern Oregon (Murray, 1971 unpublished).

According to Nelson (1980), epizootics among California ground squirrels living in dense and widespread colonies tend to be more violent and rapid than among those living in smaller aggregates. Smaller groups in montane areas sometimes do not become involved in plague among ecologically associated species. Nelson (1980) also notes that resistance to plague appears to be developing in populations in the Sierra Nevada and San Jacinto Mountains where ground squirrels have repeatedly been subject to epizootics, a phenomenon anticipated by Meyer (1942).

Montane plague cases in the Sierra-Cascade regions of California and Oregon typically are acquired during summer recreation or vacationing activities and are associated with periodic epizootics among chipmunks (*Eutamias* spp.) and golden-mantle squirrels (*S. lateralis*). Human cases usually originate from campground or vacation development areas where large populations of chipmunks and golden-mantle squirrels accumulate in response to feeding by tourists. Chipmunk feeding is an appealing activity to campers and summer residents from urban areas who otherwise see little unrestrained animal life. Campgrounds and vacation homes often provide harborage in excess of that naturally available, thus further encouraging high rodent populations. Chipmunks readily invade walls and attics of mountain residences and build nests.

The assortment of fleas on chipmunks includes *Monopsyllus eumolpi*, a chipmunk flea known to bite man readily, and a variety of other species of varying local and seasonal distributions. *Oropsylla idahoensis* and, in the Sierra Nevada Mountains, *D. montanus* are predominant on golden-mantle squirrels.

The montane biocoenoses — open coniferous forest, coniferous forest edge, and montane grasslands of the Sierra-Cascades — are ecological counterparts of biocoenoses in the Rocky Mountain region also subject to periodic epizootics more or less coincident in time. One must speculate as to why human plague cases associated with these species occur with some regularity, if low frequency, in the Sierra-Cascades and are extremely rare in the Rockies. Golden-mantle squirrels occur in both regions. The chipmunks are different species but for the most part occupy very similar niches. *S. beldingi* in the Sierra-Cascades occupies a closely similar niche to that of *S. richardsoni* in the Rockies. Epizootics in chipmunk and golden-mantle squirrel populations often are devastating in both the Sierra-Cascade and Rocky Mountains. Survival rate among rodents follow-ing epizootics tends to vary considerably from one site to another in both regions. Two principal flea species are shared between the two regions: *Oropsylla idahoensis* on golden-mantle squirrels and *Mono-psyllus eumolpi* on chipmunks. The Sierra-Cascade flea fauna, in general, is more varied than that in the Rocky Mountains, but little is known concerning species transmission capabilities or tendencies to bite man. The only observed significant difference between the two regions is that *D. montanus* is commonly and frequently found on golden-mantle squirrels in the Sierra-Cascades, but very rarely in the Rockies where golden-mantle squirrels and rock squirrels rarely occur in the same biocoenose.

The northern foci

The smaller ground squirrels, *S. beldingi, S. townsendi, S. armatus, S. richardonsi* and their relatives represent a widely distributed group in montane and northern areas that are subject to regional or localized plague epizootics. Little information is available for generalization as to the effects of plague in their populations: in some instances, we have seen populations almost eradicated; in others we have witnessed widespread decimation but little long-term impact on populations.

Eskey & Haas (1940) noted that a much greater variety of flea species is associated with the smaller ground squirrels than with their larger relatives, e.g. prairie dogs, rock squirrels, and California ground squirrels. We have confirmed their observations, but note that at any particular place and season, one or two species predominate just as on the larger squirrel species.

The human infection potential stemming from epizootic plague among this assemblage of squirrels appears to be slight. Violent and widespread epizootics among the various species have been recorded in the past (Eskey & Haas, 1940), but up to now are not known to have caused a human case. For example, in 1974–1976 a regional epizootic swept through *S. richardsoni* and *S. lateralis* populations in the Rocky Mountains and High Plains of Colorado and Wyoming. Other mammals found to be involved were several *Eutamias* species, *Sciurus aberti, Tamiasciurus hudsonicus,* and *Peromyscus maniculatus*. Serologic evidence of infection was obtained from weasels (*Mustela frenata*), coyotes (*C. latrans*), and martens (*Martes americana*). *Sciurus niger*, an eastern tree squirrel introduced and adapted to city parks in western North America, was found infected in urban Cheyenne, Wyoming, and Fort Collins, Colorado. No human cases resulted from the two-year epizootic even though animals died of plague in abundance in crowded campgrounds, city parks, golf courses, and other heavily populated areas. Large populations of *S. richardsoni* were central to the epizootic and played the major role in its amplification and geographic proliferation.

As described earlier (p. 250), extensive epizootics occurred among *S. townsendi* in the Great Basin Region of southern Idaho and northern Nevada during the 1970–1980 period, but produced no human cases. The few scattered cases reported from these foci during the past 10 years all have resulted from direct animal exposure to peripheral mammal species. As is probably the case with prairie dog fleas, the fleas of the several smaller ground squirrel species, although effective vectors among their own hosts, do not readily bite humans.

PLAGUE CONTROL

Control of commensal rodent plague

Although no human cases have been attributed to commensal rodents or their fleas in the United States since 1925, the recrudescence of rat plague in enzootic areas remains a real threat. *Y. pestis* was last isolated from rats in San Francisco in 1963 (Kartman, 1963), and surveillance activities in Tacoma, Washington, detected rat plague in 1942–1944 (Hundley & Nasi, 1944), 1954, and 1971 (Barnes, 1971). Surveillance for plague among commensal rodents and their fleas is a continuing activity in both of these cities and in Los Angeles, California, where a single *Rattus rattus* was found infected in 1981.

Rat control programs in both urban and rural areas in the western United States represent the most important barrier to the recurrence of epidemic plague, particularly in Pacific Coast cities. Most urban rodent control programs focus on environmental sanitation at both the individual and community level. Where the use of rodenticides is considered necessary, anticoagulant baits usually are used although acute rodenticides are occasionally used in sewers and other situations not accessible to people, pets, or wildlife. In some cities, anticoagulant baits may be issued to individuals rather than applied by community program personnel. Resistance to routinely used anticoagulant rodenticides has become a common phenomenon in both rural and urban situations in the United States, particularly where rodenticidal control has been used as the primary means of reducing rat populations (Brooks & Bowerman, 1974; Jackson, Brooks, Bowerman & Kaukeinen, 1973). As resistance is observed, commonly used rodenticides are being replaced by newer anticoagulant rodenticides such as Bromodialone and Brodifacoum or, in some cases, by older acute rodenticides.

Wild rodent plague

The geographic scope and the ecological complexity of wild rodent plague in western North America creates a diverse set of problems. During peak plague years, epizootics move through tens of thousands of square kilometers. Many occur in areas only sparsely inhabited by man. Mass control actions, even if technology were available, could not be economically justified by the risk of infection to man nor would they be environmentally acceptable. For these reasons, control strategies are aimed primarily at prevention of human contact with wild rodent/flea reservoirs by specific actions to reduce reservoir and vector populations in areas of human residence and activity. Because of the periodic or fluctuant expression of epizootic plague in animal populations, sound surveillance is indispensable.

Control of wild rodent fleas

Although human plague cases may be acquired via direct contact with infective animal tissues or body fluids, the bite of infective fleas remains the most frequent route of infection to man, particularly in the south-western and Pacific Coast foci where the majority of human cases occur. Insecticidal control of wild rodent fleas therefore is the most effective means of interrupting human contact with the vast, unmanaged reservoirs of infection present in the two regions.

Insecticidal control operations, conducted by local, state or Federal agencies, are precipitated by: (1) the presence of susceptible rodents and vector fleas in areas of human activity, plus the imminent possibility of their exposure as revealed by surveillance activities, (2) the observation of plague among susceptible wild rodents in areas of human activity; and (3) detection of residual plague activity among rodents and/or fleas where one or more human cases have originated. Such measures are limited to areas of actual or potential human exposure at the time of treatment. Under this regime, routine and repetitive treatments which might lead to development of insecticide resistance are avoided.

Carbaryl (Sevin[R]) in 5% or 10% dust formulations is the only insecticide presently registered for the control of wild rodent fleas in the United States. It has proven effective against *O. hirsutus* on prairie dogs (Barnes, Ogden & Campos, 1972), and *D. montanus* on rock squirrels and California ground squirrels (CDC records), and on several flea species on *P. maniculatus* (Barnes, Ogden, Archibald & Campos, 1974). In California, however, it has been reported as ineffective in the field against *Monopsyllus eumolpi* on chipmunks and *Oropsylla idahoensis* on golden-mantle squirrels during widespread epizootics in 1975–1976 (Bernard C. Nelson, pers. comm.). An additional problem with carbaryl is its short half-life; although its persistence varies widely under different field conditions, carbaryl cannot be expected to remain effective more than two weeks, thus requiring repetitive treatment of populations subject to reinfection.

A great many other insecticides have been found effective against wild rodent fleas in the United States. DDT, known to be extremely effective in field use, has been banned along with BHC and other chlorinated hydrocarbons. A number of other compounds have been found relatively effective, including diazinon, dursban, and Baygon, none of which is presently registered for wild rodent flea control. During widespread epizootics in Colorado and California in 1975 and 1976, it was necessary to obtain official exemption to use DDT rather than carbaryl: in California, because of field resistance, in Colorado because trained manpower was not available to apply the repetitive treatments that would have been necessary with carbaryl.

The means by which insecticidal dusts are delivered to the target species are at least as important as the insecticide used. In general, we use one or both of two target-directed delivery methods: burrow or runway dusting against the fleas of animals which construct burrows or trails that can be successfully located and treated, and insecticide bait stations (Kartman, 1958) against fleas or hosts that do not. Area dusting, in our experience, has been relatively ineffective except in cases where the volume of insecticide used has been so great as to endanger man and other non-target species. No conclusive statement can be made as to the best of the two methods against any specific host/flea complex. Burrow dusting is most effective against prairie dog, rock squirrel, and California ground squirrel fleas, but in some situations squirrel burrows may be hidden or inaccessible and bait stations must be used. Control failures may occur because of failure to analyse the problem; e.g. when DDT treatment of golden-mantle squirrel burrows appeared to fail in a California montane campground, we discovered that most of the animals being trapped for assessment of the treatment were attracted daily to the campground but most did not reside in it. In this case, bait stations were used to attract and treat the rodent visitors and control of fleas was quickly achieved. Dominant species and dominant individuals within species may prevent entry of other animals, thus leaving the subordinate species untreated. In montane campgrounds where rodent populations may achieve abnormally high densities, it is important that sufficient numbers of stations are used so that most of the population is treated. In some situations, the bait stations themselves may alter the structure of rodent populations by serving as an attractant, causing shifts in rodent behavior and habitat utilization.

Control of fleas on dogs and cats is of prime concern in the southwestern focus where it is believed that pets may serve as transporters of fleas from rodents to man (Weber, 1977). A much wider range of insecticidal materials is available for this purpose than for control of fleas on wild rodents and many formulations can be purchased from markets and pet shops. On Indian Reservations in the south-west, 5% carbaryl dust for flea control on pets is distributed by the Indian Health Service and the Indian tribes to individuals during plague season.

Wild rodent control

Wild rodent control for any reason is viewed with skepticism and concern by the public and environmental groups. California is the only state to consistently exert control measures against wild rodents for plague control. The program, which began early in the century with the avowed objective of "eradicating" plague by the destruction of California ground squirrels, failed to achieve this sweeping goal, but, nevertheless, has continued on a management basis in conjunction with agriculturally-oriented programs. Human plague cases have not increased significantly in the state during the past 50 years despite an eight-fold increase in human population and a continuing high level of plague activity in rodents other than California ground squirrels.

California has a background of 70 years of ground squirrel control during which the populace and public agencies have come to consider the California ground squirrel as a pest and a potential health hazard. California ground squirrel infestations rarely are tolerated around residences and in public areas. Both local and state agencies exist to assist residents in coping with the problem much as they would assist with a commensal rodent problem. The fact that California ground squirrels are well known as an agricultural pest also contributes to the public attitude toward them.

No organized control program has been undertaken against rock squirrels in the south-west despite the fact that their populations are highly amenable to environmental and rodenticidal control actions. Rock squirrels in the southern Rocky Mountain region are not considered a pest species, do little damage to agriculture, and are accepted by most residents as a normal and attractive part of the biota. Infestations near residences and public areas are viewed as part of nature even though human artifacts such as rock walls, refuse piles, and streamside riprap encourage their presence and increase local population densities. The south-western lifestyle includes a tendency to "live close to nature", thus the poorest to the wealthiest seek to live in a "natural" rather than an artificially landscaped setting. Proposals to control rock squirrels in this setting by the use of rodenticides are certain to encounter significant resistance from environmentalists. Removal of man-made harborage, accompanied by flea control and rodenticidal reduction of populations along the dry stream beds that intersect many south-western towns, could produce significant long-term reduction of human plague potential.

The effective use of rodenticides remains important to plague control. However, in recent years, the use of acutely toxic materials such as compound 1080, zinc phosphide, strychnine and cyanide has been severely restricted or banned, to be replaced by slow acting anticoagulants. Two newer, faster acting anticoagulants, Brodifacoum and Bromodialone, have shown promise against California ground squirrels and other sciurids but have not as yet been licensed for use.

It should be emphasized that control of wild rodents for plague control should be preceded or accompanied by flea control (Poland

& Barnes, 1979), otherwise the control measures themselves may release infected fleas into the environment.

Health Education

The appropriate and timely use of news media, informational circulars, pamphlets, and word-of-mouth information circulated in the form of lectures and training courses, and to members of the medical community and the public, are key elements of plague control and of health education in general.

Many wild rodent plague cases, particularly those that result from direct contact with infected animals, could be avoided, given the information and awareness of the hazard on the part of individuals. The dispensing of plague information with hunting licenses plus news releases to sportsmens' magazines and the news media may be the only effective means of reducing cases acquired by rabbit hunters in the western United States. With appropriate information and encouragement, individuals may also be convinced to avoid potentially plague-infected rodents and fleas, to dust pets for wild rodent fleas, to remove rodent harborage and to rodent-proof houses, and to carry out premises control measures against rodents and fleas on their own properties.

In New Mexico, California, and Oregon, campers and visitors to recreation areas receive plague leaflets explaining the hazard of plague, how to avoid it, and to whom to report if evidence of rodent morbidity or mortality is observed. In New Mexico and California, known plague-infected areas are posted during epizootics and in both California and Colorado, plague-infected recreation areas may be closed.

In addition to public information, the medical community is kept informed of plague surveillance information and provided with current information on case occurrence, diagnosis, and therapy.

REFERENCES

Archibald, W. S. & Kunitz, S. J. (1971). Detection of plague by testing serums of dogs on the Navajo Reservation. *HSMHA Health Rep.* **86** (5): 377–380.
Audy, J. R. (1958). The localization of disease with special reference to the zoonoses. *Trans. R. Soc. trop. Med. Hyg.* **52**: 308–328.
Baltazard, M. (1953). Sur la resistance à la peste de certaines espèces de rongeurs sauvages. I. Faits observés dans la nature. *Annls Inst. Pasteur Paris* **95**: 411–423.
Baltazard, M., Karimi, Y., Eftekhari, M., Chasma, M. & Mollaret, H. H. (1963). Interepizootic conservation of plague in its inveterate foci: a working hypothesis. *Bull. Soc. exot. Pathol.* **56**: 1230–1241.

Barnes, A. M. (1971). Tacoma rat plague situation—1971. *Proc. Int. Northw. Conf. Dis. Nature Commun. Man* **26**: 41—42.

Barnes, A. M., Ogden, L. J., Archibald, W. S. & Campos, E. G. (1974). Control of plague vectors on *Peromyscus maniculatus* by use of 2% carbaryl dust in bait stations. *J. med. Entom.* **11**: 83—87.

Barnes, A. M., Ogden, L. J. & Campos, E. G. (1972). Control of the plague vector, *Opisocrostis hirsutus*, by treatment of the prairie dog (*Cynomys ludovicianus*) burrows with 2% carbaryl dust. *J. med. Entom.* **9**: 330—333.

Brooks, J. E., & Bowerman, A. M. (1974). An analysis of the susceptibilities of several populations of *Rattus norvegicus* to warfarin. *J. Hyg. Camb.* **73**: 401—407.

Burroughs, A. L. (1947). Sylvatic plague studies. The vector efficiency of nine species of fleas compared with *Xenopsylla cheopsis*. *J. Hyg., Camb.* **45**: 371—396.

California Board of Health (Unpubl.). *Plague, Los Angeles, 1924—1925, Sacramento*, 1926.

California Department of Health (1980). California morbidity. *Weekly Rep. Calif. Dept Hlth* No. 48, Dec. 5, 1980.

Collins, R. N., Martin, A. R., Kartman, L., Brutsche, R. L., Hudson, B. W. & Doran, H. G. (1967). Plague epidemic in New Mexico, 1965. *Publ. Hlth Rep.* **82**: 1077—1099.

Cruickshank, J. G., Gordon, D. H., Taylor, P. & Naim, H. (1976). Distribution of plague in Rhodesia as demonstrated by serological methods. *Centr. Afr. J. Med.* **22**: 127—130.

Ecke, D. H. & Johnson, C. W. (1952). Plague in Colorado and Texas. I. Colorado. *Pub. Hlth Serv. Monogr.* No. 6: 1—37.

Eskey, C. R. & Haas, V. H. (1940). Plague in the western part of the United States. *Pub. Hlth Bull.* **254**: 1—83.

Hudson, B. W. & Quan, T. J. (1975). Serologic observations during an outbreak of rat-borne plague in the San Francisco Bay area of California. *J. Wildl. Dis.* **11**: 431—436.

Hundley, J. M. & Nasi, K. W. (1944). Anti-plague measures in Tacoma, Washington. *Publ. Hlth Rep.* **59**: 1239—1255.

Jackson, W. B., Brooks, J. E., Bowerman, A. M. & Kaukeinen, D. E. (1973). Anticoagulant resistance in Norway rats in U.S. cities. *Pest Contr.* **43**(4): 12, 14—16.

Kalabukhov, N. I. (1965). The structure and dynamics of natural plague foci. *J. Hyg. Epidemiol. Microbiol. Immunol.* **9**: 147—158.

Kartman, L. (1958). An insecticide-bait-box method for control of sylvatic plague vectors. *J. Hyg., Camb.* **56**: 455—465.

Kartman, L. (1963). Plague infection in *Rattus rattus* in San Francisco. *Zoonoses Res.* **2**: 131.

Kartman, L. & Prince, F. M. (1956). Studies on *Pasteurella pestis* in fleas. V. The experimental plague-vector efficiency of wild rodent fleas compared with *Xenopsylla cheopis*, together with observations on the influence of temperature. *Am. J. trop. Med. Hyg.* **5**: 1056—1078.

Kartman, L., Prince, F., Quan, S. & Stark, H. (1958). New knowledge on the ecology of sylvatic plague. *Ann. NY Acad. Sci.* **70**: 668—711.

Kartman, L., Quan, S. F. & Lechleitner, R. R. (1962). Die-off of a Gunnison's prairie dog colony in central Colorado. II. Retrospective determination of plague in flea vectors, rodents, and man. *Zoonoses Res.* **1**: 201—204.

Lechleitner, R. R., Kartman, L., Goldenberg, M. I. & Hudson, B. W. (1968). An epizootic of plague in Gunnison's prairie dogs (*Cynomys gunnisoni*) in south-central Colorado. *Ecology* **49**: 734—743.

Lechleitner, R. R., Tileston, J. V. & Kartman, L. (1962). Die-off of a Gunnison's prairie dog colony in central Colorado. I. Ecological observations and description of the epizootic. *Zoonoses Res.* **1**: 185—199.

Link, V. B. (1955). A history of plague in the United States of America. *Publ. Hlth Rep. Wash.* **70**: 1—120.

Macchiavello, A. (1954). Reservoirs and vectors of plague. *J. trop. Med. Hyg.* **57**: (part in each issue).

Mann, J., Martone, W. J., Boyce, J. M., Kaufmann, A., Barnes, A. & Weber, N. S. (1979). Endemic human plague in New Mexico: Risk factors associated with infection. *J. infect. Dis.* **140**: 397—401.

Maupin, G. O. (1970). *A survey of the Siphonaptera and ectoparasitic and inquiline Acarina associated with the black-tailed prairie dog,* Cynomys ludovicianus. M.S. Thesis: Colorado State University, Fort Collins.

McCoy, G. W. (1908). Plague in ground squirrels. *Publ. Hlth Rep.* **23**: 1289—1293.

McCoy, G. W. (1911). Studies upon plague in ground squirrels. *Publ. Hlth Bull.* **43**: 1—53.

Meyer, K. F. (1942). The ecology of plague. *Medicine* **21**: 143—174.

Miles, V. I., Kinney, A. R. & Stark, H. E. (1957). Flea-host relationships of associated *Rattus* and native wild rodents in the San Francisco Bay area of California, with special reference to plague. *Am. J. trop. Med. Hyg.* **6**: 752—760.

Miles, V. I., Wilcomb, M. J. & Irons, J. V. (1952). Plague in Colorado and Texas. II. Rodent plague in the Texas south plains. *Publ. Hlth Monogr.* No. 6: 41—43.

Murray, K. F. (1971 unpubl). *Epizootic plague in California, 1965—1968.* California Department of Health Bureau Vector Control.

Nelson, B. C. (1980). Plague studies in California—the roles of various species of sylvatic rodents in plague ecology in California. *Proc. Ninth Vertebrate Pest Conf.* 89—96. Clark, J. (Ed.). Fresno, Calif.

Pavlovsky, E. N. (1966). *Natural nidality of transmissible diseases with special reference to the landscape epidemiology of zooanthroponoses.* Urbana and London: University of Illinois Press.

Poland, J. D. & Barnes, A. M. (1979). Plague. In *CRC Handbook series in zoonoses, Section A: Bacterial, rickettsial, and mycotic diseases* **1**: 515—556. Steele, J. F. (Ed.). Boca Raton, Fla.: CRC Press.

Poland, J. D., Barnes, A. M. & Herman, J. J. (1973). Human bubonic plague from exposure to a naturally infected wild carnivore. *Am. J. Epidemiol.* **97**: 332—337.

Pollitzer, R. (1954). Plague. *W.H.O. Monogr. Ser.* **22**: 1—698.

Pollitzer, R. & Meyer, K. D. (1961). The ecology of plague. In *Studies in disease ecology*: 433—501. May, J. F. (Ed.). New York: Hafner.

Rall, Y. U. & Federov, V. N. (1960). The problem of physiographical evaluation of rodents as plague carriers and the monohostal character of natural nidi of plague. *Zh. Mikrobiol. Epidemiol. Immunobiol.* **31**: 29—35.

Rust, J. H., Cavanaugh, D. C., O'Shita, R. & Marshall, J. D. (1971). The role of domestic animals in the epidemiology of plague. I. Experimental infection of dogs and cats. *J. infect. Dis.* **124**: 522—526.

Rust, J. H., Miller, B. E., Bahmanyar, M., Marshall, J. D., Purnaveja, S., Cavanaugh, D. C. & Hla, U.S.T. (1971). The role of domestic animals in the epidemiology of plague. II. Antibody to *Yersinia pestis* in sera of dogs and cats. *J. infect. Dis.* **124**: 527–531.

Ryckman, R. E. (1971). Plague vector studies: Part II. The role of climatic factors in determining seasonal fluctuations of flea species associated with the California ground squirrel. *J. med. Entom.* **8**: 541–549.

Taylor, P., Gordon, D. H. & Isaacson, M. (1981). The status of plague in Zimbabwe. *Ann. trop. Med. Parasit.* **75**: 165–173.

Vector-Borne Diseases Division (1976). *Vector-Borne Diseases Division 1975 Report.* Plague Branch activities. Fort Collins, Colo.: U.S. Dept. Health Educ. and Welfare, Center for Disease Control.

Vector-Borne Diseases Division (1977). *Vector-Borne Diseases Division 1976 Report.* Part II: Plague activities. Fort Collins, Colo.: U.S. Dept. Health Educ. and Welfare, Center for Disease Control.

Vector-Borne Diseases Division (1979). *Vector-Borne Diseases Division 1978 Report.* Part II: Plague activities. Fort Collins, Colo. U.S. Dept. Health Educ. and Welfare, Center for Disease Control.

Velimirovic, B. (1979). Review of the global epidemiological situation of plague since the last conference of tropical medicine in 1973. In *CRC handbook series in zoonoses.* **1**: 560–596. Steele, J. F. (Ed.). Boca Raton, Fla.: CRC Press.

von Reyn, C. F., Barnes, A. M., Weber, N. S. & Hodgin, U. G. (1976). Bubonic plague from exposure to a rabbit: a documented case in the United States. *Am. J. Epidemiol.* **104**: 81–87.

von Reyn, C. F., Barnes, A. M., Weber, N. S., Quan, T. J. & Dean, W. J. (1976). Bubonic plague from direct exposure to a naturally infected wild coyote. *Am. J. trop. Med. Hyg.* **25**: 626–629.

Weber, N. S. (1977). *Plague in New Mexico.* (A special publication of the State of New Mexico Environmental Improvement Agency.)

Wheeler, C. M. & Douglas, J. R. (1945). Sylvatic plague studies. V. The determination of vector efficiency. *J. infect. Dis.* **77**: 1–12.

Wherry, W. B. (1908). Plague among the ground squirrels of California. *J. inf. Dis.* **5**: 485–506.

White, M. E., Gordon, D., Poland, J. D. & Barnes, A. M. (1980). Recommendations for the control of *Yersinia pestis* infections. *Inf. Contr.* **1**: 324--329.

White, M. E., Rosenbaum, R. J., Canfield, T. M. & Poland, J. D. (1981). Plague in a neonate. *Am. J. Dis. Child.* **135**: 418–419.

WHO (1970). Expert Committee on Plague (4th Rep.). *W.H.O. Tech. Rep. Ser.* No. 447: 5–25.

WHO (1971). International health regulations (1969) Part II, Article 4: 11–12. Geneva: WHO.

WHO (1980). Plague surveillance and control. *W.H.O. Chron.* **34**: 139–143.

WHO (1981). Human plague in 1980. *WHO Weekly Epidemiol. Rec.* **56**: 273–275.

Wolf, K. & Hudson, B. W. (1974). Paper strip blood sampling technique for the detection of antibody to the plague organism, *Yersinia pestis. Appl. Microbiol.* **28**: 323–325.

DISCUSSION

Tyrrell (Chairman) — I was wondering whether you would like to tell us a little more about the control measures. Do you only deal with things at the edge — where you've got your wild habitat and the new house is being built — or do you ever try, as I gathered the Soviet plague control authorities do, to get advance warning of the enzootic in the animals and then try to head it off before it gets to areas where other vectors could take over?

Barnes — We haven't dealt with it on a mass basis as the Russians have. Furthermore, I don't think we ever shall, because I think in many instances it has been counter-productive in the United States. For example, in California in the early days it was believed that they could eradicate plague by eradicating the California ground squirrel, which was the major epizootic host of plague in California. They attempted that. They spent literally millions of dollars at a time when the dollar was worth a great deal more, and this programme went on from about 1910 up into the 1920s — it still goes on, as a matter of fact, though on a much more sensible basis of control under specific circumstances where man has made a misjudgement of increased animal populations, and it's done on a management basis. But it was obviously an impossibility, and I think ecologically detrimental; and people in California recognized that and a great deal of conflict came about that ultimately was resolved by the sort of management control programme of today.

Nelson — I was involved in a number of epidemics of plague in east Africa with Heisch, and the last publication that came out of this work was by Heisch and Karl Meyer, showing that enormous numbers of rodents in east Africa are serologically positive to plague even when there is no epizootic. I suspected that these were false positives, and I would like to know whether the maps that you produced of plague in the United States were based on similar serological evidence.

Barnes — Let me discuss this a little bit. Both Heisch and Meyer were the early-day heroes of plague, in their various spheres, and both of them made a mistake that any of us is prone to make, namely, they used a test before it was proven, and they had inadequate controls. The passive haemagglutination test for plague will cross-react with pseudotuberculosis — you're seeing pseudotuberculosis antibodies. (We use a passive haemagglutination inhibition on all PHA-positive specimens and we can eliminate those that cross-react.) This is obvious if you take Heisch's data from east Africa, also Meyer's from some of the offshore islands and also his big survey on the island of Hawaii, which was disastrous for the whole future — because of that mistake, they spent years fighting plague when it no longer existed in Hawaii. If you make a curve on titers in any population in which a disease has occurred or is occurring, you get a normal curve; if you look at Heisch's data, or Meyer's data, you will not see a normal curve at all: you'll see a fantastic number of low titers. I'm not going to condemn what was done: I would only condemn any of us if we should pay any further attention to it.

Tyrrell — Can I just get a little bit of amplification on that answer? I gather what you are implying is that you believe that in east Africa, like most other places, the disease is fairly fatal in rodents, and you don't have widespread silent infections going on over large areas when there's no evidence of disease activity?

Barnes — From what I can see in east Africa and what's been going on recently, it's a highly focal disease there; there are then periodic or sporadic outbreaks that move across great distances. There was one recently in Zimbabwe about

1975 or 1976 that was concurrent with our outbreak in the United States — and that's another strange thing about plague, outbreaks seem to occur concurrently on a world-wide basis, we don't have any idea why. In Zimbabwe Cruickshank and Taylor[1] used dog serosurveys to follow it across Zimbabwe and could check its progress and also identify its residual focal points. It made its foray and then retreated leaving little pockets of infection to continue, and that's what I expect in east Africa as well.

Steck — Do you have any indication of what factors were involved — you said that plague was disappearing from Hawaii? What mechanism was involved there?

Barnes — It was actually lack of mechanism, I think, in Hawaii. In the first place, the only mammal native to Hawaii — not so native as we might have thought, but at least it was there when Europeans settled the island — was *Rattus exulans hawaiiensis*; subsequently *Rattus norvegicus* and *Rattus rattus* were introduced, and in the sugar cane fields of Hawaii *Rattus rattus* became fairly common in the denser parts, while *exulans* was on the edge, and for a short period all the conditions coincided to enable plague to be supported locally here and there, possibly even by traveling from one place to another. But sooner or later, if you don't have consistent conditions, if you miss even one link that's necessary for the support of the disease in nature, you're going to lose it, and that's what ultimately happened on Hawaii: it died out on an island-by-island basis. And that has happened fairly frequently in other parts of the world too: for example Java, where plague remains endemic in central Java, but has disappeared from other places, because the lack of support, year after year, in terms of physical factors in the environment or biological factors, maybe in one crucial year would have eliminated it from this place or that.

Fielding — Is there any evidence that America's political and military adventures throughout the world could provide a mechanical vector in the retransportation of plague back to the North American continent from places in the Far East such as Vietnam or the Philippines?

Barnes — You have asked a question that is of interest to me because in 1969 I assisted in the planning of what was called a retrograde cargo programme from Vietnam, which was designed to prevent the reintroduction of both agricultural pests and public health pests from Vietnam, principally to Japan because that's where most of the cargo was going back to; and what we really found was that our main problem had to do with *Aedes* mosquitoes in old tyres. Plague wasn't much of a problem. We did have a couple of introductions of *Rattus exulans* into a Californian navy base at one time, neither of which was plague-positive, and a couple of rats were intercepted in Okinawa; but it wasn't a big problem at all, probably because of the way ships are constructed and the way cargo is handled now.

Tyrrell — Presumably you're implying too that, because of the way most people live and the way rats and other vectors are controlled, without doing anything very specific about it wildlife plague doesn't sweep across the human population in the way it did in the Middle Ages.

Barnes — Or in Vietnam. In fact I worked in Vietnam with WHO and also with the South Vietnamese Ministry of Health on plague control for quite a long time during that period, while plague was epidemic in Saigon and in Da Nang and Nha Trang, and the same thing happened there that happened when plague was intro-

---

[1] Cruickshank, Gordon, Taylor & Naim (1976) — see list of references, pp. 264–267.

duced in other tropical places: it persisted in the urban environment for a period
of time. Then such things as the development of herd immunity and resistance
on the part of the rat population, perhaps, caused it to disappear from the urban
environment and it spread to rural areas; and that's where the big plague
problem was at that time, during the retrograde cargo programme.

Gordon Smith — It must be recognized that there is plenty of evidence of transportation of *Aedes aegypti*, of gonococci, of mosquitoes on aircraft, indeed of
people and animals infected with pathogens and of mosquitoes, tsetse flies,
Simulium flies etc. from one place to another. While such transportation may
be increased by military operations and the disorders associated with war, this
is a much more widespread process.

Barnes — Well, when we get to page two we have to think in terms of receptivity
once it gets there. With regard to plague, for example, if you introduce a couple
of rats to Point Magoo in Santa Barbara, California, on a sterile naval base with
no receptive rat and rat flea populations, they're not going anywhere anyway.

Symp. zool. Soc. Lond. (1982) No. 50, 271–285

Conservation in Relation to Animal Disease in Africa and Asia

D. M. JONES

Institute of Zoology, The Zoological Society of London, Regent's Park, London NW1, England

SYNOPSIS

Disease in wild animals that have become well adapted to their habitat, over often tens of thousands of years, rarely constitutes a threat to that population. In recent years the spread of man and his domestic animals has frequently provided a direct source of some infective organism or parasite to which wild animals are not accustomed. By reducing the available habitat, many indigenous species suffer from a less adequate nutrition and become more susceptible to the clinical effects of organisms which otherwise would not have affected them.

In a few cases, wild animals are the natural but clinically unaffected hosts of a number of diseases which are potentially dangerous to domestic stock and even man. Because of this, large tracts of Africa have been depleted of many of the larger wild species, on the theory that this will reduce the disease risk to those animals of interest to man. Such actions have often been based on inadequate research. A further increasing threat to wildlife has been the use of agricultural chemicals, many of which are not now used in developed countries. This chapter will be illustrated by examples of these problems.

INTRODUCTION

Where animal populations have remained undisturbed, the role of disease as a limiting factor of those populations is probably relatively insignificant. When the habitat is altered, usually by man's activities, new diseases to which the wild animals are susceptible may be introduced, overcrowding may occur, or the availability and quality of the animal's diet may deteriorate, all of which results in a greater susceptibility to infection and parasitism. This chapter outlines some of the many ways in which the wild animal populations of Africa and Asia are influenced by disease, particularly highlighting conditions which have not been discussed in the earlier more specialized contributions.

DISEASE IN UNDISTURBED POPULATIONS

In general, overt disease in populations of animals which are living in equilibrium with their environment is probably uncommon, but because the undisturbed areas also tend to be remote, few detailed studies of wildlife mortality in such areas have ever been carried out.

Helminths

Many wild animals carry heavy burdens of endoparasitic helminths, but only occasionally do these cause clinical problems. Schistosomes, for example, are found in a wide range of ungulates in tropical areas of high humidity, their main pathological effect being to cause thrombosis and mild phlebitis, usually of the veins of the liver and intestines. Many species have their own particular schistosomes. *Schistosoma leiperi* is found only in both the red lechwe (*Kobus leche*) and the Nile lechwe (*Kobus megaceros*) despite the fact that the two species are separated by 5000 km. *Schistosoma bovis* of cattle is also regularly found in a wide range of game where they have been grazing in association with domestic ruminants. McCully, Van Niekerk & Kruger (1967) in a survey of the parasitic infestations of the hippopotamus (*Hippopotamus amphibius*) noted that lesions associated with schistosomes were found in the majority of the population and it appeared that occasionally they were clinically affected by them.

Cases of cholangitis caused by the bile duct hook worms, *Monodontella giraffae* of the giraffe (*Giraffa camelopardalis*), and *Grammocephalus clathratus* of the African elephant (*Loxodonta africana*), are relatively frequent findings, particularly where the population levels of these animals are high. Juvenile animals are usually more severely affected. Ostertagiasis has also been reported to cause fatalities: *Ostertagia ostertagi* has caused deaths in eland (*Taurotragus oryx*). A wide range of other nematodes affecting wild ruminants, many of which are also found in their domestic relatives, are listed in an excellent review of game diseases in southern Africa by Basson *et al.* (1971).

Lung worms can also cause a considerable problem in wild ruminant herds. *Pneumostrongylus* and *Protostrongylus* spp. often produce interstitial pneumonia which can be either localized or widespread. This problem has been noted especially in blue wildebeest (*Connochaetes taurinus*) and impala (*Aepyceros melampus*) in which there is a relatively high incidence of the parasites.

Protozoa

Apart from the helminths, many Protozoa have been recorded as causing clinical disease in wild mammals. Howe (1971) lists a wide range of ungulate species, many of them from Africa and Asia, where erythrocytic forms of *Theileria*, many of them unidentified, have been observed in natural infections. It is now well known that *Theileria lawrencei* of the African buffalo (*Syncerus caffer*) is found in a large proportion of that species and is almost certainly the causative agent of east coast fever in domestic cattle. Most *Theileria* have very little effect on their natural hosts although *Theileria taurotragi* has caused deaths in eland (Grootenhuis, 1979). Those forms which undergo schizogony in the histiocytes are placed in the genus *Cytauxzoon*, while those which undergo schizogony in the lymphocytes retain the generic name *Theileria*. *Cytauxzoon* spp. are relatively rare in wild mammals and have only been recorded from three species, the greater kudu (*Taurotragus strepsiceros*), Grimm's duiker (*Sylvicapra grimmia*), and the eland. Nietz & Thomas (1948) reported a series of fatal infections in Grimm's duiker with what they called *Cytauxzoon sylvicaprae*. *Theileria* are found in a wide range of Asian and African animals including primates (Primates), the aardvark *Orycteropus afer* and the warthog *Phacochoerus aethiopicus*. All *Theileria* are dependent on arthropod vectors and the range of the infection is therefore dependent on the range of the relevant arthropod.

Mange Mites

Another group of parasites which have occasionally caused considerable clinical problems are the sarcoptic mange mites which are found on a number of bat species in Africa and the Far East. The mites are found on the wing membranes where the females appear to cause most damage. Frequently the movements of the parasites tend to be seasonal. Often the mites form nodular cysts which may become encapsulated and secondarily infected with bacteria. This in turn causes deterioration in the condition of the host and a septicaemia may ensue. Occasionally bats are found with the wing membranes and sometimes the surface of the body covered with these pale nodules. Although the migration of the mites may cause additional skin damage it is also probable that periodic variations in the availability of the animals' food supply will lead to changes in the bodily condition of the host which itself determines the clinical effect of the parasitic mite.

Anthrax

One bacterial disease which seems to have achieved greater significance in game animals in recent years has been anthrax. In southern Africa, where this problem has been studied extensively, a wide range of large mammal species have been affected. Antelopes, in particular hartebeeste (*Alcelaphus buselaphus (caama)*), springbok (*Antidorcas marsupialis*), black wildebeest (*Connochaetes gnou*), greater kudu and roan antelope (*Hippotragus equinus*) have suffered considerable losses in the Kruger National Park, where the spores, which have considerable longevity, are picked up mainly from the mud on the edges of waterpools. An attempt to vaccinate the animals, either by aerial spraying or by herding them into corrals for hand injection, appears to have had a moderate effect in reducing the incidence of infection.

DOMESTIC ANIMALS AND WILDLIFE

Where domestic ruminants and large wild mammals come into regular contact, two problems may ensue. First, where the available food supplies are limited, competition for the resource may lead to its excessive use and degradation of the habitat, with a resultant loss of bodily condition for all the animals involved. This frequently leads to an increasing incidence of disease in both the domestic and wild animal populations. Retrospective studies of the antibody content of sera from a wide range of African ungulates often provide evidence of their contact with diseases such as foot-and-mouth disease, bluetongue, rinderpest, contagious bovine pleuropneumonia and trypanosomiasis. Although lesions are occasionally found from which these organisms can be recovered, serious clinical cases are usually rare. In the more arid areas of Africa and Central Asia the available grazing at certain times of the year is very limited. At those times the wild animal populations which frequently migrate along the same routes as the nomadic people and domestic animals come under considerable pressure from overhunting because during this period the domestic stock are least productive and unable to provide sufficient food in the form of milk or meat for their human owners.

Exchange of Infectious Disease

The second major feature of this regular contact is the interchange of a number of diseases between the two animal populations. One population is often resistant while the other is highly susceptible. *Theileria parva* for example (which is probably the same species as *T.*

lawrencei) is carried by up to 90% of the buffalo population, but clinical cases are virtually unknown. When regular contact occurs between cattle and buffalo severe losses may be seen in the cattle which suffer from anaemia, leucopenia, oedema, diarrhoea and dyspnoea.

Foot-and-mouth disease virus can often be isolated from a range of wild ruminants and it is probable that wildlife may be a source of this infection to domestic stock although the reverse is also true. Nevertheless, in most parts of Africa, this disease is not a major factor in reducing domestic animal productivity and most ungulate populations, wild and domestic, have become resistant to it.

The case with African swine-fever is quite the reverse. Domestic pig husbandry on a large scale is comparatively rare in Africa. Where local custom and religious constraints permit the keeping of pigs, it is probable that the indigenous breeds have developed a degree of resistance to this virus. Any attempts to improve the stock by importing European breeds have usually led to an increase in the susceptibility of these animals to infections. All three wild African porcine species are carriers of the virus of African swine-fever and, providing suitable arthropod vectors are available, the disease is quickly transmitted to domestic pigs if no effort is made to keep the wild and domestic hosts well separated.

The part which young migrating blue wildebeest play in the transmission of malignant catarrhal fever virus (M.C.F.) to domestic cattle has been well documented (Plowright, Ferris & Scott, 1960). The Masai tribesmen of the Serengeti plains of east Africa have long associated the calving season of the wildebeest with an upsurge in the number of clinical cases of M.C.F. in their cattle. As a result, they traditionally take steps to ensure that their animals are grazing well away from the wildebeest herds at that time of year. A viraemia can be demonstrated in wildebeest calves at a relatively early age and it is probable that M.C.F. is one of the many factors which contributes to the relatively high losses amongst this species as well in some years.

Another organism to which antibodies can quite frequently be demonstrated in wild ungulates is *Brucella abortus*. In the Jonglei region of the southern Sudan where the Zoological Society is contributing towards a broadly-based ecological survey of the region, chronic brucellosis is a major contributory factor in the poor health of many of the cattle herds in the area. Part of the veterinary work there at present is being directed towards establishing whether the large numbers of migrating wild ruminants are likely to have a role in the transmission of this and other diseases although it is much more likely that brucellosis is a self-perpetuating problem within the groups of cattle.

New Vaccines

The study of the mechanisms of the transmission of disease between wildlife and domestic animals, apart from elucidating more accurately the role which wild animals play, has also produced other advantages. Significant amongst these has been the finding that an organism of interest in the wild animal may be somewhat dissimilar to that known for the related domestic species. As a result, it may be possible to develop new vaccines for domestic stock, based on less virulent organisms found in wild relatives. A recent example of this was the finding of *Besnoitia* cysts in the cardiovascular system of many blue wildebeest and impalas in South Africa. None of the infected antelopes had skin lesions which are frequently a typical finding in bovine besnoitosis. Experimental infection of cattle with the antelope strains of *Besnoitia* produced immunity and did not cause the disease or lead to the development of any detectable cysts. As a result of this work a tissue culture vaccine, prepared from the blue wildebeest strain, is now being used in cattle (Bigalke, Van Niekerk, Basson & McCully, 1967).

The diseases which wild animals can transmit to man's domestic animals have naturally been of more concern to him than infections which have passed in the other direction. During the last century and the early years of this century, rinderpest caused enormous losses amongst wild ruminants in Africa. In recent years, populations of the already rare gaur (*Bos gaurus*) and wild Indian buffalo (*Bos indicus*) have become threatened by rinderpest virus spread by unvaccinated domestic cattle which often enter the areas "protected" for the wild species and graze in close association with them. The gaur population, in particular that of southern India, has been reduced in number very severely from this cause. When the Nuffield Institute for Tropical Animal Ecology was active in what is now the Ruwenzori National Park, Uganda, a relatively high incidence of bovine tuberculosis was discovered in a number of wild ruminant species, especially buffalo, which had almost certainly contracted the infection from local domestic cattle in which the disease had been endemic for some time. It is interesting to note that one of the major sources of African ungulates for exhibition purposes in European and American zoos in recent years has been central and northern Uganda. It is possible that many of the cases of bovine tuberculosis which are now diagnosed regularly in the ungulates of European and American collections originated from a source of infection in Uganda.

CONTACT BETWEEN MAN AND WILDLIFE

Where some wild species come into close contact with humans, they may act as a carrier to him of a number of clinically important diseases. In recent years Lassa fever, Ebola virus and the Marburg virus amongst others have led to a few dramatic incidents with high human mortality in Africa. Although the vervet monkey (*Cercopithecus pygerythrus*) may be involved in the transmission of the Marburg agent, it is probable that all these viruses will ultimately be found to have a rodent or other small mammal as their natural host in which they have no clinical effect. Outbreaks of such diseases tend to be spectacular but localized, and because in most cases the exact source of the infection is unknown, there has been as yet no concerted effort to try to eliminate possible carrier animals; and this would probably be largely impractical anyway.

The transmission of human diseases to primates is now well known but also tends to be a local phenomenon. The dissemination of some species of Mycobacteria around the world has been closely associated with the spread of man. Primates such as macaques (*Macaca* spp.) and langurs (*Presbytis entellus*) which may come into close association with villages and urban communities in central and southern Asia, often have a much higher incidence of human tuberculosis (*Mycobacterium tuberculosis*) than animals caught in more remote areas. Suppliers of laboratory primates from Asia used to catch animals found on the fringe of human communities, knowing that they were approachable and therefore more easily caught. Very often the animals were considered to be pests and both these factors led to large numbers of primates being imported into America and Europe with a strong likelihood of tuberculous infection. As the disease in primates generally follows a similar pathological course to that in man, it is probable that a proportion of the wild primates would die of this disease whether they were trapped or not.

Measles

The trapping and movement of primates has also led to other important diseases being transmitted to them. Seven years ago a shipment of black and white colobus monkeys (*Colobus guereza*) caught in northern Tanzania all died of measles in quarantine at Regent's Park. The presenting clinical picture was of an acute pneumonia (Hime, Keymer & Baxter, 1975). This infection had almost certainly been transmitted by children living in the village where these animals were being held prior to shipment. Although in the normal course of events colobus monkeys would not associate closely with African villages, the problem of carrying human infections back to a wild population is a possibility and would be made worse if infected animals were released back into the wild.

Hepatitis

A further interesting example of cross-transmission has been the use made of dealers in west Africa of human sera which is given to young chimpanzees (*Pan troglodytes*) after capture in the hope that this will confer some passive immunity against human diseases. This trade has now largely ceased. But it has become apparent that because of the very high incidence of human hepatitis types A and B in the region many chimpanzees were given serum from carriers of the viruses. This is evident from the large number of carrier animals which have now been identified in captivity. Hepatitis A and B have no clinical effects in the chimpanzee, but another as yet unidentified virus which causes hepatitis and is also carried by man less frequently, causes acute disease in large apes and could therefore be a serious threat to life.

INDIRECT EFFECTS OF MAN ON WILDLIFE

Barriers to Movement

In many areas of Africa and Asia, man's efforts either to fence his domestic stock in, or to fence wild animals out from particular areas, have had fairly dramatic effects on animal populations. The efforts to clear the tsetse fly from large areas of central Africa also involved the maintenance of considerable lengths of game fencing which had the effect of keeping some species out of the "protected" area, but was ineffective in preventing the movement of small ungulates such as bushbuck (*Tragelaphus scriptus*) and warthog which were carriers of trypanosomes. More recently, the clearance of large tracts of woodland combined with the sometimes misguided efforts to develop western-style cattle ranches, have affected the normal movements of game animals. This has a number of effects. Many fences tend to act as traps and migratory animals forced on by instinctive pressures very often become entangled in them. More significantly, animals may be diverted into unsuitable habitat, to their detriment and occasionally to the detriment of the habitat which they originally occupied. If large, grazing ruminants are denied a particular area where they must have had a very significant influence on the maintenance of the natural vegetation, it is likely that over a period of years the vegetation will change. As many of these areas are often in "marginal land" where the ecosystems are very fragile, even minor changes can in the long term reduce the diversity of plant and animal life and ultimately its overall productivity. The land often becomes degraded and in many cases is then unsuitable for domestic livestock as well.

Many forms of agricultural development in Africa and Asia have a similar displacement effect. Apart from diverting wild animals, such schemes often trap them in small pockets where there are inadequate food supplies. Animals then die of starvation, exacerbated by disease, or are forced to move onto the agricultural schemes where they are likely to be shot. One recent example, in which the veterinary department of the Zoological Society was involved, occurred during the development of the Uda Walawe Valley in southern Sri Lanka. In making the valley available for peasant agriculture, "development" was carried out in a haphazard manner, leaving small areas of secondary forest in which relatively large numbers of elephants (*Elephas maximus*) became trapped. The animals began to lose condition and started raiding the surrounding plantations where they were shot at. As a consequence many of them died of secondarily infected wounds or of starvation. As Sri Lanka is a Buddhist country, considerable efforts were made to catch and move the survivors, some by drug immobilization and others by driving them using large numbers of people and considerable noise.

The Jonglei Canal

In the Jonglei area of the southern Sudan, a canal is under construction to link two points on the White Nile, south and north of the Sudd swamps. The area around the proposed route of the canal contains very large numbers of wild herbivores, of which the dominant species is the tiang (*Damaliscus korrigum*). It is probable that the canal itself will not prevent the east—west migration of these animals, provided that they are not harassed as they swim across it. The main threat to their seasonal movements would be extensive irrigation projects and other forms of development along the banks of the canal which might create an obstacle, not of 100 m in width, but possibly of several kilometres. At the present time, although these animals carry serological and other evidence of contact with diseases of potential significance, the population is healthy, but the situation could change dramatically if their usual migratory movements were obstructed. The significance of this resource is that up to 50% of the total protein intake of much of the human population in the area consists of game meat. It is estimated that some 15 000 tons of tiang meat is probably consumed there every year. With 450 000 tiang in the area this makes no difference to the numbers of these animals, which are readily replenished, but if that population was to be severely reduced by agricultural development, milk production from cattle would have to increase at least threefold to make up the protein loss.

Biological and Industrially Produced Toxins

Although a number of naturally produced toxic substances are implicated in the deaths of wild animals, their significance is far surpassed by the effects of the increasing use of industrial chemicals, principally those used to control plant and animal "pests". The most important of the biological toxins are those produced by *Clostridium botulinum* and these have already been discussed elsewhere in this volume. Blue–green algae are also thought to have been implicated in a number of deaths in waterfowl, and a number of toxins from moulds and fungi, notably those produced by several *Aspergillus* spp., have caused losses from time to time.

Oil and Heavy Metals

Development in Africa and Asia has brought with it the extensive use of many chemicals which are toxic to wildlife. Oil spillage, for example, is as much a problem off the coasts of southern and western Africa as it is in Europe and America and has similar effects on seabirds in particular. The discharge of industrial effluents often containing heavy metals such as mercury, chromium and cadmium has led to considerable rises in the tissue levels of these substances in marine life and these increases have become particularly evident in carnivorous Cetacea and Pinnipedia at the end of the food chain. This problem is now well known, for example, around the coasts of Japan.

In Kenya there was considerable concern in recent years about the rising levels of copper in Lake Nakuru in the Great Rift valley caused by industrial effluent from a factory which had been discharging waste products from chemical processing into the drainage catchment area of the lake. Water copper levels were rising steadily and there was concern that the metal was being taken up by the blue–green algae which formed the basic food source of the invertebrates on which many of the birds feed. Following the expression of considerable international concern for this unique resource, an enlightened Kenyan government moved the factory elsewhere and so removed the problem.

Agricultural Chemicals

On a much broader basis, the use of many agricultural chemicals, mainly herbicides and insecticides, is very widespread throughout the Third World. Many of these substances are highly effective and relatively cheap which makes them attractive to an impecunious buyer. The irony is that many of these substances are produced in and even donated by countries where their use is now illegal on conservation grounds. The more sinister wartime use of defoliants in south-east Asia has been well documented, and has undoubtedly contributed to large-scale destruction of wildlife in that area, partly by its direct effects, and partly by destruction of the habitat.

The group of substances which give the most cause for concern at the present time are the organo-chlorine insecticides. The well-publicised exhortations from some of the world's leading development agencies to eradicate the tsetse fly and other insect pests across Africa brings with it an increasing demand for such substances, and with it development money to pay for them. The principal compounds employed are DDT and Dieldrin, although there are a number of other related substances which are also in use. The great popularity of these materials, apart from their effectiveness, is that they persist for very long periods. A single spraying with DDT at the manufacturer's recommended levels may well keep an area free of insect pests for several years. The problem usually is that in the hands of inexperienced users, there is a tendency to the belief that if a little is effective, then a lot must be a great deal more effective. As a result the long-term problems that ensue are magnified. One recent example of the misuse of DDT was the attempt to halt the northward spread of the tsetse fly from Benin into southern Niger. In the "W" National Park the boundary between the two countries is formed by the Mekrou river which is a tributary of the Niger system. On both sides of the river, and particularly on the southern, Benin, side, lie some of the best stretches of typical Guinea savanna and mixed woodlands in west Africa. Until recently, the tsetse fly was only found on the Benin side, but because of the risk to a relatively small area containing cattle on the Niger side, it was decided to remove the riverine forest on both sides of the river and to spray the cleared areas and the margins of the remaining forest regularly from the air with DDT. Unfortunately the local officials responsible were not willing to listen to the advice of the development personnel who were assigned to the project and who were advocating minimal use of the substance. Relatively large quantities of DDT were dropped on the area with predictable results. Fishing in the Mekrou and considerable lengths of the Niger river where the Mekrou drains into it has had to be abandoned. Unfortunately, nobody yet has quantitatively surveyed the problem there, as the resources are not available.

A private report on the chemical contamination of the environment in Zimbabwe has recently been presented to the Director of National Parks and Wildlife Management in Salisbury and has revealed some very disturbing results in a country which has been much more conscious of this problem than most in Africa for many years (W. R. Thompson, pers. comm.). It was decided, in 1980, to collect every available egg from fish eagles (*Haliaetus vocifer*) nesting along the south side of the Zambesi river between Kariba and Victoria Falls. The residues of DDT in 33 clutches of fish eagle eggs averaged 60 parts per million. The egg shells were 16% thinner than those laid before the use of DDT in the country. It is worth remembering that the American bald eagle populations (*Haliaetus leucocephalus*) began to decline when DDT reached 30 parts per million in the eggs and egg shells became 10% thinner. Signs of reproductive failure in smaller birds of prey such as the peregrine falcon (*Falco peregrinus*) and the black sparrow-hawk (*Accipiter melanoleucus*) have already been noted in Zimbabwe and levels of DDT in zooplankton and fish from the Zambesi have risen markedly. The organo-chlorines are often said to be safe on the basis that they rarely cause immediate death. In animals in good condition these substances are stored in fat reserves and in the brain and liver where they accumulate. The species which are most likely to ingest high levels of the toxin are at the top of the food chain. At certain seasons of the year, when other environmental conditions reduce the dietary energy intake of the animal, fat reserves are mobilized and organochlorines are released into the blood stream in relatively large quantities. At this point, incoordination, ataxia and convulsions are seen and the animal dies relatively quickly.

Many national aid agencies will not now advocate the use of these substances, but prefer the quickly detoxified organophosphates. The main risk in the next decade will probably come from the small farmer who buys the materials privately without realizing the long-term effects he may be producing.

EFFECTS OF CLOSE MANAGEMENT

It has often been assumed that as development takes place and wilderness areas become diminished there is no necessity to monitor and manage the resultant smaller spaces available to wildlife. If large numbers of animals become restricted to a relatively small area, the pressure on available food supplies increases, and the likelihood of a disease epidemic or more chronic health problems results. Over the last four decades there has been considerable interest in the development of game farms in southern Africa in particular. Usually game is managed in association with domestic stock. The domestic animals use the land on which particular grasses suitable for beef and dairy farming can be grown without long-term adverse effects, while the game species use the more marginal land on which cattle would be unprofitable. Game meat in southern Africa commands considerable prices but this is not the case yet elsewhere on the continent, although as has been described above the meat constitutes an important local protein resource. Some farmers contemplating the close management of game tried to achieve more profit by assuming that the stocking rates could be increased beyond the carrying capacity of the land. In other cases, landowners who were keen to have game animals on their land were not so willing to cull them regularly and the same effects of overpressure on the available food supplies took place.

One problem which has frequently been noted in these areas has been the increase in the incidence of sarcoptic mange. It is not uncommon, for example, to see large numbers of blue wildebeest and impala with extensive chronic dermatitis which can be seen from a considerable distance as light grey or brown areas, mainly affecting the head and back. These animals are invariably in poor condition, and may also harbour large numbers of endoparasites which are probably also contributing to their decline. A regular programme of population surveys and an assessment of the available food plants and the condition of the environment are essential in deciding the carrying capacity of the land and the numbers of animals which should be culled annually.

CAPTIVE BREEDING AND REINTRODUCTION

It is now generally accepted that the zoo world has a responsibility to breed and so replace the animals which it exhibits. In the case of rare species, many zoos are now building up stocks which will become available for reintroduction schemes where these are practical. A number of projects, particularly with birds, have already been initiated with varying success. One of the main problems which the captive breeding centres encounter is that because they can maintain only relatively small numbers of the animal concerned, it is frequently necessary to move stock about, often internationally, in order to maintain genetic diversity. While the technology for moving gametes and embryos around the world is being developed, the animals themselves are having to be transported. Animal movements are complicated where the importing country is concerned about the risk from disease potentially carried by wild animals to domestic stock of importance to their agricultural industry. In Britain, for example, a number of rare ungulates are breeding well, but closely inbred. To obtain new stocks, for example of addax and scimitar horned oryx from north Africa, would be impossible at present because of the concern that bluetongue virus could destroy our sheep industry.

Many species of birds carry and are susceptible to Newcastle disease and some psittacines, for example from Africa, are known to carry forms of the virus particularly virulent for domestic poultry. The importation of birds into Britain is not as complicated a procedure as that encountered for large mammals, but nevertheless, statutory regulations to protect the farming community do have an indirect effect on captive propagation programmes for wildlife throughout the world. It is interesting to note that recently, when the first groups of Arabian oryx bred in captivity in the United States were available for reintroduction into the Middle East, Oman, which has a policy of not allowing into the country ungulates with antibodies to the bluetongue virus, had to refuse some of the Arabian oryx which had been offered to them for a new National Park because they turned out to have positive titres.

Reintroducing animals may also put another species potentially at risk from disease. In Indonesia a considerable number of young orang-utans, originally intended for illegal sale by animal dealers, have been confiscated and hand-reared by well-meaning individuals with the hope that they could be reintroduced. Fortunately, one of the research workers, interested in the ecology of this species, had had a veterinary training and discovered that a number of these young animals, particularly those confiscated in markets or from villages, had human tuberculosis. Had reintroduction been attempted in areas where there were already wild populations of orang-utans, it is quite feasible that the susceptible orangs would have contracted tuberculosis and that a new and highly dangerous disease would have been introduced to an otherwise healthy population.

REFERENCES

Basson, P. A., McCully, R. M., Kruger, S. P., Van Niekerk, J. W., Young, E., Devos, V., Keep, N. E. & Ebedes, H. (1971). Disease conditions of game in Southern Africa: recent miscellaneous findings. *Vet. Med. Rev.* 1971: 313–340.

Bigalke, R. D., van Niekerk, J. W., Basson, P. A. & McCully, R. M. (1967). Studies on the relationship between *Besnoitia* of Blue wildebeest and Impala and *Besnoitia besnoiti* of cattle. *Onderstepoort J. vet. Res.* 34: 7–28.

Grootenhuis, J. G. (1979). *Theileriosis of wild Bovidae in Kenya with special reference to the Eland.* Meppel: Krips Repro.

Hime, J. M., Keymer, I. F. & Baxter, C. J. (1975). Measles in recently imported colobus monkeys (*Colobus guereza*). *Vet. Rec.* 97: 392.

Howe, D. L. (1971). Theileriosis. In *Parasitic diseases of wild mammals*: 343–353. Davis, J. W. & Anderson, R. C. (Eds). Iowa: State University Press.

McCully, R. M., van Niekerk, J. W. & Kruger, S. P. (1967). Observations on the pathology of Bilharziasis and other parasitic infestations of *Hippopotamus amphibius* from the Kruger National Park. *Onderstepoort J. vet. Res.* 34: 563–617.

Nietz, W. O. & Thomas, A. D. (1948). *Cytauxzoon sylvicaprae* gen. nov., spec. nov., a protozoan responsible for a hitherto undescribed disease in the duiker (*Sylvicapra grimmia*). *Onderstepoort J. vet. Sci. Anim. Ind.* 23: 63–76.

Plowright, W., Ferris, R. D. & Scott, G. R. (1960). Blue wildebeest and the aetiological agent of bovine malignant catarrhal fever. *Nature, Lond.* 188: 1167–1169.

Symp. zool. Soc. Lond. (1982) No. 50, 287–297

The Control of Disease in Wildlife when a Threat to Man and Farm Livestock

W. M. HENDERSON

Yarnton Cottage, High Street, Streatley, Berkshire, England

SYNOPSIS

In the developed countries the diseases of farm livestock transmissible to man have been largely eliminated. In Great Britain it was the virtual control of tuberculosis in the early 1960s that permitted the very considerable expansion of the British dairying industry. The method used for the final elimination of infection of the major animal plagues is the slaughter of susceptible animals which are infected or which have been exposed to the risk of infection. The wildlife/domesticated animal relationship is discussed for a number of diseases in the context of the dilemma presented by the wish to avoid the destruction of wildlife when it is the wild creature that is the source or the transmitter of infection of a disease which is of economic or public health importance. In the case of foot-and-mouth disease in Africa the buffalo is a carrier of the virus, showing no clinical signs. If the contact between the buffalo and cattle is diminished by game fences and if the cattle are regularly vaccinated against foot-and-mouth disease, no sacrifice of the wildlife nor of the cattle is called for in the particular circumstances of this buffalo/cattle interaction. There are other diseases which demand more drastic action for effective control and particular attention is given to rabies and tuberculosis. The badger/cattle interaction with regard to tuberculosis in the south-west of England is discussed in some detail. It is concluded that there is irrefutable evidence that the diseased badger population is responsible for herd breakdowns to the tuberculin test. The policy of the destruction of the badger is strongly supported but with a plea for the search for a more acceptable practice. On the other hand, the current tactics of the Ministry of Agriculture, Fisheries and Food are criticized as being insufficiently vigorous and extensive.

INTRODUCTION

In discussing the transmission of disease between different species, it is convenient to consider three population groups, namely, wild animals, domesticated animals and man. One of the successes of the veterinary and medical public health services of the developed countries has been the extent to which the diseases of farm livestock transmissible to man have been largely eliminated. In terms of the demand on medical care, the three most important are usually considered to be, or to have been, tuberculosis, brucellosis and hydatidosis. The reduction in the prevalence of tuberculosis in cattle which, in Great Britain, resulted in achieving control and a great diminution in the number of reactors to the tuberculin test by the early 1960s, was essential for the subsequent very considerable expansion of the British dairying industry. The present success of the programme of eradication of brucellosis from the national herd reflects great credit on the veterinary services and on the dairy and beef sectors of the agricultural industry. Other spectacular successes have been achieved during the last 100 years, with the eradication of rinderpest in 1877, of bovine pleuropneumonia in 1898, of glanders in 1928 and swine fever in 1966, to say nothing of the long periods of freedom of which we can now speak in the case of rabies and foot-and-mouth disease. When these particular diseases affect farm livestock, the method used finally to eliminate the infection is the slaughter of susceptible animals which are infected or which have been exposed to the risk of infection. Such taking of the animal's life but anticipates the fate for which it was reared and the slaughter of the domesticated animal does not give rise to the dilemma which is of such concern that it was judged to be appropriate to hold this two-day symposium. This does not, of course, take into account the great loss caused by slaughter of flocks and herds, the quality of which has been built up by years of devoted application of good husbandry and careful sire and dam selection.

The dilemma is presented by the widely-held wish to avoid the destruction of wildlife when it is the wild creature that is the source or the transmitter of infection of a disease which is of economic or public health importance. In most societies there are some qualifications to this point of view. This regard for wildlife is not usually extended to the category of pests — things noxious, destructive or troublesome. There are not many, in general, who would weep for the rat or the biting insect. There are, in addition, specific economic or even aesthetic interests which result in certain species being considered to be pests; from bullfinches to jays and magpies, from moles to rabbits, grey squirrels and foxes.

DISEASES THAT CAN BE CONTROLLED BY ACCEPTABLE MEANS

During this symposium, the fascinating story has been told of the rinderpest cattle and wild game relationship with, perhaps, the unexpected conclusion that the game may be less of a menace as a source of infection for cattle than was generally supposed. The problem of trypanosomiasis is somewhat different. There are three approaches towards attempting to solve the problem of this insect-transmitted protozoological disease of wild animals, domesticated animals and man. One is to try to eliminate the tsetse fly or its habitat, a second is to attempt to eliminate the wildlife hosts of the trypanosomes and the third is to alleviate the effects of the infection of man and livestock by drug therapy or by immunological endeavours. In the early attempts to control trypanosomiasis, for example nagana (principally *Trypanosoma vivax, T. congolense* and *T. brucei brucei*) in Natal, Republic of South Africa, the impala and the bushbuck were slaughtered but not the warthog nor the bushpig. The problem was not solved and opposition to the destruction of wildlife was aroused. The next attack was by aerial and other spraying for destruction of the tsetse fly which proved to be successful in all but the thick bush in close proximity to rivers. Similar programmes using insecticides are being applied in the tsetse fly areas of Botswana with considerable success being claimed as a result of the application of accumulated knowledge about the breeding habits of the fly.

The disease situation and the wild animal/domesticated animal relationship in and around the Kruger National Park in South Africa provide three examples of methods by which the domesticated stock are relatively well protected from diseases of wildlife without the taking of drastic measures.

Buffalo Disease, or Corridor Disease, a Type of Theileriosis

This type of *Theileria parva* infection, so called *T. lawrencei* infection, is maintained in the buffalo. Ticks which have fed on infected buffalo may cause a fatal disease when they feed on cattle. This resembles East Coast fever except that the life-cycle of this particular form of the parasite is not completed in cattle and no trouble results if the cattle and the buffalo are kept apart. This is achieved by the complete enclosure of the Kruger National Park and of smaller parks in Natal by game fences so as to retain the buffalo within the parks from which there is a ban on them being moved. The buffalo in parks in Cape Province are outside the range of the tick vector and are free of *T. parva*. Buffalo from these parks may be moved to other parts of the country for the establishment of *Theileria*-free buffalo herds.

Malignant Catarrhal Fever

This is a herpes-type virus infection of cattle and other animals. It has long been known that the appearance of the disease in cattle is a consequence of their grazing being shared with wildebeest. The critical factor is that the infected wildebeest calf sheds cell-free virus in contrast to cell-bound virus which is typical of the infection in older animals. The control procedure, where feasible, is to keep the cattle and the wildebeest apart by fencing. It is interesting to note that an outbreak of this disease occurred in the herd of Père David's deer at Whipsnade in 1959.

Foot-and-Mouth Disease

During the last 15 years much information has been collected in a number of African countries in collaboration with the Animal Virus Research Institute, Pirbright, on the part played by wild animals in the epidemiology of foot-and-mouth disease. The first information, which was scanty and haphazardly obtained, resulted from the collection of material from lesions of the disease found on shot game. It was only when immobilization by "darting" was introduced that the samples required for the detection of antibodies and carrier virus could be collected on a significantly large scale. Positive antibody titres as evidence of foot-and-mouth disease infection have been obtained from many of the species that might be expected to be susceptible, namely, the ungulates. Three require particular mention: the buffalo, the impala and the kudu.

Of all the animals from which samples of blood have been collected for detection of antibodies, the buffalo has tended to show the highest percentage of evidence of infection with the highest antibody titres. Also, the buffalo has yielded the highest percentage of pharyngeal samples positive for the presence of carrier virus. Nevertheless, the buffalo, or to be more precise, the African wild buffalo, does not normally show clinical signs of foot-and-mouth disease. The domesticated Asian buffalo, by contrast, becomes clinically affected. Buffalo calves appear to become infected with the virus from the herd "pool" when their maternal immunity is waning, say between four and eight months of age. They may then become carriers of the virus with no clinical signs of the disease but with the demonstrable production of antibodies. From two years of age the antibody titre remains high but the samples for the detection of carrier virus become increasingly negative. There are many instances where this situation would appear to prevail but with no spread of infection from the buffalo to domesticated livestock. On the other hand, there are instances in which other wildlife, especially the impala and, to some extent, the kudu, become infected with clinical signs of the disease. Such a situation presents a hazard to domesticated livestock which may thus be exposed to a greater source of infection than from carrier animals. This was the situation in the neighbourhood of the Kruger National Park in the 1950s when it was obvious that game in the Park were the cause of many outbreaks of foot-and-mouth disease in cattle on farms on the Park's outskirts. The farmers in these areas were under continuous restrictive measures and, being unable to market their livestock, agitated for the source of the infection to be eliminated. A more conservative attitude prevailed and a six-foot-high barbed wire and steel fence was erected around this Park of some 2.15 million ha. In addition, a regular programme of vaccination of the cattle against the three prevalent types of foot-and-mouth disease was initiated and has been maintained in adjacent areas because the integrity of the fence cannot be guaranteed. This policy, which calls for no sacrifice of the wildlife nor of the domesticated animals, has been successful. This is in spite of evidence of continued infection being maintained in the buffalo population within the confines of the Park. This population is maintained at around 25—30 000 head. Culling is required to keep to this level. This is done by cropping the Park herds for buffalo meat which has a ready sale through the catering services to the visitors to the Park.

OTHER DISEASES NOT SUBJECT TO SUCH ACCEPTABLE MEASURES OF CONTROL

Rabies

Rabies is a virus disease of widespread distribution affecting many species, with dogs, foxes, bats and carnivorous wild animals being of importance as regards the hazard presented to man and his livestock. There are two generally separate situations. One is so-called urban rabies in which the infection is maintained in the dog population, especially stray dogs, with man being at considerable risk. This situation can be controlled by vaccination of dogs and elimination of strays. The other situation is when the disease is maintained in a wildlife population with the most frequent casualty among man and his animals being cattle. This is certainly so in many parts of South America where the vampire bat by infection with the rabies virus is the cause of severe losses in cattle. Control is attempted by destruction of the colonies of bats. In South Africa, the black-backed jackal is easily singled out as the animal most likely to transmit rabies infection to cattle. Destroying it becomes a high priority when livestock owners find that cattle losses because of rabies become unacceptably high. On the continent of Europe rabies has spread widely in the red fox westward from east Prussia since the end of the Second World War with cattle being the most commonly affected domesticated animal. Population density is an important factor in the spread of disease. In the United States of America it has been observed that enzootic rabies can be maintained in a fox population as sparse as one animal per square kilometre. In Denmark the disease disappeared when the fox population was reduced to 1 fox per 4—5 km^2. It is of the greatest importance that the quarantine regulations continue to be strictly enforced in the United Kingdom and that the utmost vigilance is maintained to prevent the illegal smuggling into the country of animals in prohibited categories. The problem which will next be discussed would pale into insignificance if rabies became established in our wildlife.

Tuberculosis

The high incidence of tuberculosis in cattle coupled with the consequential high risk of milk-borne infection for man was the cause of much concern in the early years of this century, but it was not until the 1930s that organized effective action began to be taken.

The importance of the problem was one of the factors that led to the creation of the Agricultural Research Council in 1931. The Council in its first Report published in 1933 stated that the most urgent subject for enquiry for the economic well-being of the British farmer was that of animal disease. One of the original members of the ARC, and President of the Royal Society, Professor Sir Frederick Gowland Hopkins, was appointed chairman of the Economic Advisory Council Committee on Cattle Diseases, the report of which was published in 1934. It was noted therein that some 40% of cows throughout the country were infected with tuberculosis. The registration of herds attested to be free of tuberculosis began in 1935 during which year 22 000 cattle had been slaughtered under the Tuberculosis Order. By the end of that year the roll of attested herds stood at 99. By 1950 there were 55 000 attested herds and only 5 000 cattle were slaughtered under the Order. In that year an Area Eradication Plan was introduced. In 1960 the whole of Great Britain became attested with 243 933 herds. Twenty-eight tuberculous cattle were seized for slaughter and the incidence of reactors to the tuberculin test was 1.9 per 1000 in a cattle population of between 10 and 11 millions. The number of reactors continued to decline except in some areas in the south-west of England where, by the end of the 1960s, there were more herd breakdowns than in the whole of the rest of the country. The routine study of each of these breakdowns failed to identify the source of infection. In June, 1971, *Mycobacterium bovis* was isolated from a badger carcass found on a farm in Gloucestershire where bovine tuberculosis existed in the cattle. This isolation of the tubercle bacillus at the Central Veterinary Laboratory, Weybridge, marked the start of the collection of evidence, which is now irrefutable, that there are areas in the south-west of England in which the badger population is infected with the bovine type of tuberculosis and that this infection is transmissible to cattle — and thus threatens the health of our herds, especially our dairy herds, which has been so dearly won during the last half century and more.

What is the evidence which justifies such a positive statement?

(1) Tuberculosis has persisted in cattle in specific areas.

(2) In these areas the badger population is infected with bovine tuberculosis.

(3) Tuberculosis is uncommon or non-existent in all wildlife species except the badger. On the few occasions when *Mycobacterium bovis* has been isolated from other species, the infection has not been of a progressive nature.

(4) Tuberculosis in the badger is a progressive disease; the kidneys and lungs are often severely affected and, in such cases, large numbers of tubercle bacilli are excreted in the urine, the bronchial pus and, to a lesser extent, in the faeces.

(5) Exposure of calves to badgers either experimentally or naturally infected with tuberculosis has resulted in the calves becoming positively reactive to the tuberculin test and developing lesions of tuberculosis.

(6) Experiments have shown that the tubercle bacillus contained in urine, bronchial pus and faeces from infected badgers may survive for at least four weeks during the winter.

(7) Within the problem areas, the frequent use of the tuberculin test in cattle has permitted the removal of reactors before they would be expected to be able to transmit the disease but cattle have continued to become infected.

(8) On a very few problem farms where the badger sets have been persistently gassed over two to three years, the situation has returned to normal with no further occurrence of tuberculin reactors and with no continued evidence of the disease persisting in that particular population of badgers.

(9) There is little or no evidence of tuberculosis in badgers outside the problem areas.

Critics of the Ministry's policy complain that the evidence for the described involvement of the badger in the transmission of tuberculosis to cattle is circumstantial and incomplete. But in no other epidemiological context — among the many diseases of wildlife which it is accepted are transmissible to domesticated animals or man, or even among the diseases of domesticated animals transmissible to man — is there much stronger evidence for the identification of a source of infection than that for the rôle of the badger in this specific situation.

The responsibility of dealing with outbreaks of a notifiable disease of livestock such as tuberculosis is that of the Ministry of Agriculture, Fisheries and Food. The investigations that were conducted after the first infected badgers were found showed that tuberculosis was present in the badger populations in many parts of Gloucestershire, in parts of Avon, Cornwall, Devon, Wiltshire and Dorset, all areas in which herd breakdowns had occurred. There was no question but that the Ministry had to take some action. Examples have been given of how a disease in wildlife can be prevented from prejudicing the health of domesticated animals by a variety of measures. These measures have, in common, two key factors. One is the effective separation of the wild species from the domesticated species and the other is the protection of the latter by vaccination, a procedure which also results in reducing the degree of exposure to the infection of the wild species. In the particular case of the badger/cattle ecology, physical separation is not yet seen to be feasible and an effective vaccine for the protection of cattle against tuberculosis has yet to be developed. Under these circumstances, there was no apparent alternative to the drastic method of the elimination of the infective hazard to cattle, namely the killing of the badger. A much more distinguished person, the President of The Zoological Society of London, has already gone over the same ground in much greater detail and if there is interest in acquiring more substance to the reasoning presented, this can be found in the Zuckerman Report published in August, 1980.

MEASURES TO CONTROL THE SPREAD OF TUBERCULOSIS FROM BADGERS TO CATTLE

There are three parts remaining to this chapter. The first part supports the killing of the badger but questions the Ministry's tactics. The second part questions whether we must be content with the Ministry's policy. The third part attempts to assess the impact of the disease on the badger population.

The premise on which the final section of this chapter is based is that the infected badgers must be kept apart from the susceptible cattle.

The Ministry's Tactics

The policy to achieve the separation of infected badgers from susceptible cattle is to eliminate the badger population. To quote from the Fifth Report (1981) issued by the Ministry on this subject:

> it was decided that, whenever a herd breakdown occurred for which no other source of infection could be found and the disease was found in badgers in the locality, the sets considered to be used by infected badgers and other badgers in contact with them should be gassed.

The effectiveness of this procedure is dependent on at least two factors — the population density of the badgers and the speed of repopulation of gassed sets. By all accounts, the population density is surprisingly high and the speed of repopulation after one gassing exercise is unfortunately rapid. Too rapid repopulation will result in the incoming badgers being exposed to residual infection in the sets and their environs. The rate of repopulation can be controlled by repeated gassing but this must impose a further toll upon the badger population. Repopulation would, presumably, be slowed down if the area cleared of badgers were much larger in extent than the particular contaminated area. In that the guidelines being used by the Ministry direct gassing only of sets considered to be used by infected badgers (see above) it could be argued that the Ministry's policy has been too restricted and insufficiently vigorous with regard both to the frequency of gassing and to extent of the area to be gassed. Such a statement is no more than a commentary as it is made without access to and detailed study of all the available information. I urge that the relevant study be pursued. What is implicit in this comment is that the frontier between the badgers and the cattle has to be pushed right back into clean areas. Otherwise, I see no end to the present unsatisfactory situation.

The Ministry's Policy

The second point is whether we should be content with the present policy — and whether we should be content with the more drastic policy which I have suggested should be considered.

I submit that we cannot be complacent about this issue. A less drastic and more acceptable solution to this harassing problem must continue to be sought. In the meantime, however, I strongly advocate the continuation of the present policy. I ask those who are more familiar with recent advances in immunology than I, whether the possibility of stimulating an active immunity to tuberculosis in cattle merits being looked at again. I ask whether it is necessary to consider increasing the overall scale and the frequency of tuberculin testing so as to permit the rapid removal of reactors. If, for example, the initial infection of badgers was from cattle, repetitions of this occurrence would be intolerable were they to be in other parts of the country not, so far, under suspicion.

The Diseased Badger Population

Whatever the initial source of infection for the badger and however it was caused, there is no doubt that, in the problem areas, tuberculosis is not a dead-end infection in the badger; it is now an established disease in these particular groups of the population. There is no way in which the tuberculous badger may be cured, there is no way in which the susceptible badger may be protected against the infection. Unfortunately, tuberculosis in the badger is a progressive disease characterized by open lesions with the discharge of massive numbers of tubercle bacilli. The results of the examination of badger carcasses picked up in fields, woods and buildings in Gloucestershire, Avon and Wiltshire, provide sufficient evidence that tuberculosis is a significant cause of death in badgers in these areas. Of approximately 200 such carcasses examined between 1972 and 1979, the cause of death in no less than over 50 of these animals (28%) was tuberculosis. Under these circumstances the only action that I can prescribe as a veterinarian is the elimination of this, so far, minor part of the country's badger population in order to safeguard the health of the major part.

This is a distressing situation but the recipe for failure is action that is too little and too late.

ACKNOWLEDGEMENTS

I acknowledge with thanks the assistance provided by the Director of the Division of Veterinary Services, Pretoria, and the Director of the Veterinary Research Institute, Onderstepoort, South Africa; the Director of the Botswana Veterinary Services; and the Chief Veterinary Officer of the Ministry of Agriculture, Fisheries and Food of Great Britain.

REFERENCE MATERIAL CONSULTED

Bovine Tuberculosis in Badgers
    Third Report (1979),
    Fourth Report (1980),
    Fifth Report (1981),
    Ministry of Agriculture, Fisheries and Food,
    H.M. Stationery Office, London.
Badgers, Cattle and Tuberculosis
    Report by Lord Zuckerman (1980).
    Ministry of Agriculture, Fisheries and Food,
    H.M. Stationery Office, London.

GENERAL DISCUSSION AND CLOSING REMARKS

Molyneux — Could I ask Sir William, and possibly through him other members of the audience — because I think he raised some very important points in relation to the badger populations — whether we know the extent of movement of badgers? If so, what would be required barrier zones? I would strongly support the suggestion about the need to be really firm in relation to protecting the other populations which are outside that zone.

Henderson — I am sure I am not the best qualified person to reply, but I will give an answer and there may well be someone in the audience who can expand on it.

From the information which I felt it necessary to acquire in studying this whole problem, there are two quite outstanding points. One is that, in these areas of high density of badger populations, the number of badgers is quite extraordinarily high. Every available bit of ground would appear to be the territory of the badgers from a set. This is well illustrated in the information contained in the reports of the Badger Panel published by the Ministry. The other thing is that the size of the territory does not appear to be particularly large. From the bait marking techniques and the radio telemetry techniques that have been used it would appear to be rather exceptional for the odd badger to extend much beyond the territory of the set and, if that happens, it does not appear to go further afield than perhaps the neighbouring two or three sets.

I was very impressed with the amount of information that has been collected, because I well remember some 35 years ago investigating an outbreak of foot-and-mouth disease in which the hedgehog population became infected, and nobody could tell me how far a hedgehog wandered in its normal activities. The badger information appears to be extremely good.

Thompson — We have been making a study, as part of the Ministry's research into badger ecology, since 1975 on the movements of badgers — particularly in one small area of Gloucestershire where there is some bovine tuberculosis in surrounding herds. Added to that rather long-term ecological study, we are also accruing information on badger movements with the recolonization of a cleared 1200-ha area in Dorset and the ten times larger area of Thornbury in Gloucestershire. In both these areas, by removing the infected badger population the disease has died out in cattle also and has been kept clear for a period of three years in each case, and now the badgers are being allowed to come back again. The evidence we have so far — which is not very great at the moment — is that the badgers are travelling longer distances in recolonizing than they travel in areas where they are firmly established and very dense, as in the study site in Gloucestershire.

I should like, if I may, to make a few comments on what Sir William was saying and on some of the papers yesterday. Obviously there are many examples of wildlife reservoirs of diseases which afffect humans and domesticated animals, but there is little work actually being done on the management of the wildlife involved. Sir William discussed the alternatives available. Many of us feel that population management is a fairly large undertaking. We are very interested in wildlife management in the Ministry because so many of the problems which farmers have with wild mammals and birds entail modifying husbandry practices, actual killing of animals in some cases, and in others the use of repellents or other techniques of control, but we have never tried to manage a population of an animal like the badger.

We thought initially that tuberculosis in badgers was very local and discontinuous in its distribution. This was so in Dorset and that was a very neat example of how we could eradicate the disease, perhaps temporarily but possibly fairly permanently, in both the badgers and the cattle. If the disease is more widely distributed among the badger population than we have thought hitherto, it does seem that we are into a much bigger scheme of research and control altogether. I think that we need to approach this fairly cautiously and try to find some way in between, say, complete eradication of badgers (which all conservationists deplore) and, on the other hand, trying to fence in disease in the way that has been done in the Kruger Park.

So one of the things that various research staff within the Ministry have been concerned with over the past year or so is to study the distribution and progression of the disease in badgers by a refined method of faeces sampling and by periodical clinical examination of badgers from known groups which have been studied carefully by radio tracking and bait marking, so that these groups are known intimately; their movements are known with some precision and, if a member of the group is found with tuberculosis, we can find out what happens to it, where it goes, the risks it might pose to neighbouring cattle, and the extent to which infected badgers are, in fact, contaminating the environment. Following on from that, we can investigate the extent to which that contamination is linked with the movements of the cattle.

I think we are all convinced that we need a study of badger behaviour and cattle behaviour from which we can try to elucidate the ecological aspects of this disease, because not much research throughout the world has really been done on the ecology of wildlife disease. I think that we might be breaking some new ground here to very good purpose in pursuing this approach.

The only question I would like to append to these comments is this: can you, from your very wide knowledge of animal disease, think of any aspect that we may have missed out in the approach that we are making now?

Henderson — It is most encouraging to learn of the extent of the effort that has been put into the solution of this problem. As I said in my paper, I can see no alternative than to continue the present policy. Without being happy about doing so, I see no alternative in the meantime. The possibilities that have been applied with success in other conditions are not feasible. People may ask why I suggested that there had to be a vaccine to protect cattle against tuberculosis — is there not a vaccine called BCG? Well, there is, but it does not protect animals. It may lessen the development of the disease in the animal, but it is not protecting in the way that the examples of the use of vaccine which I cited are doing.

Rees (Chairman) — Mr Jones, in your experience in the African projects, do you see any parallel with our problem that we can draw from?

D. M. Jones — Not really, I am afraid. I think that this is a unique example. The methods that have been used for getting rid of disease problems in Africa do not bear any comparison with the problems we have here with badgers.

Rees — Mention was made earlier of the problem in some of the African countries where the buffalo is a reservoir of inapparent foot-and-mouth disease infection, and in many cases now in Zimbabwe where buffalo came on to some of the ranches they have had to separate them and have had to slaughter all the buffalo to protect the domestic cattle. There they had to separate, but by slaughter, which is something similar to what we have to do. In other instances, say in the Kruger Park, they have developed the fencing system, albeit they then have to keep the wildlife on one side of the fence and the cattle within their vaccinated zone on the other side of the fence.

Zuckerman — May I add a word to what Mr Thompson has said about the badger? The amount of attention given by the media to what is actually being done by the Ministry and by other responsible people to increase our knowledge of the ways of the animal — its movements in relation to cattle and so on — is in inverse proportion to the quantity of nonsense that has appeared in the Press about the subject. Mr Thompson will know that in the south-west many people who are concerned about the welfare of badgers simply do not understand the problem. I would therefore suggest that the Ministry take urgent steps to inform the general public and, in particular, the concerned public in the south-west, about what it is doing and why it has to be done. I do not believe that the Ministry should undertake work on the badger which is of little scientific interest — even if it satisfies the interests of one or two naturalists. Half a column in *The Times* the other day was devoted to a synopsis of a study, presumably published in *Nature*, saying that a main item of the diet of badgers is the earthworm and that the animals eat as many earthworms at night as they do by day. That badgers eat earthworms is common knowledge. This kind of study, put out in the name of science, misleads the public about the true nature of the work that the Ministry is now undertaking.

Rees — Thank you, Lord Zuckerman. I entirely agree with your points.

Davies — I am sometimes known, unfortunately, by the people to whom Lord Zuckerman was referring just now, as "Mr Badger" so I am one of those who has to wear armour. One of the difficulties always in speaking to people of that ilk is actually to get over the dimension of this problem. That is one thing that I would like to make clear to this audience now. We are talking about a problem that obviously is very serious, but it is serious in very small patches indeed. If you think in terms of the south-west region having a cattle population of 2.5 million, last year we had 635 reactors to the tuberculin test, and this year we anticipate having rather fewer.

When we talk about badgers and how much infection they have, it is true that at one stage there was 28% of infection in the badgers that we happened to catch over a certain period, or that were presented to us. For the last two years that figure has been of the order of 9%. So it may well be that the infection rate in the badger itself is going down.

Also, the map of the south-west region that was shown is in itself misleading because there were a tremendous number of white dots on that map — for instance the toe of Cornwall that we know as West Penwith was solidly white — but in fact our operations over the whole region in respect of badgers do not occupy more than some 2–3% of the total land area.

So I would like to get that point across — yes, it is a serious problem, but it is in very small, isolated pockets that it is in fact serious.

Rees — I think we must realize, though, that there are many people who have some disquiet about the action that we have to carry out, and so do we. I am quite sure that there is no one in the Ministry who enjoys doing this work, but we feel that it is necessary. If there are people in the audience who have this disquiet, I hope that they will now either ask questions or express their opinions. The purpose of this kind of symposium is that people do have an opportunity to express their opinions.

Kingdon — It seems to me that we use the word "conservation" very loosely. You have invited comparisons with Mr Jones' work in Africa, but I do not think that in the badger situation in England we are attaching the same meaning to the word "conservation". After all, the badger is essentially an animal that is parasitic on the farmer, and it should be regarded as a farm problem rather than a conservation problem.

Ormerod — Earlier in this symposium I put forward the view that "disease" did not occur in wild populations under balanced ecological conditions. Examples of unbalanced conditions have already been mentioned, for instance: the introduction of myxomatosis into a non-immune rabbit population produced recognizable disease; similarly the disease seen in the fox *Lycalopex* affected by *Leishmania donovani chagasi*, as described by Dr Lainson, probably indicates that the disease organism is not normally harboured by this host (Deane & Deane, 1962)[1] and may even have been introduced from elsewhere. The disease suffered by badgers in Britain when infected with bovine tuberculosis suggests that they also may not be a true reservoir of the pathogen, but that introduction of the infection has been recent and that adaptation by selection has not yet occurred.

Gallagher — Tuberculosis in badgers does produce a progressive disease which ultimately ends in death. Exactly how many badgers die of tuberculosis we do not know, but certainly when we look at badgers which have been found dead through natural causes, in fields, by sets and so on — as opposed to found dead due to road traffic accidents — about 40% died as a result of tuberculosis in the Gloucestershire area. It is difficult to work out the annual toll, but it would appear that in some areas it may be of the order of 10%, possibly more. It is very difficult to assess and this is a "guesstimate".

Plowright — I find it somewhat strange, having spent some years in Africa, that people can affirm that disease is not of importance in regulating the populations of large ungulates. Leaving aside the rinderpest story, I will give you a very shortened version of another incident, which is mentioned again in Sinclair's (1977)[2] book, of an infection which involved a herpes virus producing a generalized skin disease and which was associated with a very considerable mortality in young buffalo just north of the Serengeti Park. In that context Sinclair affirmed — and I assume that this applied to the whole 70 000 or so population of buffalo in the Serengeti area at the time — that 10 000 calves disappeared in the first year of life and that it was very unlikely that any of them would be found in a state in which they would be suitable for examination by a veterinarian. I cannot believe that there are not many other such incidents in these tremendous populations where proper investigation carried out at the right time would reveal a very significant involvement of microbial pathogens. In this case there were also Protozoa and ticks, and there was starvation involved, but I do not think that that gets us away from the point that, in the end, there is often an important pathogen to contend with, in that case a herpes virus.

A Speaker — Can I add my own comment to Dr Plowright's point? Surely it is permissible to argue by analogy from domesticated animals — take liver fluke infections in cattle and sheep, for example — that one of the effects these things seem to have is a serious effect on the productivity of flocks and herds. So perhaps it is wrong to think about regulation occurring through the death of infected animals; regulation might rather occur through a reduction in the

---

[1] Deane, L. M. & Deane, M. P. (1962). Visceral leishmaniasis in Brazil: geographical distribution and transmission. *Rev. Inst. med. trop. São Paulo* 4: 198–212.
[2] See list of references, pp. 25–27.

number of births, or in the ability of infected animals to sustain their young. It is quite clear in sheep, and also in cattle, that there is reduced productivity, there is a reduced conception rate, there is a reduced live weight gain, and milk yield and quality also fall. I think it is quite easy to argue from that analogy that pathogenic organisms in wild populations probably have a similar kind of effect on the productivity of those flocks, herds, populations of whatever animals you are considering, and therefore will reduce the numbers of animals that we observe below that which could be attained in the absence of those pathogens.

Kaplan — The obvious analogy to the badger story is what Professor Steck was discussing yesterday, and that is fox control in the reduction of rabies in Central Europe and other countries. Also, in a similar way, when jackals are found to be the vectors of rabies, especially in the Middle East and parts of Africa and southeast Asia, the only way to deal with them is to thin them out and destroy as many as you can. In the rabies field, finally, there is the question of *Desmodus rotundus*, the vampire bat, which transmits rabies to cattle in South America. A way of trying to achieve control of this disease in cattle populations where there is a heavy population of *Desmodus* is to try to destroy as many bats as possible. There are two approaches, as many of you may know. The first is the use of Warfarin, an anticoagulant, in a grease base. You just smear this on a certain number of the bats, reintroduce them to their own flock and, by grooming each other, they spread this poison; and it does reduce them. Another one is to feed the cattle an anticoagulant, and when the vampire bats feed on the cattle they ingest this poison. It is not too effective, but some studies have shown that reduction even of a difficult population such as bats has been successful. The dynamiting of the bat caves has not been so successful because they manage to find other places. It is a continuing problem and one that requires attention to a wildlife vector that will have to be pursued for many years to come.

Henderson — I think it is often not appreciated, the extent of the problem of bats spreading rabies in South America. I always remember my first "exposure" to this situation which must have been in 1957 when I visited for the first time the state of Rio Grande do Sul in Brazil, which is the main high-quality cattle-producing state in Brazil. At that time the state veterinary service was very well organized. They had 60 veterinary posts throughout the state, and the personnel manning these posts included a veterinarian, who was in charge, a clerk, two vaccinators and two bat killers. Their job in life was to identify where the bats were living, in hollow trees, caves, holes in the ground — all sorts of locations — and then attempt to kill them, usually with explosives. Subsequently, the Pan-American Zoonosis Centre of the Pan-American Health Organization in Buenos Aires developed the application of Warfarin which was received with high hopes. However, my contact with that Centre has been lost for the last year or two and I have not heard that very much has developed out of it.

I feel that a more rational approach is protection of the cattle by vaccination. There are many greatly improved vaccines against rabies, but, so often, it is a question of cost and interest.

A Speaker — I think that we ought perhaps to consider discussing a point which has not come to the fore in this Symposium over the last two days, and that is the relation between wildlife and disease which Drs Ormerod and Plowright spoke about earlier. Under natural conditions, wildlife is infected not with one pathogen but with very many, and perhaps we ought to concentrate our attention for a short time in this discussion on the interaction between organisms in wildlife. Anything that tips the balance one way or the other — for example, an organism that is known to immuno-suppress — will have a drastic effect on the outcome of any other infections which are present in that host. I think that this is a factor which we have hitherto neglected. At the risk of grinding my own axe, I would like to point out that vampire bats also transmit trypanosomiasis in South America and this may be a biological control agent for the bats, because it kills them.

Rees — Perhaps I could bring in Dr Little on this question of the interaction of various diseases. He has studied the badger, so perhaps we could ask whether he sees any application to the badger problem, or is it a major problem of tuberculosis and there are no other side issues?

Little — I think not. If we consider what Dr Gordon Smith was saying we find a lot of useful guidance. That is to consider bovine tuberculosis in terms of a maintenance host, which is cattle, and an accidental host, the badger. It would appear to me that the badger is becoming a maintenance host for this disease. However, the severity of it in certain instances would indicate that it probably is a recent disease, it has not had a long evolutionary exposure and, as such, has not reached the situation that we get with many other organisms such as *Leptospira* in a natural host where there is an exquisite balanced relationship between the parasite and the host. After all, this is basically what parasitism is all about: it is a balanced relationship without a disastrous epizootic killing. This is the situation we are seeing in the badger.

Another analogous situation occurs in New Zealand in opossums. There we can actually put dates and places to the first occurrence of the disease. The opossum was introduced into New Zealand from Australia and has recently been found to have extensive tuberculosis; there is now a widespread epidemic of tuberculosis, caused by *M. bovis*, in the opossum population in New Zealand. No such tuberculosis has ever yet been discovered in the opossum population in its native country, Australia. The opossum has been released at known dates in New Zealand and it is more than probable that it has only been exposed to this organism for perhaps 40—50 years. The disease is disastrous, it will kill populations, and yet the animal is starting to adapt and is becoming a maintenance population in New Zealand. There we have a situation which is an exact analogy to that which occurs in England. In New Zealand a wildlife reservoir of *M. bovis* in opossums exists which is causing a problem in cattle, and they are now faced with exactly the same control proposition that we have. That is a mammal with no predators, that can multiply, and the only possible control at this time is a destruction exercise. They are very fortunate in that the opossum is, in many parts of New Zealand, a declared pest and there is not a strong conservation element which is causing them to have to sit back and think a little more about what they are doing.

There may well be interactions which suppress the immune system and produce disease, but what we see with *M. bovis* is a maintenance host with accidental hosts. It may well be that sometimes an accidental host can eventually become a reservoir host.

A Speaker — I was not considering this in the context only of the badger, but over a much wider spectrum of wildlife hosts.

Steck — I think that when we talk about the ecological equilibrium between animal species and a certain disease, we have to differentiate between endogenous and exogenous disease. The first is a disease which is prevalent in a particular region, where a certain equilibrium has been reached between genetic resistance, passive and active immunity and virulence; there the population has settled to a certain way of coping with the disease. Of course, this equilibrium may not be economically acceptable in domestic animals but it could keep wild animal populations at a certain level. Exogenous diseases may appear in sweeping epidemics — just as rabies is entering some countries, and rinderpest is exogenous to some extent for certain areas — and there is for a given population no passive or active immunity, so that the effect of the infection may be devastating. However, after a while they start settling into a balance again.

People have suggested that rabies might be controlled by sarcoptic mange. The trouble is, you can control sarcoptic mange by rabies, but not the other way around because the cut-off level is different.

Gallagher — Continuing the debate that Dr Little started, I take a different viewpoint. I think that tuberculosis has probably been in our badgers for some time — I would suggest half a century, a century, possibly longer. The reason I say this is that we see tuberculosis in different manifestations. In Gloucestershire we see a particularly malignant form of the disease — a highly progressive, destructive disease — but in other parts of the south-west the disease is far more benign. Indeed, with control measures the reduction in the prevalence of tuberculosis, even in Gloucestershire, has led to a change. It seems that a more benign disease is now developing in the badger. We see more cases of, perhaps, lymph node infection, and far less of the highly destructive lung disease.

I see that our real problem in Gloucestershire, as has been indicated, is not just that of tuberculosis; it is the problem of massive overcrowding. We have a population which is almost out of control in some parts of the Cotswold escarpment. I would suggest that it is probably this tremendously high density of population which is our real problem in Gloucestershire. We have the highest recorded population density for the European badger in Gloucestershire. I suggest that this is a factor influencing the manifestation of this disease.

A Speaker — Could I ask whether anyone from the Ministry has taken badgers from outside the south-west and experimentally infected them with bovine tuberculosis from Gloucestershire to see what the outcome would be in other populations?

Little — We have infected badgers from Hampshire and Surrey with *Mycobacterium bovis* and the disease was of the severe type that Dr Gallagher has described.

Ormerod — The type of disease might be related to the dose of the infecting organism obtained at any one time, the acute miliary lesions being caused by a heavy initial dose, while the chronic may be caused by a small initial dose which allowed immunity to be developed as in human infections with *Mycobacterium tuberculosis*.

Gallagher — Yes, there certainly is some evidence for this. In Gloucestershire we see a manifestation of disease which is not seen with any frequency in other parts of the south-west. We see tuberculosis due to bites. A badger with infected lungs will cough up infected sputum, which will then contaminate the mouth in general, and bites from an animal such as this can introduce infection intramuscularly or subcutaneously and result in very early generalization and in fulminating disease the like of which we do not see in other parts of the country where the population density is much lower. So, to some extent, yes.

Chapman — Sir William mentioned two possible methods of control or containment of diseases, one being separation of domesticated animals from wild animals and the other being the vaccination of the domesticated species. Whilst agreeing that neither offers any hope for the eradication of bovine tuberculosis in badgers, I wonder if I could ask him to extend the argument. Would he comment on whether or not it might be of any value, if not of complete usefulness? I believe that there was a paper earlier this year in the *Journal of Zoology* (Cheeseman & Mallinson, 1981)[1] commenting on the incidence of tuberculosis in herds where badgers had access to buildings. I wonder if there is some possibility of reducing the contact between badgers and cattle, not necessarily by complete separation. Secondly, I wonder what his comments would be on vaccination, both of badgers and of cattle. I know that Mr Jones talked about the vaccination of wild animals in Africa earlier this afternoon.

Henderson — There is obviously quite a lot in what you say. One way of confining cattle and making the exposure to badgers less is, in certain circumstances, to keep them housed. However, this is quite against the traditional practice of dairy farming in that part of the country. I can give you an example of where this works. David Jones mentioned African swine fever as being a problem that really prevented any exploitation of domesticated pigs in certain areas of Africa. In the Republic of South Africa they have found that they can rear pigs quite successfully in areas inhabited by warthog, a carrier of African swine fever, provided they double-fence the domesticated pigs' environment, have concrete floors to the sties, and have walls not less than about $3'6''$ high. So what you are suggesting would probably work, but it is not very practicable in this particular situation.

As far as vaccination of badgers is concerned — this is begging the question: we do not have a vaccine for cattle nor for badgers, but if we did have — I found absolutely fascinating the story from Professor Steck of the success of an oral route with an attenuated virus being effective. This must be one of the very few examples of effective vaccination of wildlife without actually getting hold of the animals and administering a vaccine parenterally.

Chapman — I was not suggesting total separation. I believe the paper was talking about badgers actually getting into farm buildings and one impression was that the badgers were going for the concentrates, either in buildings or even on the ground by feed troughs. It was just extending the idea from that. If you put the feed in high troughs you might not solve the problem, but you might reduce it.

Rees — It is not only a problem of contact in buildings, obviously; it is also contamination of grassland and ingestion that way. The actual building contact is probably much lower than the danger we see in outlying fields, where the young stock are the first to show infection. Perhaps Mr Jones would comment on this from his Cornish experience?

G. W. Jones — So far as the interaction between cattle and the badger is concerned, Mr Thompson has already touched on the research work that is being done in this field. The contact between badgers in the field and their coming into contact with cattle under certain circumstances, such as when they are attracted to buildings where concentrates are being stored, is the subject of research being carried out into the behavioural interaction between the two species. As far as observations in the field are concerned, observed contacts are fairly few in comparison to the contact that you would find in just field contact of badgers and cattle feeding on the same pasture at the same time.

Thompson — I should like to follow that one up a little, Mr Chairman. I am not

---

[1] Cheeseman, C. L. & Mallinson, P. J. (1981). Behaviour of badgers (*Meles meles*) infected with bovine tuberculosis. *J. Zool., Lond.* **194**: 284–287.

rising to Lord Zuckerman's complaint about working on earthworms, because we are not, but badgers are of course going to the pastures to feed on earthworms and they are spending more time, from the Gloucestershire studies, on pasture than on any other part of their habitat. So the likelihood that there is close contact between badgers and cattle in some situations and at some times of the year is very strong. What we want to do is actually to find out what it is.

If I could just take up another point from Dr Little's comments on the opossum in New Zealand, of course he is right, it is a very close analogy. However, there are differences and one that he pointed out is that the opossum, being an introduced species, is not popular and, since it causes tree damage, is regarded as a pest. On the other hand, the skin is valuable and it is regarded as an asset by some people. So they have a problem on their hands of mixed interests involved in relation to this animal which is very successful. The badger is used a little for fur skins in this country, but very little. Our problem is really quite different — the badger is a very popular animal, as you well know.

Rees — As we have such an array of talent dealing with other diseases, perhaps we should take full advantage of them. In Sir William's paper, he said that he felt our tactics at present in the Ministry were not correct in that we were not gassing sufficiently extensively. I wonder whether any of the previous speakers would like to comment on this, as to whether they feel that limited gassing is likely to fail and that more extensive gassing is necessary before we can get ahead of the disease and break the cycle?

Steck — I think it is very important to do a very thorough survey of how far tuberculosis has gone in badgers, and not just take the cattle as indicators of how far it has gone, so that you can appropriately adjust whatever measures you take for controlling the badger population. I understand from the information we have received that this is a disease which also kills badgers and has an influence on the badger population. If we look at it in the long run, it would probably be beneficial for the badger population if one could get rid of it in the badger population. However, I think that any action should be based on good surveys.

Henderson — Could I ask a question in connection with this? I entirely agree; I tried to put it across that I was making recommendations, including that of gathering more information. I should like to ask the Chief Veterinary Officer, is it true that, if you find infected badgers and there has up to that point been no positive tuberculin reaction in the cattle, you have not got authority to destroy the badgers?

Rees — Sir William, you are asking me a very pointed question now. What you are saying is absolutely true, that up to this stage we have tried to limit our activities and we have only taken action where there has been a relationship between tuberculous badgers and reacting cattle. In those circumstances, we have gassed the sets which contain the infected badgers. We have not gone to the next step which would involve dealing with the infected sets before we had reactors in the contact cattle. This is quite true.

Henderson — It seems an unnecessary constraint, does it not, Professor Steck? From your experience with rabies, is this not so?

Steck — I think so. You may just have to wait and, if you wait, you may then reach a stage where tuberculosis will spread to the cattle again. Your contacts with cattle are probably much less than the actual spread of the disease among the badgers. However, I am not here to give advice, but it does not seem to me that one should just go out and gas vast areas where tuberculosis is not in the badgers. In our country we got rid of tuberculosis in cattle; we had apparently some cases among badgers, but to our knowledge we did not have the problem that foci of tuberculosis would remain in badgers and then spread back into cattle again.

Rees — I can assure you that we do not gas where there is no actual evidence of tuberculosis in badgers. The point Sir William was making earlier was that we should go ahead, into the clear area, and get to clean country before stopping. At the moment we only gas where there is actual evidence of infection in the badger population and also related disease in cattle.

Molyneux — I was going to support Sir William's idea. Can I ask those in the Ministry who are responsible for these activities a question? How much monitoring of the possible spread of tuberculosis in badgers is actually being carried out outside the known area of its distribution? That is the first thing. Because I believe that you have to start on the periphery and then move in, as a tactic in order to control what is clearly a serious disease problem. It seems to me to be the most appropriate strategy without any regard to what is in the cattle population.

Henderson — I was really making two points. The point I made in my paper was to suggest that you were not pushing back far enough the barrier between the diseased badgers and the cattle, and you should go to the extent where you were into the clean badger population. The other point I was making was that in your monitoring of tuberculosis in badgers, when you find tuberculosis in the badger which has not yet become evident as having spread the infection to cattle, it is a bit late to wait until you get that demonstration of infection in cattle before you take any action.

Barr — A number of points are arising in the discussion and I think one of the most important things to come out is, how do you monitor whether the badger population is healthy or diseased? We have no blood test that is efficient in any way, there is no skin test that we can use either, so we are compelled to resort to two possibilities: one, we kill badgers and culture samples from them; two, we examine the faeces. If you have a very low percentage of tuberculosis in badgers, what number of the population do you have to kill before you get a meaningful result? If you have 5% infection in badgers, a statistician may tell me that I have to kill the entire colony before I can be sure that that colony is clean. When it comes to faeces, in the past we have had some very disappointing results from culturing badger faeces. Where we have a known population we still sometimes get results like 0.5% of the faeces samples examined giving us cultures of *M. bovis*. This is when we are doing it under the best system possible, when we are collecting the badger faeces less than 24 h old and culturing immediately. In other areas, though, we are getting up to 10% positive faeces — at least it is looking as if it is going to go that way — and this could be much more promising in the future. I think it is rather too early to say exactly how we can monitor a clean or a diseased badger population, but this is the ideal — find the extent of the infection and work inwards.

On the other point, where we have experience of infected badgers and no infection in cattle, we have not come across this in many parts. Where we do have this, as we have at one position in the south of England, we are using it as an experimental study area and hoping to gain as much information as we possibly can as to the infection, or absence of it, in the rest of the wildlife population and also, perhaps, to determine under which circumstances infection does or does not spread from the badger to cattle.

Henderson — On that last point, Mr Chairman, it would be very valuable information to know the point to which you have to drop the prevalence of the disease for it to disappear. We have had information from Professor Steck on rabies, that the infection is no longer maintained at 0.3 foxes per square kilometre. The information I had was from Denmark, where the infection was no longer maintained when there was one fox to four or five square kilometres.

Lainson — I can almost hear the badger complaining bitterly about the domestic cat. After all, we know the cat distributes its oocysts of *Toxoplasma* not only in the domestic situation but also on pastures. As a result, there are sometimes outbreaks of toxoplasmosis among sheep and other domesticated animals, and these in turn act as an important source of infection for man. I suppose the badger is in the unfortunate position of not being a domestic pet; we never talk of gassing or killing off the domestic cat!

Omerod — I would like to add my support to the view expressed by Dr Lainson. Bovine tuberculosis is essentially a residual problem; pasteurization has for practical purposes abolished it from man in Britain and human tuberculosis has also been greatly reduced. Many serious diseases of man throughout the world are acquired by infection from animals; our preoccupation with the transmission of tuberculosis from badgers to bovines in the absence of human disease, may be out of proportion to its significance.

Rees — Could I now call upon Professor Payne to draw the threads together on the whole Symposium?

Payne — Mr Chairman, Ladies and Gentlemen, I should like to return to Sir Andrew Huxley's opening remarks. He referred to the potential conflict between animal disease and conservation. This has been central to our deliberations. We have tried to keep the conference on a scientific level and have concentrated on accurate knowledge and information, as opposed to myths which some, in all good faith, would substitute for truth. I am reminded of the late Professor Oakley, who once said that a hypothesis would be accepted whether or not it was true, provided it made life appear less brutish than it was before. Professor Oakley made this point in a paper called "Virgins and he-goats". It was a remarkable document. He related the myth that, in the Harz Mountains on Walpurgis night, if you confronted a he-goat with a virgin, the he-goat would turn into a youth of surpassing beauty — a lovely hypothesis in which evil, when confronted by innocence and goodness, will disappear. Oakley tells us that this was actually tried under the glare of television lighting, on top of one of the Harz Mountains, and a certain young lady was presented to a he-goat — and nothing happened. Now here is a beautiful myth we would love to be true. It even goes back to childhood myths and fairytales. Could it still be true in spite of the evidence? Professor Oakley suggested further tests using an enormous "gramophone record" on which a succession of virgins and he-goats would be presented to each other in turn — after all, perhaps the original lady was not what she pretended to be, or perhaps the goat was not of the right strain, or breed — but how far can you really go in proving a null hypothesis? This could be our situation on certain aspects of wildlife and conservation. We hang on to myths and sometimes no amount of scientific evidence will refute them.

In this conference we have been scientific. In fact such a wealth of evidence has been presented that it cannot be summarized in detail.

I have often stated that there are only four ways to prevent disease. We can eradicate, or we can vaccinate, or we can use our knowledge of epidemiology to cut the cycles of infection, we can breed superior animals, or — and this is probably the most important point — we might decide to do nothing. Indeed, the story of myxomatosis in the rabbit, for which no control was attempted, carries an important lesson, which I shall mention later.

I would now like to make a few remarks on each paper in turn.

The first, on *rinderpest*, was a major success story. In the face of potential catastrophe, the use of vaccine eliminated the problem in cattle throughout most of Africa. Not only that, it was probably eradicated in the wildlife alongside. Dr Plowright, who gave this paper, was modest in omitting to mention that his own vaccine was very much involved!

With *trypanosomiasis* we had a good example of a very complex problem where we could have over-reacted in schemes to eliminate wildlife to get rid of the vector tsetse flies, but this was not carried to extremes and common sense prevailed. However, we did have considerable success in local areas by the elimination of the fly and by the sensible use of precautions. One wonders how far we can still go and whether we did enough, fast enough.

Much came out of the paper on *rabies*. The figure to emphasize is the 0.3 foxes per square kilometre, vital to maintain spread of infection. Indeed there have been limited successes by eradicating foxes in certain areas, for instance in Denmark, but vaccine offers the best hope. There is certainly no shortage of the vehicle for the vaccine, chicken's heads, but are we being over-cautious in the use of live attenuated vaccine? Few examples exist of reversion to a virulent strain.

Myxomatosis in rabbits presents a salutary warning. Here was an example of what happens when you let a disease loose without attempts at control. Say we let this happen with other diseases such as badger tuberculosis? Would we finish up with a situation like myxomatosis? We certainly could and it would take many years, centuries in fact, to put matters right. In fact the rabbit has advantages over the badger in being able to breed faster and select out resistant strains of animals.

The paper on *influenza* presented a most interesting study. The concern to control the big pandemics is obviously in everyone's mind and new biotechnology may come to our aid to develop, store and mass-produce new vaccines. This is potentially a success story for the future.

When we came to *leishmaniasis* we had a fascinating display of colour photographs. We were treated to a series of pictures showing what ugly lesions could develop on a charming young lady's face if the disease continued unchecked. This filled us with an awareness of the need for very urgent measures for better control. We rely at present on treatment and upon very simple safeguards.

When we come to *trichinosis* and *hydatidosis* a similar story applies. Here our background knowledge of epidemiology and control is good, but it has to be applied. How can this be done when people like the Turkanas insist on disposing of their dead by leaving them out in the open for jackals to eat?

Later in the meeting we became increasingly impressed with the complexity of present knowledge. Dr Gordon Smith gave an extraordinary wide-ranging review of the extent of our present knowledge. He referred repeatedly to the existence of even more expertise in the audience. How can this be brought together, because we do need a whole view of our subject? Prediction of likely trends in disease incidence so that we could develop control measures well in advance of need might follow if we could computerize all the information. Is this a far-off possibility?

Dr Barnes told us about *plague*, the disease we all fear for historical reasons. His photographs were superb, showing horrible lesions associated with this disease and linked with historical reference. I was impressed to hear how Dr

Barnes used wild animals as sentinels, for giving advance warning of plague in humans. The value of local eradication of fleas was mentioned. I was most impressed to see the number of fleas involved! Also, he warned us how sometimes we can make misjudgments and try to eradicate wildlife where it is not helpful so to do.

We heard from Dr Smith of the dramatic incidence of *botulism* in wildfowl, with photographs showing thousands of animals dead from this disease. We heard, too, of the possibility of prevention following on from research. For instance, what exists in Camargue muds that prevents growth of *Clostridium botulinum*?

Mr Jones introduced us to the view of a practical veterinarian in controlling disease of wildlife. I thought his first-hand appreciation of animal movements around water-holes and his down-to-earth knowledge of disease control especially valuable. This will help proper planning when new bores and water holes are built. Mr Jones also has clear ideas on priority — some diseases, such as tuberculosis, brucellosis and parasite infestation, require most effort.

Sir William Henderson looked back on the long history of disease prevention, drawing on his wide experience. He gave a justification for our present policies, and emphasized the importance of properly planned strategies, carefully put into application with monitoring to check results. I was most impressed with the concept of the fence around the National Park in South Africa which surrounded such an enormous area. Sometimes we have to go to extreme lengths to keep wildlife disease separate from our domesticated animals.

Sir William gave an important and authoritative contribution on tuberculosis in badgers. He emphasized that eradication of infected badgers was essential. Clearly this is not only in the interests of domesticated animals and man, but also in the interest of disease control in the badger population itself.

I come finally to a number of questions for discussion. First, we began our symposium by fearing conflicts between agricultural development and conservation. However, are these conflicts more imagined than real? We have not yet identified a priority list for especial action.

Secondly, looking at the available preventive methods, are there major gaps? What more research is needed? Finally, there has been an explosion of knowledge in this area. Is this sufficiently easy to retrieve from the literature? Indeed much important knowledge may be unpublished and spread only by verbal contact. Certainly much of it is not in the scientific literature, but left in the form of personal anecdotes and experiences. Can we look into the future and devise a data bank for storing this information? I am already involved in planning a data bank on zoonosis for the European Commission at Luxembourg and this could be extended.

Rees — Thank you, Professor Payne, for that excellent summing-up and for leaving us with those questions. I am sure that everyone here who is engaged in these activities must always bear them in mind. I can assure you that the Symposium was thrown open to all participants, both he-goats and virgins were invited, but we were not really expecting miracles — just an exposure to the truth and the facts.

I think that we have all enjoyed the papers over the two days. You will realize that the whole Symposium stems out of the inspiration of Lord Zuckerman and we owe him a great debt of gratitude for organizing such a conference. So I would now like to invite Lord Zuckerman to say a few words and to close the Symposium.

Zuckerman — Mr Chairman, it is not you but we who owe whatever debt of gratitude is called for on this occasion. I can assure you that the Zoological Society owes a great deal to those who have come together to take part in this Symposium, a Symposium which has realized all our best expectations. I am not a pathologist nor a veterinarian, but a biologist, with a wide interest in a number of subjects. The main lesson that I shall carry away from these past two days is that the interaction of a variety of pathogens in the animal kingdom implies a series of symbiotic relationships which are one aspect of evolutionary adaptation. The interactions are very complex. One of our speakers suggested that his words might imply an interest which favoured the life of his particular pathogen rather than that of the host on which it preyed. But host and pathogen appear to me to be one so far as the evolutionary cycle is concerned. What we have been talking about also has a great deal to do with the control of animal population. I found myself asking, what indeed is a "healthy animal"? Is a healthy animal one that has no protozoan or helminthic infections? Viruses and whatever? Has such a creature ever existed? Professor Payne has just referred to the explosion of knowledge that has occurred in the subjects we have been discussing. Undoubtedly this Symposium has opened up wide horizons of enquiry and knowledge.

Sir Andrew Huxley's researches, like my own, lie outside the field we have been discussing. But it was a revelation to both of us to discover how much magnificent science has been revealed in these past few years by students of parasitic and other animal diseases. My first research post was in this Society more than 50 years ago. At the time there was a pathologist on the staff who, to my dismay, was soon appointed Chief Medical Officer to the then Colonial Office. For a time I then had to do the post-mortems of animals that died in the Gardens. It was bad news in those days if one heard about foot-and-mouth or anthrax. A couple of elephants developed anthrax in the 1920s. The infection spread, and we lost a keeper, and two others were severely ill with the disease.

When I learnt from Sir William Henderson's paper that foot-and-mouth disease is something which no longer need be dealt with in the way it always has been, I realized how far we have come. That is only one illustration of the many there are of animal diseases which have been conquered, in which symbiotic chains have been broken.

There are many other diseases that could be eradicated. There are certainly other victories ahead in your field. But we do not live in a scientific vacuum. We live in a social world, an economic world, where whatever we do may run into barriers of prejudice due to lack of understanding by the public. We have experienced no conflict at this meeting, but what we have said and what we do does occasion conflict. Having listened to all the papers of this meeting, I wish that I had delayed my response to the request made of me a couple of years ago by the Minister of Agriculture to carry out an independent enquiry into the badger/TB/cattle problem. When I embarked on it, I was totally unaware of the fact that half the general public which had views on the subject simply did not understand, indeed could not understand, the nature of the scientific evidence that has been discussed at this meeting.

There is much that could be said about the need to reconcile policies of disease eradication with what is politically necessary, and what is politically possible. Unless I am wrong, the Government would be acting outside the law if it were to turn a blind eye to measures that could be set in train to assist in the eradication of tuberculosis in cattle. If we were not vigorous in the way the tuberculin test is applied, we might find ourselves in the same position as is the Irish Government now — and called to task by the authorities in Brussels. Education is necessary to make the public at large understand why some of the things that are done in suppressing disease in animals have to be done.

It is not just the general public that needs to be educated, or at least informed generally, about the significance of disease in domesticated and wild animals. Some well-meaning but narrowly-educated young people who belong to a modern generation of ecologists — I refer to those who work on their emotions rather than on the basis of a good scientific education — also need to become better informed scientifically. David Jones has told the story of the Arabian oryx; of the captive herd that was bred in Arizona with the object of reintroducing the animal to the wild. But he also told us that further animals from the Arizona stock cannot now be moved because the herd has developed blue-tongue. He referred to the release of tuberculous orang-utans into the wild. Matters such as these are important. We have got to know what it is we are doing when we try to conserve; what it is we are doing when we reintroduce animals to the wild. These things are not always understood by the amateur conservationist or ecologist.

I am sorry that Prince Philip has had to leave. His interest was apparent to everybody. As President of the World Wildlife Fund and as one deeply concerned in the conservation movement, and so anxious to make it a more effective movement than it now is, he suggested to me when he left that the proceedings of this Symposium should be very widely distributed. I too hope they will be. He suggested that we should get in touch with the central office of the WWF in Geneva when the volume appears — which I hope will be without much delay. Like Professor Payne, I feel that the information that has been exchanged at this meeting must be disseminated widely, not only because of its substance, but also to indicate the enormous promise and challenge and scientific excitement that lies in the field which has been represented here.

Finally, just a word of thanks. When Sir William reminded us that it was Gowland Hopkins, the President of the Royal Society, who in the 1930s led the enquiry into tuberculosis in cattle, I suddenly realized that since the beginning of the nineteenth century, when two successive presidents of the Royal Society took a hand in founding this Society, we have not had one of their successors here until this meeting, with Sir Andrew Huxley as overall chairman. We are very grateful to him.

The reason why he too has had to leave before the close is that he was due to appear at the Royal College of Science at a meeting to mark the centenary of T. H. Huxley's — his grandfather's — death. T. H. Huxley first served on the Council of this Society in 1861, 120 years ago. In 1865 he expressed his dissatisfaction with the use to which the anatomical material of the Gardens was being put, and he saw to it that a committee was appointed — I have just looked this up — "to consider the mode by which the animals dying in the Gardens may be disposed of with most benefit to zoological science". His grandson has now been learning of the amazing advances that have been made since then in our knowledge of animal disease. Andrew's half-brother, Julian Huxley, was Secretary of the Society in the 1930s. He, too, saw to it that the Society lived up to its purpose as set out in its Royal Charter: "The advancement of Zoology and Animal Physiology and the introduction of new and curious subjects of the Animal Kingdom". That remains our objective.

The Society owes thanks to you, Mr Chairman, and to the other Chairmen —

Sir Cyril Clarke and Dr Tyrrell — for all you have done to make this meeting a success. We also owe our thanks to all who participated in the Symposium.

The published volume will, I believe, be the fiftieth in the series of Symposia that was started 20 years ago. It will certainly prove to be one of the best, since it has provided a model by which to design other symposia that deal with matters of wide-ranging interest. And if we can revert, in a further symposium, to any of the issues that have been revealed to us by the distinguished body of men here today, we shall be only too ready to do so. Thank you all very much.

Rees — Lord Zuckerman, once again I am sure I echo the thoughts of all the speakers and all the participants here in thanking you most sincerely for your efforts.

Author Index

Numbers in italics refer to pages in the References at the end of each article

Woodford, M. H., 191, *197*
Wooding, W. L., 130, *134*
Woodroofe, G. M., 78, 79, 81, *90*
Wooff, W. R., 40, *52*
Work, T. H., 214, *233*
Wright, G. G., 98, *111*

Y

Yakhno, M., 129, 130, *134, 135*
Yakimoff, W. L., 149, *174*
Yamane, N., 126, *135*
Yarinsky, A., 98, *111*
Yegiazaryan, K. K., 224, *231*
Young, A. M., 103, *115*
Young, E., 185, 191, *197*, 272, *285*
Young, J., 122, *134*

Z

Zdanov, V. M., 130, *134*
Zeledon, R., 39, *52*
Zen-Yoji, H., 103, *114*
Zeuner, F. E., 204, *204*
Zillman, U., 30, 32, 33, *49, 50*
Zlotnik, I., 210, *229*
Zuckerman, S., 294, *297*
Zuelzer, M., 217, *232*

Subject Index

Numbers in italics refer to figures

A

Acacia, 45
Acarines, 219
Accipiter melanoleucus, 282
Adenota kob, 6
Aedes, 269
    aegypti, 209, 212, 221, 222, 270
    albopictus, 209
    taeniorhynchus, 211
Aepyceros melampus, *6*, 272
Africa, 1—28, 181—205
    conservation, animal disease, 271—285
African horse sickness, 228
African swine fever, 219, 275
Agouti, 189
Agricultural chemicals, animal disease, 280
Akodon, 155
Albizzia, 45
Alcelaphus buselaphus, 274
Alcelaphus spp., *6*
Alopex lagopus, 59
Amastigotes, *Leishmania*, 139
Amazonian visceral leishmaniasis, *160*
American leishmanias, *160—163*
Amphibians, 37, 138
Anas platyrhyncos, 105
Animal plague, geographic distribution, 249
Animal hosts
    hybatid disease, Africa, 190
    trichinosis, Africa, 184
Anomiopsyllus spp., 254
Anopheles gambiae, 222
Anteaters, 162, 175
Antelopes, 4, 40, 42, *63*, 182, 185, 189—192, 227
Antidorcas marsupialis, *6*, 274
Anthrax, 216
    undisturbed populations, disease, conservation, 274
Apodemus, *63*
Aquatic areas, botulism, *102*
Arboviruses, 210, 219
Armadillos, 157

Arthropod transmission, spread of infections, 219
Artiodactyla, 1, 5
Arvicola, *63*
Asia, conservation, animal disease, 271—285
Aspergillus spp., 280
Australian strains, *82*
Avians, influenza, 127

B

Baboons, 130, 187
Bacillus leprae, 211
Bacillus spp., 102, 117
Bacterial multiplication, *in vivo*, botulism, 109
Badgers, 62, 293—306
Barriers to movement, man/wildlife effects, 278
Base-line epidemiological information, *Leishmania* hosts, 159
Bassariscus astutus, 245
Bats, *58*, *59*, 61, 68, 73, 160, 202, 208, 226, 273, 291—302
Bears, 188
Beauvaria bassiana, 210
Besnoitia, 276
Birds, 37, 41, *63*, 99, 100, 102, 108, 131, 135, 138, 208, 227, 233, 280, 284
    influenza, 127
    influenza virus strains, *128*, *129*
Biting between vertebrates, spread of infections, 225
Blesbok, 4
Blue-green algae, 280
Bluetongue, 228, 274
Bongo, 4, 5
Bontebok, 4
Boocerus euryceros, *6*
Bos gaurus, 276
Bos indicus, 276
Botulism, 97—119, 309
Bovid populations, east Africa, 23
Bovine ephemeral fever, 228
Bovine pleuropneumonia, 274, 288